Experimental Investigation and Numerical Modeling of Rock Brittle Failure Behavior under High Stress Conditions

Experimental Investigation and Numerical Modeling of Rock Brittle Failure Behavior under High Stress Conditions

Editors

Zhengzhao Liang
Nuwen Xu
Bei Jiang

MDPI

Basel • Beijing • Wuhan • Barcelona • Belgrade • Novi Sad • Cluj • Manchester

Editors

Zhengzhao Liang
Dalian University of Technology
Dalian, China

Nuwen Xu
Sichuan University
Chengdu, China

Bei Jiang
China University of Mining and Technology (Beijing)
Beijing, China

Editorial Office
MDPI
St. Alban-Anlage 66
4052 Basel, Switzerland

This is a reprint of articles from the Special Issue published online in the open access journal *Applied Sciences* (ISSN 2076-3417) (available at: https://www.mdpi.com/journal/applsci/special_issues/Rock_Failure_Behavior).

For citation purposes, cite each article independently as indicated on the article page online and as indicated below:

Lastname, A.A.; Lastname, B.B. Article Title. *Journal Name* **Year**, *Volume Number*, Page Range.

ISBN 978-3-0365-8700-4 (Hbk)
ISBN 978-3-0365-8701-1 (PDF)
doi.org/10.3390/books978-3-0365-8701-1

Contents

About the Editors

Zhengzhao Liang

Zhengzhao Liang is a professor at Dalian University of Technology. His main research interests include the following: (1) multi-scale modeling and fracture failure mechanisms of rocks; (2) rock mechanics methods based on digital image technology; and (3) rock dynamics and impact breakage problems. He has developed a numerical method for rock failure process analysis, and has developed a three-dimensional parallel simulation system for rock engineering disaster prediction and warning based on microseismic damage models. Professor Liang won the Rocha Medal from the ISRM in 2008 and the Young Scientist Gold Medal from the China Rock Mechanics and Engineering Society in 2014. Professor Liang has hosted more than 30 projects, including the National Basic Research Program (973 Program), the National Natural Science Foundation of China, and the Ministry of Education's special fund.

Nuwen Xu

Nuwen Xu is a professor at Sichuan University, selected by the National Young Talent Project. Prof. Xu is mainly engaged in research on the disaster mechanism, warning, prevention and control of rock mass in deep underground engineering. His main research interests are as follows: (1) numerical simulation analysis of rock engineering; (2) research on microseismic monitoring technology and analysis methods for rock engineering; (3) research on the disaster mechanism, monitoring and early warning signs of rock engineering dynamic hazards; and (4) intelligent engineering and big data analytics. He has published more than 100 papers in authoritative journals, and has presided over more than 10 vertical projects, e.g., the National Natural Science Foundation of China, the Outstanding Youth Fund of Sichuan Province, and the sub-projects of the National Key Research and Development Programme. He has received eight awards, e.g., the First Prize of Scientific and Technological Progress from the Ministry of Education of China, the First Prize of Scientific and Technological Progress from the Chinese Society of Rock Mechanics and Engineering, and the "Qian Qihu Award". He has been consecutively selected as one of the Elsevier China Highly Cited Scholars (2020, 2021), and as within the World's Top 2% Scientists 2020 by Stanford University. He is also a member of the Board of Directors of the Chinese Society of Rock Mechanics and Engineering, assistant to the President, and serves on the editorial boards of several journals.

Bei Jiang

Bei Jiang is currently an associate professor at China University of Mining and Technology (Beijing Campus), She is the Deputy Secretary of the State Key Laboratory of Deep Geotechnics and Underground Engineering, and a Director of the Party and Government Office. Associate Prof. Jiang is mainly concerned with the theory and technology of the stability control of underground engineering in civil engineering and transport, energy and mining industries. She has presided over seven national and provincial vertical projects, such as the National Natural Science Foundation of China (NSFC) Programme, Key Research Development Programme in Shandong Province, Excellent Youth Innovation Team Project of Shandong Higher Education Institutions, and China Postdoctoral Special Funding Programme, and nine other major scientific research projects entrusted by other enterprises and institutions. Moreover, she has published more than 70 academic papers, of which she is cited as the first/corresponding author of more than 20. She is also the executive director of the Soft Rock Engineering and Deep Disaster Control Branch of the Chinese Society of Rock Mechanics and Engineering, and the director of the Underground Engineering Branch of the Chinese Society of Rock Mechanics and Engineering. At present, she has won eight awards, e.g., the First Prize of Scientific and Technological Progress of Shandong Province and the First Prize of Scientific and Technological Progress of China Rock Mechanics and Engineering Society.

applied sciences

MDPI

Editorial

Special Issue on Experimental Investigation and Numerical Modeling of Rock Brittle Failure Behavior under High Stress Conditions

Zhengzhao Liang [1,*], Na Wu [1], Nuwen Xu [2] and Jiang Bei [3]

1 State Key Laboratory of Coastal and Offshore Engineering, Dalian University of Technology, Dalian 116024, China; n.wu@dlut.edu.cn
2 State Key Laboratory of Hydraulics and Mountain River Engineering, Sichuan University, Chengdu 610065, China; xunuwen@scu.edu.cn
3 State Key Laboratory for GeoMechanics and Deep Underground Engineering, China University of Mining and Technology-Beijing, Beijing 100083, China; jiangbei519@163.com
* Correspondence: LiangZZ@dlut.edu.cn

Citation: Liang, Z.; Wu, N.; Xu, N.; Bei, J. Special Issue on Experimental Investigation and Numerical Modeling of Rock Brittle Failure Behavior under High Stress Conditions. *Appl. Sci.* **2023**, *13*, 8429. https://doi.org/10.3390/app13148429

Received: 11 July 2023
Accepted: 12 July 2023
Published: 21 July 2023

1. Introduction

To meet the demands of the mining, hydropower, and transportation industries, deep rock mass engineering in China has rapidly developed. As the excavation depth increases, the stress environment in which the rock mass is located becomes particularly complex and diverse. Correspondingly, problems, such as excavation disturbance, blasting, seepage, and water heat coupling, can lead to increasingly serious deformation and damage accidents in rock engineering, thereby posing a threat to normal operations. The lack of comprehensive understanding of the brittle failure behavior and mechanical properties of deep rock masses seriously hinders the effective prevention and control of deep catastrophic accidents. Conducting research on the brittle failure and mechanical behavior of rock masses under high-level stress is still a comprehensive challenge, and it is of great significance for the stability assessment of deep rock mass engineering. This was the reasoning behind publishing this special feature on Experimental Investigation and Numerical Modeling of Rock Brittle Failure Behavior under High-level Stress Conditions.

This Special Issue was introduced to collect the latest research on related topics. More importantly, the failure mechanism and mechanical behavior of rock mass under high-level stress conditions were investigated to provide a reference for the design and construction of deep rock mass engineering. Thirty-one papers were submitted to this Special Issue, and thirteen papers were accepted (i.e., 42% acceptance rate). Looking back on the Special Issues, various topics were addressed, mainly on the failure behavior and mechanical characteristics of microscopic, macroscopic and engineering rock masses under statics, dynamics, seepage and high-temperature conditions.

2. Failure Behavior of Microscopic and Macroscopic Rock Masses

Due to the emergence and development of CT scanning low-field nuclear magnetic resonance (LF-NMR) and scanning electron microscopy (SEM), research on the microscopic and macroscopic mechanical properties and failure mechanisms of rock masses has dramatically developed. The damage evolution law and failure mechanism of mudstone under uniaxial compression were carried out by Duan et al. [1] using high-resolution CT scanning equipment. Combining digital core technology and the digital volume image correlation method, the 3D characterization of the mesostructure and the evolution process of the localized damage of mudstone were analyzed. The results showed that the aggregated mineral zone was prone to local deformation under loading, and the final propagation shape of cracks was very consistent with the propagation shape of the mineral zone. The progressive fracture process and mechanical behavior of multi-scale rocks, considering the microstructure, mesostructure, and macro joints, were estimated using a failure process analysis (RFPA) simulator and digital image processing technology [2]. The effects of shale fractures, mineral spatial distribution, and joints on the fracture formation process were studied for Brazilian splitting and joined rock masses, respectively. The results showed that the multi-scale, micro-to-macro failure process of brittle materials be estimated using the finite element parallel computing simulation method and digital images. The effects of saturation and impact loading on the dynamic mechanical behavior of frozen red sandstone were investigated using a low-temperature split Hopkinson pressure bar system (LT-SHPB) [3]. The dynamic strength, elastic modulus, and brittleness index of the frozen sandstone under impact loading tended to increase, and then decreased with saturation. However, the ultimate deformation capacity of the frozen sandstone showed the opposite trend. Additionally, the energy dissipation capacity of the frozen sample first increased, and then decreased with increasing saturation. The dynamic strength, elastic modulus, and peak strain of the frozen sandstone showed a significant strengthening effect, while its brittleness index gradually decreased at full saturation as the impact load increased. The dynamic evolution of the microstructure of the frozen sandstone due to changes in saturation was estimated using low-field nuclear magnetic resonance (LF-NMR) and scanning electron microscopy (SEM).

3. Failure Behavior and Mechanical Characteristics of Rock Mass under Complex Stress Conditions

To understand the tensile strength of macroscopic rock masses, three respective scholars studied the failure behavior and mechanical properties of conventional, frozen, and temperature–water coupling rock samples. The effect of layer orientation on strength and failure patterns of the layered rocks under direct and indirect tension via experimental and numerical tests were discussed by Gao et al. [4]. The results showed that the tensile strength, failure patterns, and progressive deformation of the layered rocks were significantly affected by the dip angle of the bedding planes. Moreover, the tensile strength obtained in the direct tension test was more accurate than that of the Brazilian disc test. The modified Single Plane of Weakness (MSPW) failure criterion was proposed to predict the tensile strength of layered rocks based on the failure modes of direct tension. The effectiveness of the failure criterion was verified. The original rock specimens were mined and processed in Yulong Copper Mine, and artificially frozen fractured marble specimens were made. Based on physical experiments and numerical simulations, the failure mechanism and mechanical properties of the Brazilian splitting of frozen rocks were studied by Wang et al. [5]. The test results indicated that frozen rock samples exhibited typical brittle failure characteristics. The tensile strength of frozen rock gradually decreased with the increase in the width and length of the ice-filled crack. It first decreased, and then increased with the increase in the ice-filled crack angle. The dynamic tensile properties of annular sandstone specimens under the influence of temperature and water were studied by Ping et al. [6]. The variation in the mass, volume growth rates, density growth rate, dynamic compressive strength, average strain rate, and peak strain with water temperature was evaluated. The results showed that the changes in the dynamic properties of sandstone specimens were not due to their own material composition, but due to the damage to their structure caused by the temperature–water coupling effect.

Furthermore, the seepage characteristics, failure mechanism, and mechanical behavior of rock masses under seepage and dynamic loads were analyzed. Firstly, the seepage characteristics during the stress–strain process of limestone under high water pressure were assessed in an experimental test by Bao et al. [7]. A tool to predict the formation of the seepage channel, namely, acoustic emission positioning technology, was proposed. The results suggested that the sudden drop in the stress–strain curve after its peak indicated the full formation of shear fractures and seepage paths. The dynamic and real-time development of microfractures can be monitored with acoustic emission technology and via seepage monitoring. Then, experiment research was conducted on soft siltstone specimens under a combination of dynamic and static loads by Dong et al. [8]. The mechanical properties and acoustic emission characteristics of soft rocks were quantitatively revealed using a creep disturbance impact loading system and acoustic emission system. The results indicated that the deformation of siltstone specimen increased with the increase in the initial average stress. As the initial average stress increased, the maximum load first decreased, then increased, and finally decreased. Moreover, the influence of the elastic modulus of each loading step on the damage evolution of the specimen under dynamic disturbance was analyzed using RFPA. The wave form characteristics during the damage of the specimens were analyzed by extracting signals at the key points. Finally, the influence of the angle on the dynamic crack propagation behavior under stress wave action was investigated by Zhang et al. [9]. The dynamic propagation behavior of cracks and new cracks around the blast hole were analyzed using AUTODYN's numerical analysis software. The stress wave theory was successfully used to analyze and predict the dangerous area after failure, proving that cracks ranging from 45° to 90° will induce characteristic cracks, which will accelerate the instability of the model.

4. Discussion on Failure Behavior and Mechanical Problems of Engineering Rock Mass

Currently, the reduction of crustal stress, stability analysis, and control technologies for engineering rock masses are hot research topics. The intelligent inverse method combining a particle swarm optimization (PSO) algorithm with a back propagation (BP) neural network was applied to the inverse analysis of crustal stress by Yan et al. [10]. The application was carried out using an underground powerhouse of Shuangjiangkou Hydropower Station as an example. The results indicated that the method improved the stability and accuracy of the inversion results. Referring to existing tunnel projects, nonelectronic equipment and electronic detonators were used for blasting tests and determining the impact of blasting construction on the range of rock loosening zones and the degree of rock fragmentation, respectively [11]. The results indicated that the range of loose rock circles around the tunnel, the decrease in wave velocity of loose rock masses, and the degree of rock fragmentation caused by normal blasting with nonelectronic detonators were larger than those caused by electronic detonators. Then, the surrounding rock stability of the excavated geologically weak section of the #2 diversion tunnel in the Xulong Hydropower Station was assessed by Qian et al. [12] using a numerical model. The results indicated that during the excavation process, the surrounding rock of the tunnel section was most significantly damaged near the arch crown and corner of the side wall with tensile failure. As the excavation progressed, the number of microseismic events at the corners of the arch and side walls gradually increased, and the energy of acoustic emissions steadily accumulated. Therefore, the likelihood of a collapse and rock burst in the area increased. Furthermore, the effect of the fracture geometric parameters on the failure mechanism and mechanical parameters of engineering rock mass in the dam site area of Lianghekou Hydropower Station were investigated using the ShapeMetriX^{3D} system and RFPA^{3D} in an uniaxial compression test [13]. The results showed that the fractured rock mass mainly represented a compressive–shear composite. Moreover, the influence of the fracture geometric parameters on the uniaxial compressive strength was greater than that of the elastic modulus.

5. Future Development

Although the Special Issue is closed, researchers still require a deeper understanding of the failure mechanism of deep rock masses during construction. In this case, this Special Issue will receive more attention, facilitate in-depth discussions, and encourage outstanding research.

Acknowledgments: This issue would not have been possible without the contributions of brilliant authors, professional reviewers, and the dedicated Applied Sciences Editorial Team. Additionally, congratulations to all the authors. Regardless of the final decision to submit the manuscripts, the suggestions and comments from the reviewers and editors helped the authors to improve their papers. We would like to take this opportunity to record our sincere gratefulness to all reviewers and editors. Finally, all authors have read and agreed to the published version of the manuscript.

Conflicts of Interest: The authors declare no conflict of interest.

References

1. Duan, D.; Chen, X.; Feng, X.; Wang, X.; Sun, J. Study on Failure Mechanism of Mudstone Based on Digital Core and Digital Volume Correlation Method. *Appl. Sci.* **2022**, *12*, 7933. [CrossRef]
2. Liu, X.; Liang, Z.; Meng, S.; Tang, C.; Tao, J. Numerical simulation study of brittle rock materials from micro to macro scales using digital image processing and parallel computing. *Appl. Sci.* **2022**, *12*, 3864. [CrossRef]
3. Xu, J.; Pu, H.; Sha, Z. Dynamic Mechanical Behavior of the Frozen Red Sandstone under Coupling of Saturation and Impact Loading. *Appl. Sci.* **2022**, *12*, 7767. [CrossRef]
4. Gao, M.; Liang, Z.; Jia, S.; Zou, J. Tensile Properties and Tensile Failure Criteria of Layered Rocks. *Appl. Sci.* **2022**, *12*, 6063. [CrossRef]
5. Wang, T.; Li, P.; Tang, C.; Zhang, B.; Yu, J.; Geng, T. Tensile Characteristics and Fracture Mode of Frozen Fractured Rock Mass Based on Brazilian Splitting Test. *Appl. Sci.* **2022**, *12*, 11788. [CrossRef]
6. Ping, Q.; Gao, Q.; Wu, Y.; Wang, C.; Shen, K.; Wang, S.; Xu, Y. Study on the Dynamic Splitting Mechanical Properties of Annular Sandstone Specimens with Temperature–Water Coupling in a Coal Mine. *Appl. Sci.* **2022**, *12*, 4608. [CrossRef]
7. Bao, C.; Yin, Y.; Tang, S.; Jiang, A.; Li, H. Seepage Characteristics and Failure Prediction during the Complete Stress–Strain Process of Limestone under High Water Pressure. *Appl. Sci.* **2022**, *12*, 6041. [CrossRef]
8. Dong, L.; Wang, D.; Sun, X.; Jiang, Y.; Luan, H.; Zhang, G.; Liang, B. Experimental Study on the Mechanical Properties and Acoustic Emission Characteristics of Deep Soft Rocks under Low-Frequency Dynamic Disturbance. *Appl. Sci.* **2023**, *13*, 6399. [CrossRef]
9. Zhang, J.; Wan, D.; Gao, W.; Zhou, L.; Wang, M. Numerical Study on the Dynamic Propagation Model of Cracks from Different Angles under the Effect of Circular Hole Explosion. *Appl. Sci.* **2023**, *13*, 7955. [CrossRef]
10. Yan, H.; Liu, H.; Li, Y.; Zhuo, L.; Xiao, M.; Chen, K.; Pei, J. Inversion Analysis of the In Situ Stress Field around Underground Caverns Based on Particle Swarm Optimization Optimized Back Propagation Neural Network. *Appl. Sci.* **2023**, *13*, 4697. [CrossRef]
11. Fu, H.; Guan, X.; Chen, C.; Wu, J.; Nie, Q.; Yang, N.; Liu, J. Formation Mechanism and Control Technology of an Excavation Damage Zone in Tunnel-Surrounding Rock. *Appl. Sci.* **2023**, *13*, 1006. [CrossRef]
12. Qian, H.; Tan, Z.; Li, B. Stability Analysis of Surrounding Rock in the Diversion Tunnel at the Xulong Hydropower Station based on RFPA3D and Microseismic Monitoring. *Appl. Sci.* **2022**, *12*, 9939. [CrossRef]
13. Wu, N.; Liang, Z.; Tao, Y.; Ai, T.; Li, G. Sensitivity Analysis of Fracture Geometry Parameters on the Mechanical Behavior of Rock Mass with an Embedded Three-Dimensional Fracture Network. *Appl. Sci.* **2022**, *12*, 9284. [CrossRef]

applied sciences

Article

Study on Failure Mechanism of Mudstone Based on Digital Core and Digital Volume Correlation Method

Dong Duan [1,2,*], Xiaoyu Chen [1,2], Xiaojing Feng [1,2], Xin Wang [1,2] and Jiang Sun [1,2]

1 College of Mining Engineering, Taiyuan University of Technology, Taiyuan 030024, China
2 Key Laboratory of In-Situ Property-Improving Mining of Ministry of Education, Taiyuan University of Technology, Taiyuan 030024, China
* Correspondence: duandong@tyut.edu.cn; Tel.: +86-151-3515-3856

Abstract: In order to study the damage evolution law and failure mechanism of mudstone under different stress states, with the help of high-resolution CT scanning equipment, in situ CT scanning experiments of mudstone under uniaxial compression were carried out. Combined with digital core technology and the digital volume image correlation method, the 3D characterization of meso-structure and the evolution process of localized damage in mudstone were analyzed. The research shows that brittle minerals such as quartz in mudstone often exist in the form of agglomerated strips, resulting in the formation of weak structural planes at the contact surfaces of different minerals. There are a large number of primary intergranular pores near the mineral accumulation zone. With the increase in axial load, the connectivity of pores will gradually increase, cracks will gradually emerge, internal pores will develop abnormally, and rocks will reach the critical state of failure; at this time, the throat number and coordination number of pores increase obviously. There was no obvious difference found in the distribution of mineral particles of different sizes, and the slip between mineral zones was mainly dominated by small particles. The accumulated mineral zone was able to easily form a weak surface, and the aggregated mineral zone under loading was easily able to produce local deformation, which is related to the mechanical properties of the mineral zone and its surrounding rock matrix, with the rock failure easily occurring along the junction of the two minerals. The displacement in the polymeric mineral zone was small, the deformation displacement of the rock skeleton dominated by clay minerals near the quartz mineral zone was larger, and the stronger quartz minerals restrained the rock skeleton deformation in the region.

Keywords: uniaxial compression; CT scan; microscopic destruction; digital volume image correlation

Citation: Duan, D.; Chen, X.; Feng, X.; Wang, X.; Sun, J. Study on Failure Mechanism of Mudstone Based on Digital Core and Digital Volume Correlation Method. *Appl. Sci.* **2022**, *12*, 7933. https://doi.org/10.3390/app12157933

Academic Editor: Daniel Dias

Received: 20 July 2022
Accepted: 6 August 2022
Published: 8 August 2022

1. Introduction

Rock is a kind of heterogeneous material in which a large number of irregular primary pores, fractures, and mineral particles are distributed, which directly affect the physical and mechanical properties of rock. Therefore, the focus of the study of the macro-mechanical properties of rock is to clarify the microstructure characteristics of the rock. In this paper, the uniaxial failure mechanism of mudstone is studied on the basis of digital core and digital volume image correlation method.

The numerical reconstruction method is a widely used digital core construction method, one that is based on core slice images. The main construction principle is to reconstruct 3D digital cores by process simulation method. The pore network model is established by using the maximal ball algorithm, lattice Boltzmann method, image processing and threshold segmentation, fractal theory, and mathematical statistics to realize the quantitative representation of micro pore structure [1–3]. Yin Duohao et al. [4] and Lin Chengyan et al. [5] evaluated the characteristics of the pore network model through parameters such as pore throat radius, coordination number, and permeability. It was found that the increase in throat length and pore coordination number were the main factors causing meso-damage. Zhang Huimei et al. [6] studied the freeze-thaw rock damage variables of mesoscopic pores.

The physical experiment method is to use a high-power optical microscope, electron microscope, and CT to directly scan the rock samples to construct digital cores. Yang Zhifeng et al. [7] found the relationship between three kinds of pores and mineral composition by electron microscope and CT scanning. Mikko Voutilainen et al. [8] also conducted related studies. Many scholars obtained the law of fracture changing with stress and found that the CT scanning technique can better evaluate the spatial attribute, distribution, and connectivity of macropores [9–11]. On the basis of a high-resolution 3D CT scanning system, it is widely used to automatically identify the main mineral composition in rock and analyze the law of fracture propagation in rock [12–14]. The influence of mineral composition and microstructure on rock mechanical properties is discussed by constructing digital cores [15–19].

It is impossible to analyze the internal dynamic strain of rock only with the help of digital core. The digital volume image correlation method provides the possibility for this. Guo Weina et al. [20] found that DIC technology can effectively characterize the strain evolution and crack formation and propagation of materials. Bay et al. [21] extended the digital image correlation method and proposed the continuous field 3D digital volume image correlation method DVC. Mao Lingtao et al. [22] revealed the strain evolution characteristics and failure mechanism of glass fiber composites by the DVC method.

The failure of rock has an extremely important relationship with its internal microstructure, main diagenetic composition, and spatial distribution of diagenetic composition. In order to study the failure mechanism of rock, it is necessary to explore the process of crack propagation and development and stress failure zone distribution from macro and micro angles. It is well known that rock is a kind of complex heterogeneous material. The composition and distribution of main components and pore space distribution in rock have an important influence on the mechanical properties of rock. The way in which to accurately describe the composition and spatial distribution of the main diagenetic minerals and the spatial distribution of primary pores and cracks in the rock, as well as the way to establish the diagenetic composition and the relationship between cracks and rock failure is the focus of this article.

In this paper, the influence of brittle mineral composition on rock failure was studied by establishing a pore network model based on digital core by using the CT scanning method in a physical experiment, combined with XRD and SEM analysis results. The quantitative analysis of minerals and pores from the aspects of content and distribution characteristics can comprehensively master the microstructure information of rock, so as to indirectly evaluate the mechanical properties and failure mechanism of rock.

2. Real-Time Loading CT Scanning Experimental System and Experimental Scheme

2.1. Real-Time Loading CT Scanning Experimental System

The experimental system is mainly composed of two parts:

(1) CT scanning system: The minimum focus size of the CT experimental equipment is 3 μm, which can distinguish the pore with 0.5 μm aperture and the crack with 1 μm width. After the loading equipment is fixed on the high-precision micro-CT mechanical rotation platform, the sample is in the best scanning position through the movement in X, Y, and Z directions.

(2) Loading system: The loading system is a uniaxial loading equipment independently developed by the research group. The main structure includes an oil pump, motor, hydraulic cylinder, accumulator, tubing, displacement sensor, loading push rod, aluminum–magnesium alloy casing, spherical seat, and so on. The maximum output force of the cylinder of the loading device: Fmax = 40 KN, the cylinder stroke: S = 40 mm, the effective range of the displacement sensor: 40 mm, and the accuracy of the displacement sensor: 0.001 mm. After fixing it on the μCT225kVFCB CT mechanical turntable, the CT scanning of rock samples under the condition of real-time

loading and steady pressure can be realized, and the pressure control system of the loading system can realize digital pressure control, as shown in Figure 1.

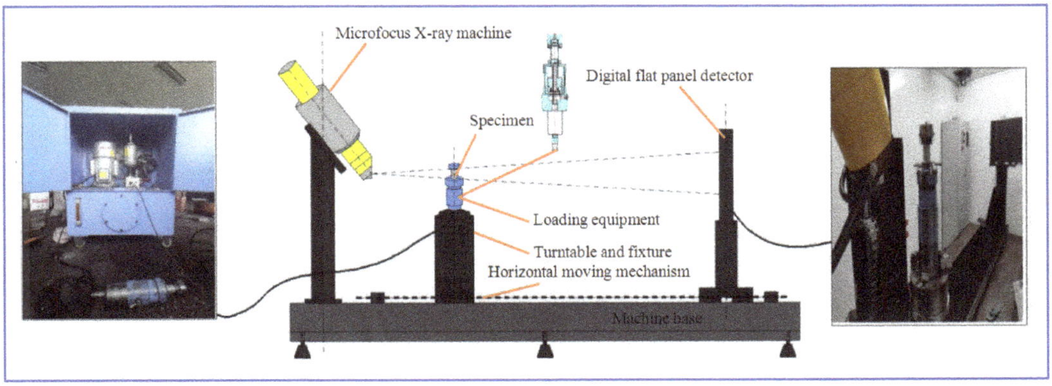

Figure 1. CT platform experimental system.

2.2. Experimental Scheme

The intact mudstone without obvious cracks was selected. Considering the resolution requirement of CT scanning, the samples were machined into φ 20 mm × 40 mm cylinders by stone processing machine, and 10 samples were made. Six of them were tested for mechanical properties, and the average peak load was 67.2 MPa, as shown in Figure 2.

Figure 2. Part of mudstone test piece.

Then, real-time pressurized CT scanning experiments were carried out on the remaining four samples. During the experiment, the sample was pressurized by operating the pressurization system outside the CT room. After the pressure was loaded to the expected pressure, the pressurization was stopped to stabilize the loading equipment, and then CT scanning was carried out. The voltage and current of CT scanning were 200 kV and 240 μA, respectively. The magnification of the rock scanning area was 12.8, that is, the minimum visible crack width was 15.4 μm. Through each scan, 1500 layers of two-dimensional CT images could be obtained, that is, the thickness of each layer was 27 μm.

3. Construction of the Digital Core Model

3.1. Composition Analysis of Mudstone

According to the results of the XRD experiments, the main components of the samples were quartz and clay minerals, accompanied by a small amount of pyrite and plagioclase. The content of quartz was 45.1%, and the content of illite was 43.74% (Figure 3, Table 1).

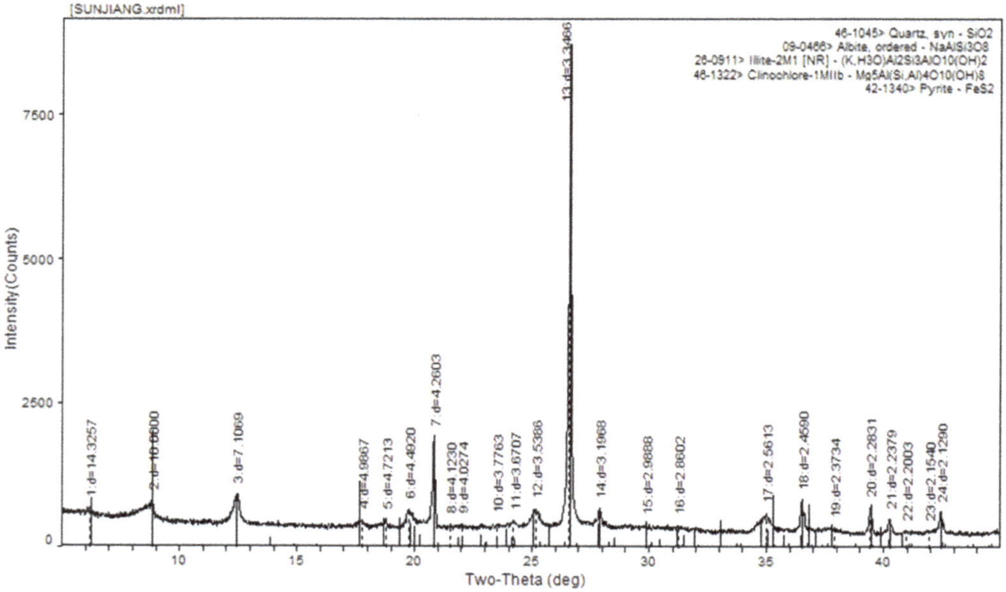

Figure 3. XRD test results of mudstone samples.

Table 1. Main chemical composition of mudstone (%).

Category	Quartz	Illite	Chlorite	Plagioclase	Pyrite	Others
Content (%)	45.1	43.74	4.86	4.6	0.2	1.5

According to the SEM experimental results, when the sample was magnified to 30 times (Figure 4b), the mudstone contained a large number of quartz particles, with clay minerals mostly filling in the large particles and being attached to the edge of the large particles; the surface of the mudstone was relatively flat, the overall structure was relatively dense, and a small number of cracks were developed near the large quartz particles. When the sample was enlarged to 1300 times (Figure 4c), there were a large number of flake aggregates and microcracks in the field of view. According to the results of XRD analysis, these flake materials were clay minerals such as illite.

According to the results of XRD and SEM analysis, the quartz content of the mudstone was high; the other materials were mainly clay minerals; and the accumulation mineral zone was quartz mineral zone, which was embedded in the clay mineral. The distribution of the two minerals in the whole mudstone sample was extremely uneven, and the quartz mineral zone in the upper right part of the sample was denser, with the distribution range being larger in the lower left part, but more discrete. To sum up, brittle minerals such as quartz are beneficial to the development of pores in rocks and the preservation of primary pores. There may be a large number of intergranular pores between brittle minerals and clay minerals. The initiation location of cracks is likely to develop from these pores and produce macroscopic cracks.

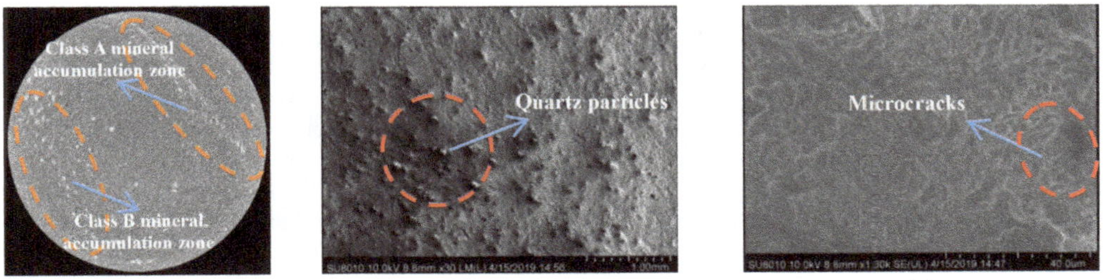

(a) (b) (c)

Figure 4. Microcracks and pores near mineral particles. (**a**) Mineral accumulation zone. (**b**) Quartz particles. (**c**) Microcracks near mineral particles.

3.2. Construction of Digital Core Model

The CT scanning images of different loading stages under uniaxial in situ loading were reconstructed. By extracting the information of porosity, mineral spatial distribution, and mineral content in different locations, a digital core model with mineral composition, mineral spatial distribution, and fracture information was established. According to the results of XRD and SEM, the extensive and easy-to-capture parts were extracted by the maximum inter-class method, and the part of microstructure information that was difficult to extract was based on the combination of watershed threshold segmentation and TopHat algorithm fine segmentation, so as to ensure that the feature elements of the scanned samples can be extracted completely and provide the basis for subsequent digital analysis and numerical calculation. A complete digital core model was obtained by adding mineral composition, rock skeleton, and pores. The construction process of digital core is shown in Figure 5.

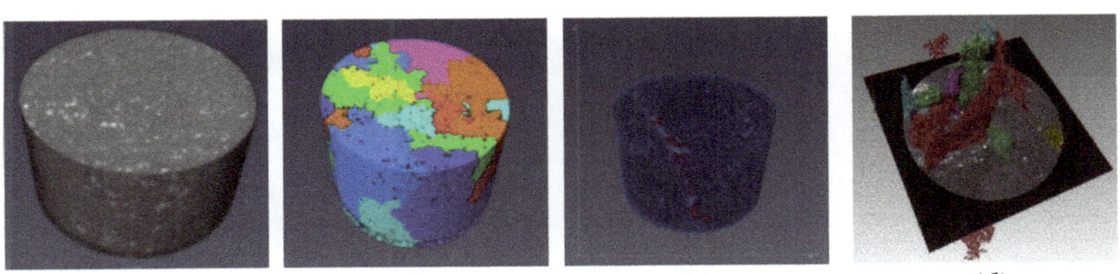

(a) (b) (c) (d)

Figure 5. Gray image segmentation process. (**a**) Primitive mudstone. (**b**) Sub-block statistical feature. (**c**) Calculation of connectivity of mineral particles. (**d**) Extraction of aggregate mineral zone.

4. Evolution Characteristics of Digital Core Fracture Model

According to the origin and distribution of pores, pores mainly include organic pores, inorganic pores, and primary microcracks. The minimum resolution of scanning in this paper was 15.4 μm/pixel; the mineral composition of the mudstone studied was relatively simple; and the distribution of cracks was mainly concentrated in the mineral zone, with emphasis on the intergranular pores and intragranular pores between mineral particles.

4.1. Distribution and Volume Change of Pores and Fractures

The change of internal pores in the process of rock loading can clearly reflect the time and space variation characteristics of rock instability and failure. Pore throat is a narrow channel in which pores are connected with each other in rock mass or soil, and pores are the space between solid mineral particles in rock and soil. In this paper, the porosity of CT data in the process of real-time in situ loading was analyzed, the pore connectivity in the process of rock failure was located and quantitatively counted, the pore evolution data obtained from experiments were analyzed, and the crack propagation trend and change characteristics were studied.

Figure 6 is a 3D distribution figure of pores and pore throats in mudstone at different loading stages. Through statistical analysis, we found 805 pores and 2 pore throats in mudstone in the initial stage, and the coordination number of pores was basically zero. No matter the spatial distribution and connectivity or the results of numerical statistics, the pore connectivity in the initial stage was poor. The initial pore volume was 0.804 mm^3 (the size of the pore ball is plotted according to the proportion of the equivalent diameter of the pore).

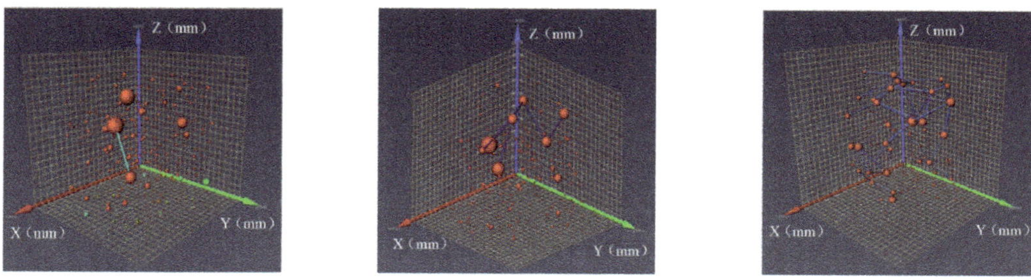

(a) (b) (c)

Figure 6. Pores and pore throats distribution of mudstone in the third loading stage. (**a**) First loading stage. (**b**) Second loading stage. (**c**) Third loading stage.

With the progress of loading, the number of pore throats in the 3D pore network model increased, and the number of pores in the first loading stage (18.5 MPa) decreased. At this stage, the mudstone was in the compaction stage; a small number of pores were closed and compacted under the action of external load; and a small number of pores were connected, which was characterized by the enhancement of pore connectivity, and the fracture volume was 1.035 mm^3.

When the loading continued, the pore connectivity in the second loading stage (37.5 MPa) was further enhanced, indicating that new pores were generated and the old pores were further developed and connected, and a macroscopic crack appeared on the surface of the sample. In the pore network model inside the sample, it can be seen that the pore network model was mainly developed into two clusters, and the positions of the two pore connecting clusters were very close to the distribution of the mineral zone. At this stage, the rock damage increased gradually, the number of pore development and penetration was also much higher than that in the previous loading stage, and the fracture volume also increased to 2.32 mm^3.

With the continuous development of pores and the accumulation of penetration, the local microcracks gradually expanded into macroscopic cracks and further penetrated, reaching the maximum load limit of mudstone, and the damage of rock reached the peak. In the third loading stage, the fracture volume reached 180.75 mm^3, and there was an obvious stress drop, indicating that in the previous stage, the pores in the mudstone had developed to a critical state of failure, which was extremely prone to instability failure, and the adjacent pores penetrated into cracks. The mudstone was completely destroyed, as shown in Figure 7.

Figure 7. Complete failure result of mudstone and 3D image of fractures.

Table 2 shows the parameters of fracture development in different loading stages of mudstone.

Table 2. Fracture development parameters of mudstone at different loading stages.

Loading Phase	Fracture Volume (mm^3)	Number of Pores	Number of Pore Throat	Number of Macro Cracks
Initial stage	0.804	805	2	0
The first stage	1.035	810	26	0
The second stage	2.32	1286	264	2
The third phase	180.75	-	-	5
The fourth stage	210.56	-	-	7

4.2. The Changing Law of Pore Coordination Number

The pore coordination number represents the pore throat number connected between pores and pores, and the pore coordination number can be directly related to the connectivity of the pore network model. The wide distribution of the coordination number can be understood as the rich development of micro-joints in this area. The coordination number is often used to evaluate the reservoir transport capacity of porous media.

The pore network model visually shows the change of the pore network at different loading stages (Figure 8). In the initial stage, most of the pores were isolated pores, which were dispersed in the rock. With the loading, the connected throat appeared between the adjacent pores and further expanded into a network. The pore-connected throat initially appeared near the mineral zone and gradually developed into a pore network in the two mineral accumulation areas. In the initial stage, the average coordination value of primary pores in mudstone was close to 0, indicating that the connectivity of primary pores was poor. At the first loading stage, the average coordination value of pores did not change much from that of the previous stage, and the pore compaction stage in the corresponding stress–strain curve was basically consistent with the pore change characteristics of this stage. At the second loading stage, the increase in the average coordination number of pores corresponded to the pore connectivity in the 3D pore network model. At the stage before the peak stress of loading failure, the average coordination number of pores increased rapidly, which may have been due to the fact that the pores in this stage had developed to the critical state of failure.

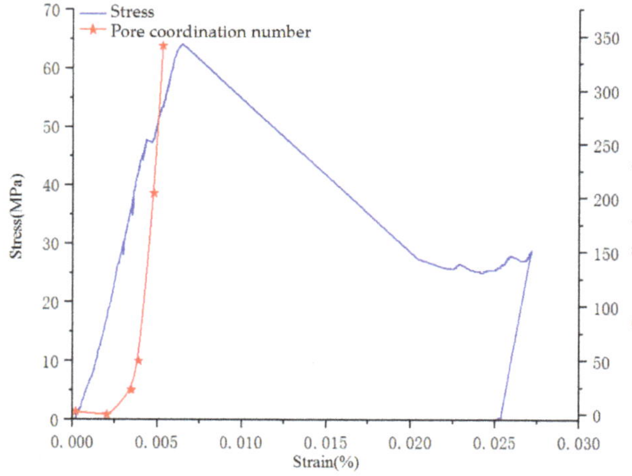

Figure 8. Change of pore coordination number during mudstone loading.

4.3. *Spatial Distribution of Minerals in Rock Failure Process*

Quartz is one of the main minerals of mudstone, and the existence of this mineral has a significant influence on the mechanical properties of mudstone.

$$D_{eq} = \sqrt[3]{\frac{6V_{3d}}{\pi}} \tag{1}$$

The calibration analysis algorithm can be used to calculate the particle volume distribution parameters in the extracted mineral particle model, and then the equivalent diameter distribution can be calculated by using the equivalent diameter Formula (1), and the diameter of each particle can be counted to establish the particle equivalent diameter distribution histogram, as shown in Figure 9 and Table 3. According to the results of calculation and statistics, the diameter of quartz mineral particles in the mudstone sample was mainly distributed in the range of 25–100 pixels, and the scanning resolution was 15.4 μm/pixel. According to the equivalent diameter formula, the diameter of quartz particles in mudstone samples was distributed between 0.38 and 1.54 mm, and the proportion of quartz particles with 0.38 mm diameter was the highest, accounting for 64.7% of all quartz particles.

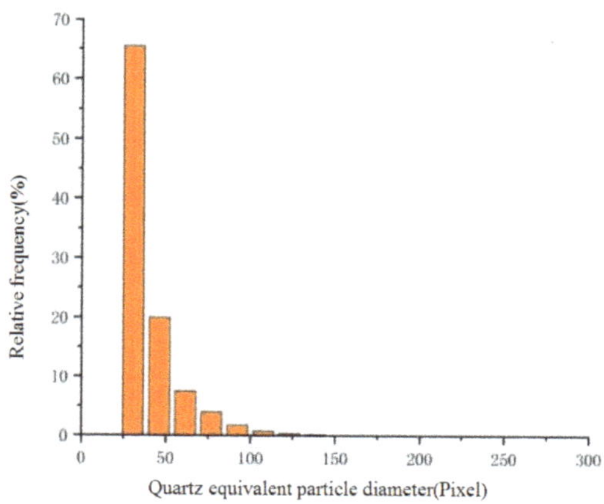

Figure 9. Histogram of equivalent diameter distribution of quartz particles.

Table 3. Particle equivalent diameter distribution parameter table.

Mean/Pixel	Standard Deviation/Pixel	Max/Pixel	Min/Pixel	Quantity
38.93	34.68	328.98	22.35	19,961

The particles with different equivalent diameters were characterized in space, and the particles in different positions were statistically analyzed, with the influence of the distribution of quartz particles with different particle sizes on the damage and its change law analyzed in depth.

Figures 10 and 11 show the spatial distribution of quartz particles in the initial unstressed state of mudstone specimens. A total of 19,962 quartz particles were collected by CT scanning. The equivalent diameter of mineral particles larger than 75 voxels was divided into large particles, and those with equivalent diameters less than 75 voxels were divided into small particles. Among them, there were 656 large particles, accounting for 3.28% of the total, and 19,302 small particles, accounting for 96.72% of the total. The aggregate quartz particles were 1487, accounting for 7.42% of the total. In order to show the spatial distribution of quartz mineral particles more simply and clearly with different particle sizes, as well as to further show the relationship between particles and fracture development, the display and observation were carried out from the perspective of 45° and overlooking, respectively. Each surface observed was consistent with the position of digital core modeling and CT scanning of mudstone sample, as shown in Figure 12.

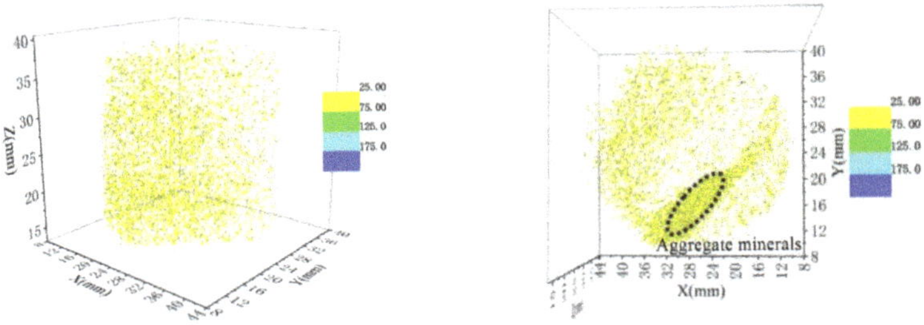

Figure 10. Spatial distribution of small size mineral particles.

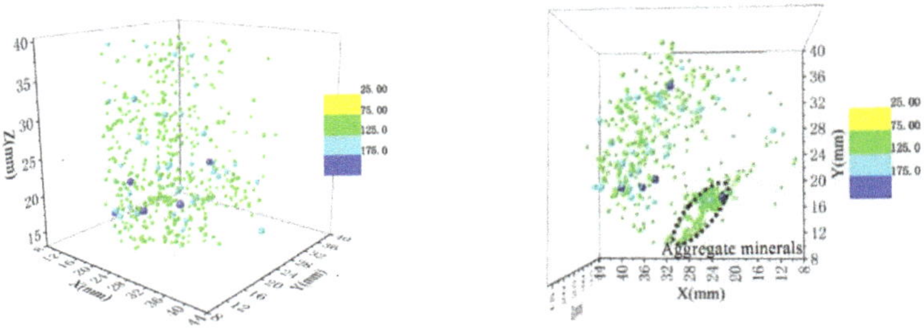

Figure 11. Spatial distribution of large-size mineral particles.

From the spatial distribution of small particles, the small quartz particles were basically uniformly distributed in the whole body of the mudstone sample; the number was relatively large; and the distribution was relatively concentrated in the XOY section, forming an accumulation zone that ran through the whole body of the sample, with the mineral zone on the other side being relatively scattered, and there being no obvious aggregation law. The occurrence relationship between them was essentially consistent with the macroscopic analysis of mineral composition, having a good corresponding relationship. By comparing the spatial distribution of quartz particles with different equivalent diameters, we found that there was no special aggregation of quartz particles in mudstone samples, and the spatial distribution rules of quartz particles of the two sizes were very similar. There was no obvious relationship between mineral accumulation zone and quartz particle size, so it is necessary to further analyze the quartz particles in the digital core in the loading stage.

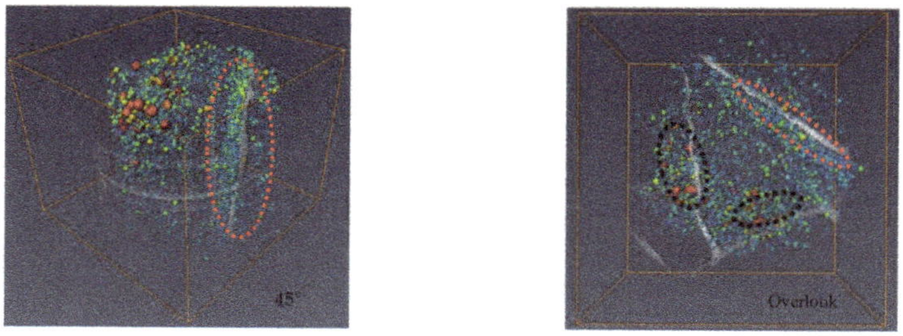

Figure 12. Relationship between final failure cracks and quartz particle distribution.

From the relationship between the distribution form of quartz particles and the distribution of cracks in the failure stage (Figure 12), macroscopic cracks were mainly distributed in the quartz mineral zone (Figure 4a). The quartz mineral zones of A and B aggregation types led to different initiation modes of cracks—the A mineral zone was denser, and the shape and location of cracks were consistent with these kind of mineral zones. The type B mineral zone was more discrete, but there were mineral clusters with large equivalent diameter, and the changes of crack shape and pore position must have been deformed by the influence of nearby space compression. Under the action of axial load, mudstone samples mainly showed along weak plane shear slip failure and along weak plane splitting failure, mainly because quartz mineral strength is larger; clay mineral strength is small; and under axial pressure, slip failure occurs between two kinds of minerals with different mechanical properties. On the other hand, the intergranular pores around the gathered quartz minerals continued to develop and expand under the action of external forces, thus forming macroscopic cracks.

5. Analysis of Mudstone Failure Process by Digital Volume Correlation Method

When using CT image to analyze damage, it will be affected by the scanning resolution of the CT system, and the scales studied by different analysis methods will have their own applicability. The digital core and pore network model is suitable for the analysis, which is larger than the resolution scale. The digital volume image correlation method can be used to analyze the local deformation and evolution process caused by the micro-crack, which is smaller than the CT resolution scale before the occurrence of the CT resolution scale crack. The result of the CT scan was taken as the reference image of the DVC computation, and the scan image of the later loading stage was taken as the computation image, and so on, until the final destruction of the image as the computation image and the computation phase being completed. The range of the DVC computational area is shown in Figure 13.

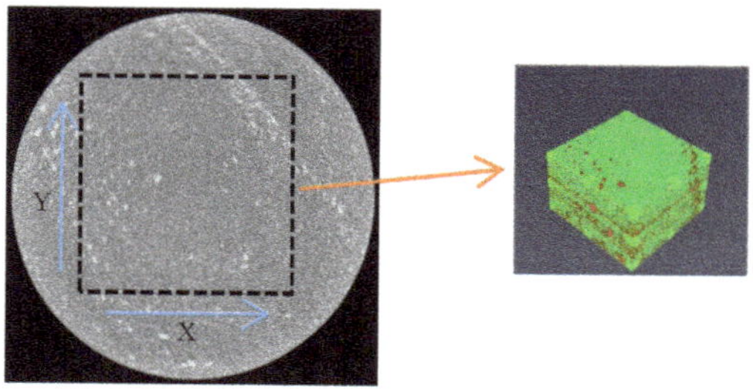

Figure 13. DVC computational area.

5.1. Analysis of the Evolution Process of Displacement Field

On the basis of the displacement distribution histogram calculated by DVC, the displacement field of mudstone sample under initial loading was analyzed. The X-axis displacement was uniformly distributed between −5 voxels and 5 voxels, and the frequency was concentrated at about 20. The displacement distribution of the Y-axis was close to that of the X-axis; there was only negative displacement in the Z-axis, which indicates that the mudstone sample was compressed and expanded in the X-axis direction. Moreover, a small abrupt increase in frequency appeared in the range of −5 voxels displacement, indicating that there were a large number of quartz mineral particles, micropores, and microcracks in the rock. At the initial stage of loading, microcracks were affected by external loads, and microcracks and micropores were gradually compressed and closed, which was the reason for voxel negative value in the numerical computation. On the other hand, a large number of quartz particles in the mudstone samples were hard and widely distributed in space. Under the influence of external load, the partially agglomerated quartz particles expanded and slid towards the X-axis and Y-axis after longitudinal extrusion, resulting in both negative and positive displacement in the X-axis and Y-axis. At the same time, in the state of uniaxial compression, at the initial stage of loading, the mudstone sample was mainly affected by the external load in the loading direction; the longitudinal microcracks were closed; and the hard mineral particles slipped in the plane, resulting in the expansion of the sample in the macroscopic range (Figure 14).

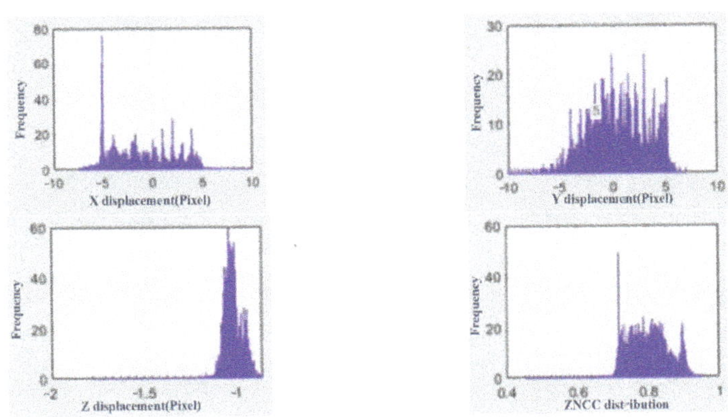

Figure 14. DVC computation of x, y, and z direction displacement and ZNCC distribution histogram.

According to the analysis of the evolution process of the displacement fields u, v, and w in the x, y, and z directions of mudstone (in which u = u(x, y, z), v = v(x, y, z), w = w(x, y, z)) (Figure 15), we can know that the sample is mainly subjected to axial vertical compression, and because there were a large number of mineral particles in the sample, the sample showed obvious heterogeneity with the overall deformation being extremely uneven. In the first loading stage, the internal pores of mudstone samples were mostly isolated pores, and the pore connectivity was poor. The second loading stage (37.5 MPa) was the starting point of the localized deformation of the sample. The displacement value of the sample near the mineral zone was larger, and the obvious local deformation had appeared inside the sample. With the increase in the load, the overall element displacement of the specimen was gradually increasing. There were few primary cracks and micropores in the rock, and the porosity of the rock sample was low. At this time, a macroscopic crack appeared on the surface of the sample, and the pore network model was mainly developed into two clusters, with the position of the two pore connecting clusters being very close to the distribution of the mineral zone. In this stage, the rock failure increased gradually, and the number of pores and penetration was much higher than that in the previous loading stage. The large displacement area corresponded to the mineral distribution area in the 3D computational area shown in Figure 13 and gradually expanded, because in the first and second loading stages, the mudstone was in the elastic stage, and the lithology of the mudstone was more compact and the pores were compacted. In the third stage of loading, when the stress value reached 48.5 MPa, the internal microcracks developed rapidly, showing obvious positive and negative displacement zones, indicating that some areas had obvious local deformation, and these areas expanded further with the loading. The rapid growth of the average coordination number of pores may have been due to the fact that the pores at this stage had developed to the critical state of failure. At the peak load 67.5 MPa, there was a large range of large displacement regions in the sample as a whole, and the large displacement regions changed strongly compared with the previous stage, being redistributed and concentrated. There was an obvious boundary between the large displacement region and the low displacement region, and there was an obvious delamination phenomenon in each displacement region of the sample. The regions with smaller displacement values essentially appeared at the junction of the mineral zone, which was consistent with the final failure form of the sample. At this time, the internal microcracks and pores of the sample expanded and penetrated, and the sample was at the critical point of instability and failure.

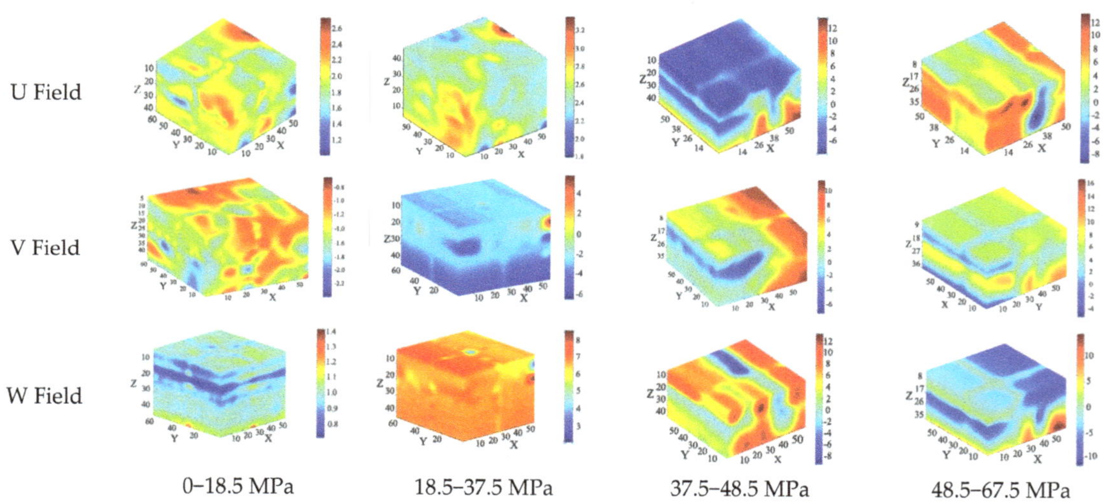

Figure 15. Evolution process of displacement field u, v, and w in x, y, and z directions.

5.2. Analysis on the Evolution Process of Strain Field

The strain field can better reflect the evolution process of local deformation in rock, six strain components of Cauchy strain tensor were solved by local least square fitting method, and the equivalent strain and volume strain under different loads can be calculated.

It can be seen from the equivalent strain nephogram (Figure 16) and the 3D volumetric strain nephogram (Figure 17) of the mudstone sample during full loading, at the initial stage of loading, the overall strain of the sample was small, and the high strain area was mainly concentrated in the XOZ plane and YOZ plane. The reduction to the real sample was in the quartz dense accumulation area of the mudstone sample, and the high strain area was scattered near the quartz particles. When loading to 18.5 MPa, there was a higher equivalent strain in the lower left corner of the sample, and the maximum equivalent effect of the sample was 0.01; according to the strain and the ultimate failure strain of mudstone, it can be inferred that there should be small cracks in this area of the sample at this time. With the progress of loading, the main strain region gradually shifted to the right side, showing the phenomenon of localized zone, which reflected the closure and change of primary pores, cracks, and other structures in the rock at the initial stage of loading. The equivalent strain in the mudstone went through a process of first dispersion and then concentration, which was because the microcrack near the mineral zone was closed at the initial stage of loading, resulting in a small range of strain concentration in the elastic stage, resulting in the rapid increase in the main strain of this part; with the increase in loading, the deformation of the sample tended to be gentle, and the equivalent strain entered a quiet period. When loading to 37.5 MPa, the sample began to show local deformation, and the maximum volume strain inside the specimen was 0.3, indicating that when the load is low, the volume near the mineral zone changes greatly, so it is easy to produce local deformation in the first step. When loading to 48.5 MPa, the microcracks and pores in the sample continued to develop and penetrate, forming new cracks and large-scale local deformation. The equivalent strain in the sample was rapidly redistributed with the new deformation region, forming an obvious boundary between the two mineral zones.

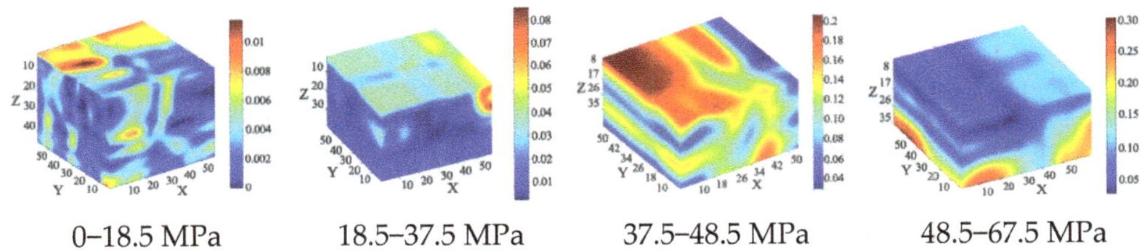

Figure 16. Evolution process of equivariant strain nephogram.

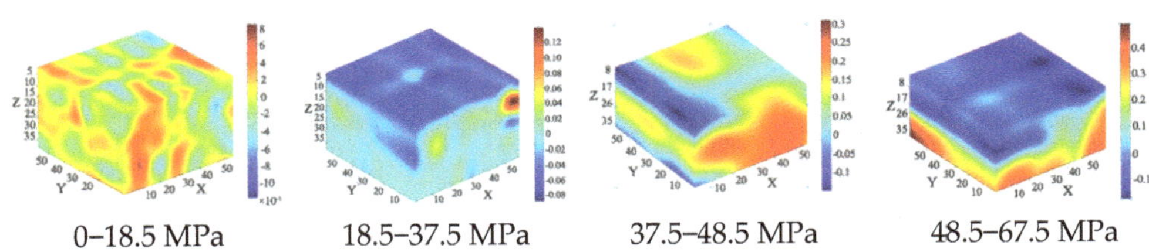

Figure 17. Evolution process of volume strain nephogram.

Figure 18 shows the variation curve of volumetric strain and equivalent strain with axial stress in the process of uniaxial loading of mudstone. From the curve in the graph, it can be seen that there was a certain correlation between them. According to the change of the curve, it can be divided into three stages:

(1) The stage of compaction (a–b). During the loading process (0–18.5 MPa), the volume strain of the sample was negative, and the equivalent effect increased slowly, indicating that the sample was affected by axial load and the sample was compressed. Combined with the equivalent strain distribution nephogram, the strain region was mainly distributed in the mineral zone, indicating that the intergranular pores near the mineral particles were squeezed, the pore connectivity increased, and the overall porosity showed an increasing trend. The number of pore throats between isolated pores increased, the coordination number of pores increased, and deformation occurred.

(2) The stage of microcrack development (b–c). During the loading process (18.5 MPa–48.5 MPa), the volumetric strain reached the minimum rapidly, and with the increase in load, the volumetric strain began to rebound and increased to 0. At this stage, the equivalent strain was redistributed from the local discrete strain region to form a large range of strain region, and the high strain region was gradually connected, maintaining a rapid growth trend, while the porosity continued to increase, and the connectivity of the pore network increased. The number of connected pore throats between pores increased, and the coordination number of pores increased. It showed

that the sample began to expand, and a large number of micro-hole cracks began to develop.

(3) The stage of microcrack propagation (c–d). During the loading process (48.5 MPa–67.5 MPa), the change rate of volumetric strain and equivalent strain began to slow down, and the equivalent strain region was connected, with there being a clear boundary between the high strain region and the low strain region, indicating that cracks that were higher than the identification range of CT resolution were generated, and that the micro hole crack had expanded into a macroscopic crack. The stress corresponding to point c can be used as the critical point of the macroscopic crack, which was 62.9% of the peak load. The increasing trend of porosity, pore connectivity, and pore coordination number were essentially the same as that of the previous stage, and there was a crack on the surface of the sample, with a trend of continuous growth.

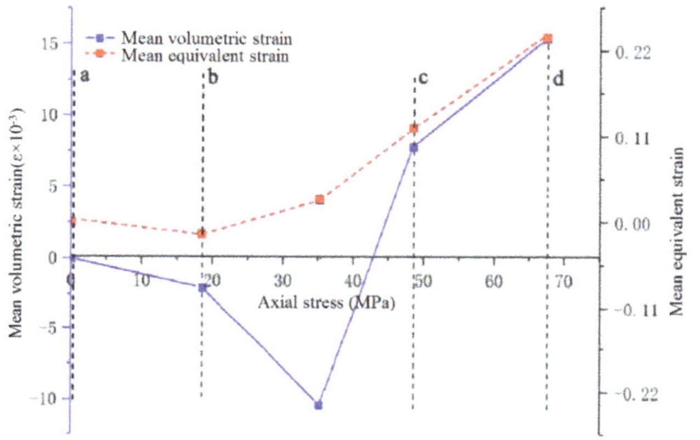

Figure 18. Variation graph of volume strain mean and equivalent strain mean. (**a**) 0MPa. (**b**) 18.5MPa. (**c**) 48.5MPa. (**d**) 67.5MPa.

6. Conclusions

In this paper, mudstone was taken as the research object, and the failure mechanism of mudstone was studied according to rock microscopic physical property analysis, uniaxial in situ CT scanning, digital core reconstruction, pore network extraction, mineral composition, and pore structure. The following conclusions were drawn:

(1) The diagenetic mineral composition of mudstone was relatively simple, being mainly composed of brittle materials such as quartz and clay minerals, and the proportion of brittle minerals was relatively high, which was the main reason for the high strength of mudstone. Moreover, through the scanning electron microscope, it was found that the brittle mineral quartz and so on often exist in the form of agglomerated strips in the mudstone, resulting in the formation of structural weak surfaces on different mineral interfaces. These results are similar to those reported by Zhao et al. (2013).

(2) The digital core models of mudstone at different loading stages were established by using a uniaxial loading in situ CT scanning test system. There were a large number of primary intergranular pores near the mineral accumulation zone, and the connectivity of pores will gradually increase with the increase in axial load. Macroscopically, the porosity of the rock increased, and cracks gradually appeared, with the internal pores being abnormally developed. When the rock reached the critical state of failure, the throat number and coordination number of pores clearly increased.

(3) There was no obvious difference in the distribution of mineral particles of different sizes, and the interior of mudstone was mainly affected by a large number of small size mineral particles, indicating that the slip between mineral zones was mainly dominated by small-size minerals.

(4) The final propagation shape of the crack was very consistent with that of the mineral zone, indicating that it was easy to form a weak surface in the aggregated mineral zone, and the aggregated mineral zone under loading could easily produce local deformation, which is related to the mechanical properties of the mineral zone and its surrounding rock matrix. When rock breaks, it is easy to produce along the junction of the two kinds of minerals. The results are in good agreement with similar study by Yin et al. (2019).

(5) The displacement in the polymeric mineral zone was small, the deformation displacement of the rock skeleton dominated by clay minerals near the quartz mineral zone was larger, and the stronger quartz minerals restrained the rock skeleton deformation in the region.

Author Contributions: Methodology, D.D.; validation, D.D., X.C. and X.W.; formal analysis, X.W.; investigation, X.C.; resources, J.S.; data curation, D.D.; writing—original draft preparation, D.D. and X.C.; writing—review and editing, X.F.; visualization, X.C.; supervision, X.F.; project administration, D.D.; funding acquisition, D.D. All authors have read and agreed to the published version of the manuscript.

Funding: This research was funded by the Chinese National Natural Science Foundation (no. 51304143) and the Chinese National Natural Science Foundation (no. 51604184).

Informed Consent Statement: Informed consent was obtained from all subjects involved in the study.

Conflicts of Interest: The authors declare no conflict of interest.

References

1. Dong, H.M.; Sun, J.M.; Lin, Z.Z.; Chui, L.K.; Yan, W.C. Quantitative characterization and characteristics analysis of microscopic pore structure in natural gas hydrate based on CT scanning. *J. China Univ. Pet. Ed. Nat. Sci.* **2018**, *42*, 45–54.
2. Ehab, M.K.K.; Gerard, P.C.J.B.; Massart, T.J. Modelling stress-induced permeability alterations in sandstones using CT scan-based representations of the pore space morphology. *Int. J. Rock Mech. Min.* **2022**, *150*, 104998. [CrossRef]
3. Zhang, Z.H.; WEI, W.; Zhan, J.; Jia, H.B. Determining method of multiscale fractal dimension of red bed sandstone pores based on CT scanning. *Bull. Geol. Sci. Technol.* **2022**, *41*, 254–263.
4. Yin, Z.H.; Xu, Q.J. Microscopic damage detection of sandstone using digital core technology. *J. Hydr. Eng.* **2021**, *40*, 169–178.
5. Lin, C.Y.; Wang, Y.; Yang, S.; Ren, L.H.; You, C.M.; Wu, S.T.; Wu, Y.Q.; Zhang, Y.M. 3D Modeling of digital core based on X-ray computed tomography. *J. Jilin Univ. Earth. Sci. Ed.* **2018**, *48*, 307–317.
6. Zhang, H.M.; Mu, N.N. Study on Meso-Damage of Freeze-Thaw rocks based on 3D reconstruction. *Mech. Eng.* **2021**, *43*, 687–694.
7. Yang, Z.F.; Zeng, J.H.; Feng, X.; Feng, S.; Zhang, Z.D.; Qiao, J.C. Small-scaled heterogeneity of tight sandstone reservoirs and oil accumulation characteristics. *J. China. U. Min. Techno.* **2016**, *45*, 119–127.
8. Voutilainen, M.; Miettinen, A.; Sardini, P.; Parkkonen, J.; Sammaljarvi, J.; Gylling, B.; Selroos, J.O.; Yli-Kaila, M.; Koskinen, L.; Siitari-Kauppi, M. Characterization of spatial porosity and mineral distribution of crystalline rock using X-ray micro computed tomography. C-14-PMMA autoradiography and scanning electron microscopy. *Appl. Geochem.* **2019**, *101*, 50–61. [CrossRef]
9. Wang, D.K.; Zhang, P.; Wei, J.P.; Wu, Y.; Zeng, F.C. Dynamic evolution characteristics of fractures in gas-bearing coal under the influence of gas pressure using industrial CT scanning technology. *J. China. Coal. Soc.* **2021**, *46*, 3550–3564. [CrossRef]
10. Larmagnat, S.; Des Roches, M.; Daigle, L.F.; Francus, P.; Lavoie, D.; Raymond, J.; Malo, M.; Aubiès-Trouilh, A. Continuous porosity characterization: Metric-scale intervals in heterogeneous sedimentary rocks using medical CT-scanner. *Mar. Pet. Geol.* **2019**, *109*, 361–380. [CrossRef]
11. Farokhian, D.; Azin, R.; Ranjbar, A. Application of medical and dental CT-Scan technologies for determining porosity distribution of the Persian Gulf coastal zone and Zagros basin core samples. *J. Afr. Earth. Sci.* **2019**, *150*, 96–106. [CrossRef]
12. Mao, W.Z.; Lv, Q.; Zheng, J.; Peng, Y.; Liu, J. Analysis of mineral composition and Meso-Structure of ruanite using ctimages. *J. Eng. Geol.* **2022**, *30*, 216–222.
13. Cao, A.; Long, W.; Du, H.B.; Chen, F.F.; Zhao, H.; Wang, G.Q.; Yuan, P. Study on the influence of asphaltene recognition on rock physical parameters based on micro CT technology. *Contemp. Chem. Ind.* **2021**, *50*, 2869–2872, 2876.
14. Kumari, W.G.P.; Ranjtth, P.G.; Perera, M.S.A.; Li, X.; Li, L.H.; Chen, B.K.; Isaka, B.L.A.; De Silva, V.R.S. Hydraulic fracturing under high temperature and pressure conditions with micro CT applications: Geothermal energy from hot dry rocks. *Fuel* **2018**, *230*, 138–154. [CrossRef]
15. Yang, Y.M.; Ju, Y.; Liu, H.B.; Wang, H.J. Infilence of porous structrue properties on mechanical performances of rock. *Chin. J. Rock. Mech. Eng.* **2009**, *28*, 2031–2038.
16. Zhao, B.; Wang, Z.Y.; Wu, J.P. Relation between mineralogical composition and microstructure to the mechanical properties of rock materials. *Coal. Geol. Explor.* **2013**, *41*, 59–63, 67.

17. Yin, X.M.; Yan, E.C.; Wang, L.N.; Chen, L. Quantitative microstructure information extraction and microscopic morphology analysis of anisotropic schist. *Rock. Soil. Mech.* **2019**, *40*, 2617–2627, 2729.
18. Li, J.; Liu, C.; Liu, H.M.; Wang, J.D.; Zeng, Z.P.; Xie, Y.T. Research on the Mesoscopic Damage Mechanism of Shale Reservoir Based on Digital Core. *Chin. J. Rock. Mech. Eng.* **2022**, *41*, 1103–1113.
19. Wu, N.; Liang, Z.Z.; Zhang, Z.H.; Li, S.H.; Lang, Y.X. Development and verification of three-dimensional equivalent discrete fracture network modelling based on the finite element method. *Eng. Geol.* **2022**, *306*, 106759. [CrossRef]
20. Guo, W.N.; Bao, J.W.; Zhang, P.; Sun, Y.Q.; Ma, Y.X.; Zhao, T.J. Mechanical Properties and Deformation Characteristics of Hybrid Fiber Strain Hardening Cementitious Composites Based on Digital Image Method. *J. Chin. Ceram. Soc.* **2022**, *50*, 1401–1409.
21. Bay, B.K.; Smith, T.S.; Fyhrie, D.P. Saad M Digital volume correlation: Three-dimensional strain mapping using X-ray tomography. *Exp. Mech.* **1999**, *39*, 217–226. [CrossRef]
22. Mao, L.T.; Wei, K.; Zhu, Z.Y.; Liu, H.Z. Measurement and analysis of 3D deformation in glass fiber reinforced composites by using computed tomography. *J.Reinf. Plast. Compos.* **2018**, *9*, 5–10.

applied sciences

MDPI

Article

Numerical Simulation Study of Brittle Rock Materials from Micro to Macro Scales Using Digital Image Processing and Parallel Computing

Xin Liu [1], Zhengzhao Liang [1], Siwei Meng [2,*], Chunan Tang [1] and Jiaping Tao [2]

[1] State Key Laboratory of Coastal and Offshore Engineering, Dalian University of Technology, Dalian 116024, China; liuxin_net@163.com (X.L.); LiangZZ@dlut.edu.cn (Z.L.); tca@mail.neu.edu.cn (C.T.)

[2] Research Institute of Petroleum Exploration Development, China National Petroleum Corporation, Beijing 100083, China; taojiaping93@sina.com

* Correspondence: mengsw@petrochina.com.cn; Tel.: +86-15901478803

Abstract: The multi-scale, high-resolution and accurate structural modeling of rocks is a powerful means to reveal the complex failure mechanisms of rocks and evaluate rock engineering safety. Due to the non-uniformity and opacity of rocks, describing their internal microstructure, mesostructure and macro joints accurately, and how to model their progressive fracture process, is a significant challenge. This paper aims to build a numerical method that can take into account real spatial structures of rocks and be applied to the study of crack propagation and failure in different scales of rocks. By combining the failure process analysis (RFPA) simulator with digital image processing technology, large-scale finite element models of multi-scale rocks, considering microstructure, mesostructure, and macro joints, were created to study mechanical and fracture behaviors on a cloud computing platform. The Windows-Linux interactive method was used for digital image processing and parallel computing. The simulation results show that the combination of a parallel RFPA solver and digital image modeling technology can achieve high-resolution structural modeling and high-efficiency calculation. In microscopic simulations, the influence of shale fractures and mineral spatial distribution on the fracture formation process can be revealed. In the mesostructure simulation, it can be seen that the spatial distribution of minerals has an impact on the splitting mode of the Brazilian splitting model. In the simulation of a joined rock mass, the progressive failure process can be effectively simulated. According to the results, it seems that the finite element parallel computing simulation method based on digital images can simulate the multi-scale failure process of brittle materials from micro to macro scales. Primarily, efficient parallel computing based on a cloud platform allows for the multi-scale, high-resolution and realistic modeling and analysis of rock materials.

Keywords: multi-scale; digital images; parallel computation; 3D numerical modeling; mechanical and fracture properties

Citation: Liu, X.; Liang, Z.; Meng, S.; Tang, C.; Tao, J. Numerical Simulation Study of Brittle Rock Materials from Micro to Macro Scales Using Digital Image Processing and Parallel Computing. *Appl. Sci.* **2022**, *12*, 3864. https://doi.org/10.3390/app12083864

Academic Editor: Arcady Dyskin

Received: 29 December 2021
Accepted: 7 April 2022
Published: 11 April 2022

1. Introduction

The mechanical properties and structural safety of rocks are relevant to many fields, including mining engineering, civil engineering and petroleum engineering. Many accidents occur due to the strength of rocks exceeding their ultimate bearing capacity, and rocks' instantaneous failure often leads to catastrophic consequences. Due to the heterogeneity and opacity of brittle materials, it is difficult to accurately visualize their internal structure and characterize the fracture process [1–3].

Traditional fracture mechanism research mainly focuses on macro-mechanical tests, such as the uniaxial/triaxial compression test [4–8] and indoor fracturing model experiments [9–12]. However, in these studies, it is impossible to predict or determine which components of the sample will interact with the cracks before testing. Numerical methods have been widely used [13,14] to solve the above problem. However, in most cases, to simplify the numerical calculation, the inhomogeneity of the material is ignored, which will cause a deviation between the simulated crack morphology and the actual crack morphology [15]. Therefore, to consider the influence of heterogeneity in the model, many scholars have adopted different statistical methods to incorporate heterogeneity into the numerical model, and to assign different mechanical properties to the units in the numerical model [16–19]. For example, heterogeneity was introduced into materialhrough the Weibull statistical model [14].

The above methods are used to characterize the heterogeneity of rock materials; they are random-level models that cannot entirely reflect the actual geometric changes taking place in the microstructures of the materials [20]. Additionally, few documents describe how the virtual structure generated by this method reflects the natural structure of the rocks [21]. The internal microstructure and composition determine the macroscopic mechanical properties of rocks, and the contours of different components determine their macroscopic fracture mechanisms [22]. Therefore, the actual internal structure of rocks should be considered as much as possible in mechanical analysis and fracture research [23].

Reconstructing digital models based on digital images has been applied in fields such as petroleum engineering [15,24]. Previous studies have shown that digital image data, such as mineral and fracture data, can be used to characterize the heterogeneity of materials quantitatively [25]. Digital image processing (DIP) technology can accurately introduce the heterogeneity of materials into numerical models, and depict the physical qualities and microstructures of actual materials more accurately [26–28]. These investigations have shown that models that take into account real microscopic heterogeneity may better predict materials' mechanical behavior and fracture reactions. However, these studies are limited to the establishment and research of two-dimensional (2D) numerical models.

The accuracy and efficiency of numerical simulation research depend on the computing performance of numerical codes. This method has reduced advantages in three-dimensional (3D) numerical research due to the limitations of computational performance. The coarse mesh of the 3D dimensional model fails to characterize the actual structure of rocks, so this method is not widely used in mechanical and fracture behaviors research. For instance, Yu et al. [28] reconstructed a 3D finite element (FE) numerical model of a jointed rock mass based on X-ray microcomputed tomography (micro-CT), and studied its mechanical behavior. The pixel size of the micro-CT image is 30 μm, and the element size of the reconstructed finite element numerical model is 2/3 mm × 2/3 mm × 2/3 mm. The element size of the reconstructed finite element model is about 22 times larger than the pixel size of the original image, which affects the accuracy of the reconstructed finite element model. Ref. [2] used the same method to reconstruct the numerical model of volcanic rock, and studied the failure behavior of volcanic rock through uniaxial compression and Brazilian splitting numerical experiments. The numerical model replicated the pore microstructure of natural volcanic rock. Some mechanical parameters, such as the strength and elastic modulus of the constituent minerals, are not reflected in the numerical simulation because of the limitations of calculation scale and model accuracy. Therefore, it is necessary to develop a more effective method to eliminate these problems, and to improve the 3D numerical model's calculation scale and efficiency.

In this paper, multi-scale three-dimensional (3D) finite element numerical models characterizing the actual structures of rocks were reconstructed based on digital image processing techniques and high-performance computing methods to study their mechanical behavior and failure processes. First, different scanning techniques were used to obtain digital images characterizing the actual structures of rocks at different scales. Microscale digital images were obtained from FIB-SEM (Focused Ion Beam Scanning Electron Microscopes), mesoscale digital images were obtained via the micro-CT scanning technique, and macroscale digital images were obtained from a video camera. A numerical code Parallel Rock Failure Process Analysis (RFPA3D) simulator was used to reconstruct multiple-resolution 3D numerical models considering natural structure, such as microscopic shale models, mesoscale models, and CJRM (Columnar jointed rock mass) models, using the digital images. Finally, the numerical models were simulated and analyzed for their failure processes under external loadings.

2. Methodology

2.1. Brief Description of RFPA3D

Through continuum mechanics, RFPA3D is programmed to simulate discontinuous mechanical problems by reducing parameters after element failure. This code reproduces nonlinear deformation characteristics by introducing the heterogeneity of material properties into the model and assuming that the mechanical parameters conform to the distribution statistics (Weibull distribution function, distribution function, or normal distribution function). This paper assumes that the strength and elastic modulus follow the Weibull distribution function [14,19,29,30], as:

$$P(u) = \frac{m}{u_0}\left(\frac{u}{u_0}\right)^{m-1} \exp\left(\frac{u}{u_0}\right)^{m}, \tag{1}$$

where u_0 is the mean value of the elements (such as strength or elastic modulus) put into the numerical simulation program, and u is the mechanical parameter of the element. In addition, the parameter m describes the shape of the distribution function and reflects the degree of homogeneity of the material; it is called the homogeneity index.

It is initially linearly elastic until the specified damage threshold is reached. This study has calculated the damage thresholds using the maximum tensile strain criterion and the Mohr–Coulomb criterion. As expressed in Equation (2), the maximum tensile strain criterion primarily determines whether a crack has been initiated. The second damage threshold, defined in Equation (3), is the Mohr–Coulomb criterion, which determines whether an element is damaged in the shear mode in the absence of tensile damage.

The maximum tensile strain criterion is:

$$\varepsilon_3 \leq \frac{\sigma_{t0}}{E}, \tag{2}$$

The Mohr–Coulomb criterion is:

$$F = \sigma_1 - \sigma_3 \frac{1+\sin\phi}{1-\sin\phi} \geq \sigma_{c0}, \tag{3}$$

Until the element stress satisfies the above criteria, the elastic modulus remains constant and the same as before loading. According to the strain equivalence assumption, the elastic modulus of an element degrades gradually with damage progress, as follows:

$$E = (1-D)E_0, \tag{4}$$

It is characterized by an elastic–brittle damage constitutive relation, with a given residual strength (illustrated in Figure 1). As shown in the negative half of the axis of Figure 1, there are three kinds of elements determined by the damage variable, namely, elastic elements ($D = 0$), damaged elements ($0 < D < 1$), and complete failure elements ($D = 1$). Under multiaxial stress states, we can express the parameter D of an element as [31–33]:

$$D = \begin{cases} 0 & \bar{\varepsilon} > \varepsilon_{t0} \\ 1 - \frac{\sigma_{rt}}{\bar{\varepsilon}E_0} & \varepsilon_{tu} < \bar{\varepsilon} \leq \varepsilon_{t0} \\ 1 & \bar{\varepsilon} \leq \varepsilon_{tu} \end{cases}, \tag{5}$$

$$\varepsilon_{tu} = \eta\varepsilon_{t0}, \tag{6}$$

where $\bar{\varepsilon}$ is the equivalent principal strain. It is defined as follows:

$$\bar{\varepsilon} = -\sqrt{\langle -\varepsilon_1\rangle^2 + \langle -\varepsilon_2\rangle^2 + \langle -\varepsilon_3\rangle^2}, \tag{7}$$

where ε_1, ε_2, and ε_3 are the three principal strains and $\langle\rangle$ is a function defined as follows:

$$\langle x \rangle = \begin{cases} x & x \geq 0 \\ 0 & x < 0, \end{cases} \tag{8}$$

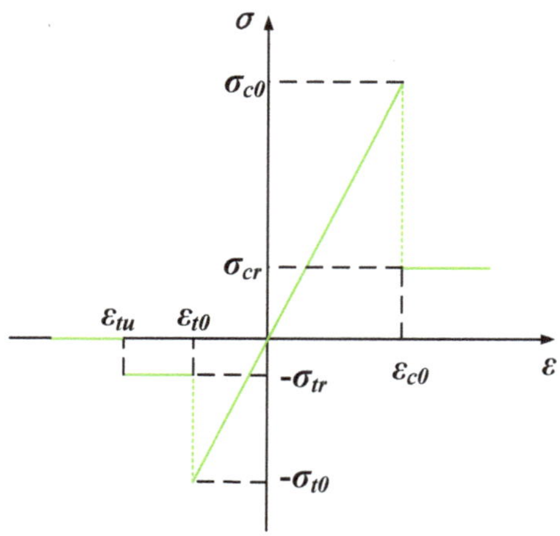

Figure 1. Elastic–brittle damage constitutive law of elements.

As shown in the positive half of the axis in Figure 1, for a shear mode element under multiaxial stress states, the damage variable D can be described as follows [31–33]:

$$D = \begin{cases} 0 & \varepsilon_1 < \varepsilon_{c0} \\ 1 - \frac{\sigma_{rc}}{\varepsilon_1 E_0} & \varepsilon_{c0} \leq \varepsilon_1, \end{cases} \tag{9}$$

According to the constitutive law of elastic damage, the element gradually deteriorates. It is important to emphasize that the finite element analysis will fail if the modulus equals zero. Therefore, the limit of the elastic modulus is specified as a relatively small number, i.e., 10^{-5}.

2.2. Parallelization Strategy

The RFPA3D system consists of pre-processing, elastic stress analysis, failure analysis, and post-processing. The RFPA3D procedures are completed with coordinated processing on an interactive Windows-Linux platform. The linear elastic stress analysis module is implemented on the Linux operating system cluster, and the other modules are completed on the Windows platform. After the data of the Windows platform are prepared, they are completely transferred to the Linux cluster, and the finite element parallel computing program on the Linux cluster is automatically started to complete the parallel computing. After the parallel calculation, it sends the results back to windows to view the results in real-time. A Hybrid OpenMP/MPI (message passing interface) parallel programming model is used to solve the finite element problem [34]. First, the whole domain is separated into local data sets scattered across the domain. A processing node is given to each partition, and MPI is used to communicate between processors and interconnect computing nodes between distributed memories. It is the most time-consuming process in the entire analysis process. It is also the core of the numerical realization, because each loading step requires a stiffness calculation and the resolving of the system equation. Parallelism and coarse-grained domain decomposition can improve the computational scale and speed of the master–slave solution.

This parallel computing structure uses a master–slave programming model, dividing the computational process into several slave processes (see Figure 2). It is a control program that does not participate in calculations, but is responsible for obtaining global machine numbers, initializing data, and processing data before receiving it. The data include unit information, node information, and material properties. According to the domain decomposition algorithm (See Figure 3, where the grid nodes in the I partition are divided into two types: internal nodes and boundary nodes. When solving the physical quantity of the partition boundary node, it is also necessary to exchange information with the adjacent area of the I partition to obtain the information of the node connected to the boundary of the I partition, that is, the information of the external node. The internal nodes and boundary nodes are the nodes to be solved in the I partition. When solving the boundary nodes, the external nodes are auxiliary nodes that provide boundary information), the program automatically decomposes the entire solution region into relatively independent sub-regions, and distributes the sub-region data into the corresponding processors. When the calculation of each slave process is completed, the master process retrieves the results of each sub-region from each slave process. It outputs the results of the entire solution region. Each slave process is a calculation program responsible for completing subtask calculations. Each slave process receives the master process data and individually completes the unit stiffness matrix, the total stiffness matrix synthesis, and the load vector assembly of each sub-region. The sub-region system equations are solved iteratively using the Krylov subspace preconditioner biconjugate gradient stabilized method. It is necessary to exchange data with adjacent sub-regions to update the data during solving. The sub-area solution is completed via the slave process, and the physical quantities of the stress and strain waiting to be solved are output into the master process.

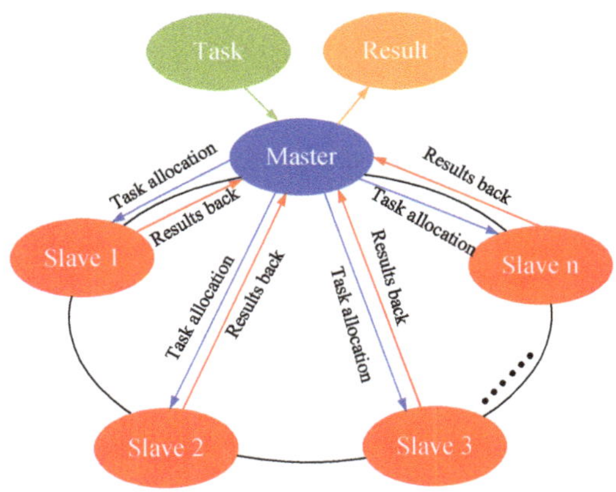

Figure 2. Flowchart of the parallel FEM solver.

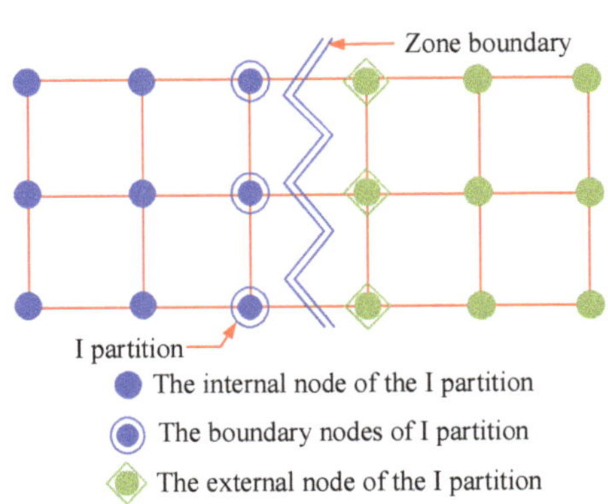

Figure 3. Example of the partitioning of a finite element grid.

The calculations in this study are completed on the Int Cloud computing platform, which integrates the resources of nearly 20 mainstream cloud service providers and supercomputing centers in China as the infrastructure, which can provide parallel computing for 10,000-core clusters. At present, a single cloud workstation node can provide more than 100 core Xeon platinum processors (Intel Xeon platinum 8260 series, main frequency 3.1 Ghz, L3 cache 36 MB), 8 Tesla V100 computing cards, 40,960 core CUDA (Compute Unified Device Architecture) computing cores, 736 GB memory, 256 GB video memory, 10 Gbps internal network bandwidth and 100 Mbps public network bandwidth for each user. The current comprehensive theoretical calculation peak of the platform exceeds 500 TFlps, and the storage capacity exceeds 100 PB. The underlying server virtualization system is the KVM (Keyboard Video Mouse) architecture. The platform uses the CentOS (Community Enterprise Operating System) system. The user cloud operating system is Windows Server 2016 Datacenter version.

2.3. Three-Dimensional Numerical Model Reconstruction by Digital Image Processing

Using the DIP technology, the actual microstructure of heterogeneous rocks could be precisely extracted from images and incorporated into numerical models. Pixels make up a digital image on a computer. Each pixel consists of the intersection area on either side of horizontal and vertical scan lines of equal width, resulting in a square element with a width of w. Each pixel is represented by three primary colors, namely, red (R), green (G), and blue (B). The reconstruction of 3D numerical samples was divided into four steps based on the digital image. As a first step, we collected high-resolution pictures of the research object, which provided information about its microstructure—numerical models of different scales with different image sources. The shale images at the microscale in Section 3.1 were obtained using Focused Ion Beam Scanning Electron Microscopes technology (Figure 4a). Mesoscale shale images were obtained using the micro-CT scanning technique in Section 3.2. The columnar jointed rock mass image of Section 3.3 was obtained using a video camera [35]. Second, the image vectorization processing used DIP technology to convert the image into vectorized data that could be recognized by numeric codes. In order to construct a 3D model, the extracted digital microstructures were stacked layer by layer in a particular sequence and at specific intervals based on variation within a certain thickness t (Figure 4c,d). The 3D FE meshes were simultaneously established (Figure 4c). A pixel's size determines the dimensions of the elements, which were all hexahedral quads. The thickness of each element was equal to the distance between layers. That is, the length and width of each element mesh are w, and the height is h. Third, we determined the segmentation threshold to distinguish and segment the constructed 3D microstructure of the finite element model, with microcracks, cracks, voids, and mineral particles. Finally, the DIP method was applied to assign the corresponding mechanical properties to the elements of the components (Figure 4d). In the following section, we describe the third step in detail.

In order to accurately characterize the microscopic and macroscopic structures of the models, the image processing methods and segmentation thresholding methods were slightly different for each model. The following briefly introduces the method of determining the segmentation threshold, taking the microscale shale in Section 3.1 as an example. The gray histogram threshold method was adopted to segment microstructural segmentation [2] from the solid matrix to determine the segmentation threshold. Figure 4b shows a microscopic image of the shale and the corresponding grayscale histograms of the I values. I represents the average of R, G, and B, while the *X*-axis represents the level of grey and the *Y*-axis represents the pixel count. The gray level is similar in the same material, but there are significant differences among different materials. The peaks reflect this difference in the histogram. Therefore, the different rock components can be classified by identifying the grayscale at the start and end of each peak. That is, the gray value at the location of each trough is the segmentation threshold. After this, we artificially adjusted the repeated segmentation threshold until the derived model could accurately reflect the reality of the microstructural and mineral distribution inside the shale. The final segmentation thresholds of this model are 25, 75, and 225. Here, the I value in the range of 0–25 corresponds to fractures, the I value in the range of 25–75 corresponds to organic matters, the I value in the range of 75–225 corresponds to the carbonate mineral matrix dominated by calcite, and the I value in the range of 225–255 corresponds to pyrite.

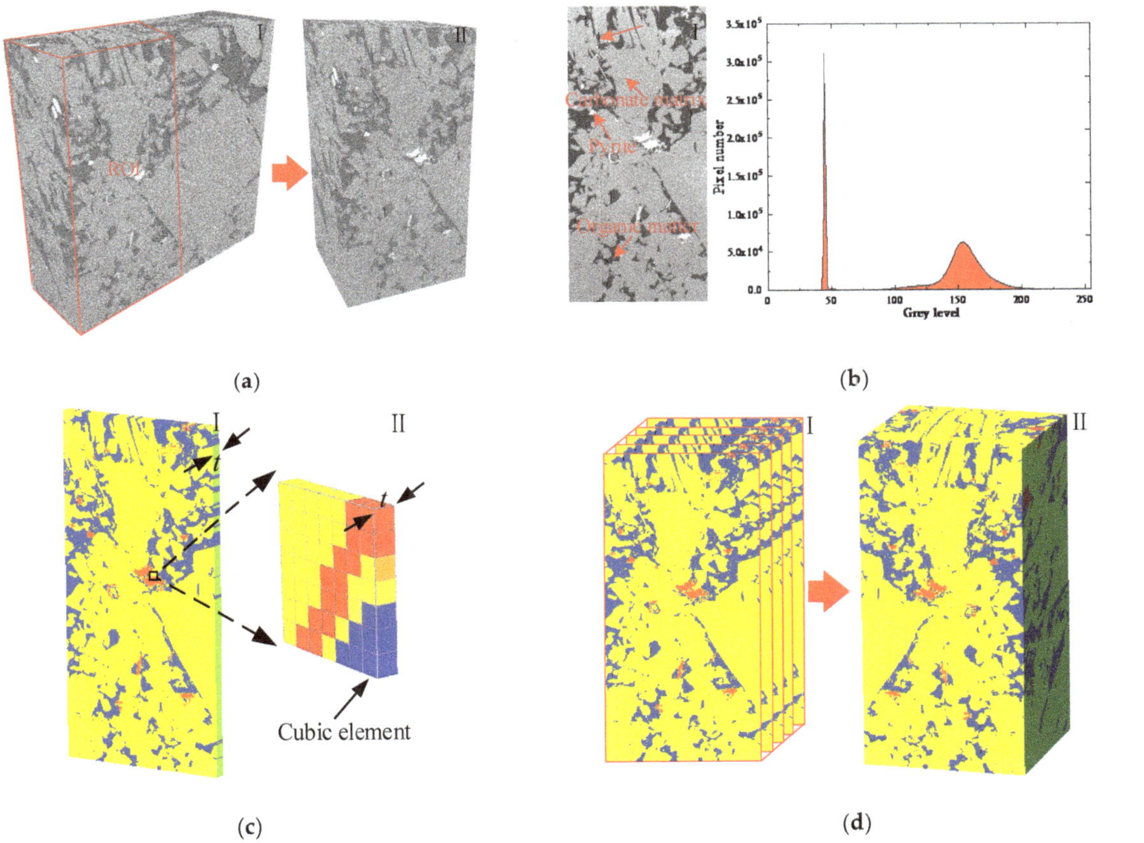

Figure 4. Reconstruction of 3D numerical models by digital image processing. (**a**) Getting the picture of the study model. (**b**) Determination of segmentation threshold. (**c**) A layer generated with a vertical extension of a slice image (Vectorization). (**d**) The 3D image stacking and numerical reconstruction model of shale specimens.

3. Multi-Scale Numerical Modeling of Rock Failure Process under Compression

3.1. Microscale Simulation

Shale is a natural composite material composed of various inorganic minerals and organic matter [36,37], and its macro-mechanical behavior depends on its microstructure and mineral composition [2,38]. During the fracturing process, the microstructure of shale will affect fracture propagation and fracture morphology, resulting in more tortuous and complex fractures [12,39]. Therefore, it is vital to study the mechanical properties and crack propagations of shale at the microscale. However, it is difficult to directly or indirectly observe and record the interaction between fractures and different mineral components of shale, and describe the crack propagation path, at the microscopic scale. Understanding of the shale failure mechanism on the microscale is not complete as yet [40]. Numerical experimental methods provide an effective way to solve the above problems. While the heterogeneity index can introduce the heterogeneity of the rock matrix, along with the statistical distribution, it cannot accurately represent the heterogeneity of the actual material [20]. The actual spatial distribution of the microscopic heterogeneity and some mechanical parameters of the mineral composition of the shale should be considered during numerical modeling [2].

3.1.1. Microscale Numerical Model

FIB-SEM has been widely used to characterize shale's nanoscale structural features [41,42]. Based on the images obtained by FIB-SEM, this section reconstructed an actual 3D multi-mineral component shale numerical model to study the influence of mineral distribution on the shale fracture process at the microscopic scale. In this way, the interaction between cracks and different mineral components can be observed, and the propagation process of cracks at the microscopic scale can be presented. The shale samples in this section were collected from the Yanchang Formation in the Ordos Basin, Gansu, China. The experimental equipment used in this section was the FIB-SEM HELIOS NANOLAB650 from the Research Institute of Petroleum Exploration and Development of China, and the acquisition pixel was about 10 nm.

Figure 4a shows shale sample 3D FIB-SEM images, and the single image resolution is 1500 pixels by 1500 pixels, with a pixel size of 10 nm $\times$ 10 nm. The interval of longitudinal cutting was 10 nm for each layer, and 750 layers were cut. Therefore, the size of the 3D FIB-SEM image of the shale sample obtained was about 15 $\times$ 15 $\times$ 7.5 μm (Figure 4a(I)). In our research, the FIB-SEM image analysis results show that the white particles are pyrite, the light gray minerals are silicate minerals, the dark gray minerals are carbonate minerals, the gray-black minerals are organic matter, and the black is fractures (Figure 4a,b).

The objective of this section (Figure 4a(II)) was to select the region of interest from Figure 4a(I)), the size of which was about 15 $\times$ 7.5 $\times$ 7.5 μm. In an attempt to increase the calculation speed without affecting the calculation accuracy of the model, when constructing the numerical model, the pixels of the image were reduced five-fold, and one image was taken at intervals (the size of the element is 50 $\times$ 50 $\times$ 50 nm). The method introduced in Section 2.3 reconstructs the three-dimensional numerical model of shale at the microscopic scale, as shown in Figure 5c,d ((c) front view, (d) back view), represented by the elastic modulus diagram. Comparing Figure 5a,b and Figure 5c,d, it can be found that the reconstructed numerical model (Figure 5c,d) reproduces the real microstructure of the shale well, and the introduction of inhomogeneous coefficients characterizes the inhomogeneity of single minerals more realistically, which makes the reconstructed numerical model more natural and realistic. The total number of finite element elements in the specimen section is 6.75 million. The blue mineral is pyrite, the yellow mineral is the carbonate matrix dominated by calcite, and the red mineral is organic matter.

In contrast, the transparent mineral is fractured, with contents of 1.09, 8.54, 18.18, and 0.19%. The shale numerical specimen was then built using the mechanical input parameters in Table 1 [43–45]. The uniaxial compressive simulation was carried out with an external displacement at a constant rate of 2.25 $\times$ 10^{-4} μm/step in the vertical direction.

Table 1. Material properties of the shale FE model.

Materials	E_L(MPa) [43–45].	P_L(MPa)	m	ν	ϕ (°)	C/T
Organic matter	20,000	60	2.5	0.34	35	10
Clay	59,000	155	2.8	0.32	35	10
Carbonates (calcite)	70,000	190	3	0.31	35	15
pyrite	90,000	310	5	0.19	35	15

3.1.2. Microscale Numerical Modeling Results

The spatial distribution of shale minerals at the microscopic scale affects the crack initiation position, propagation path, and final failure pattern. Despite the small amount of pyrite, cracks tend to begin in the organic matter around pyrite. Pyrite has a decisive effect on where cracks emerge. The crack propagates easily along the organic lamina, and the distribution of the organic lamina also determines the failure pattern. The simulation system can reproduce the typical characteristics of the failure process of microscale multi-mineral rock, and the failure characteristics obtained by the numerical simulation are in reasonable agreement with the physical test results [46].

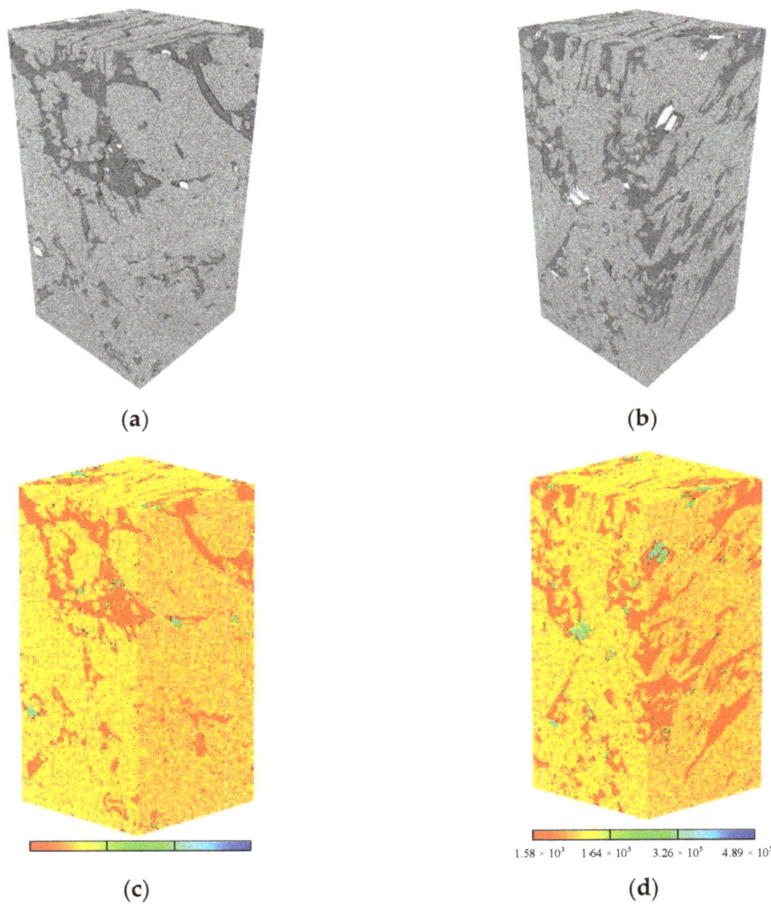

Figure 5. Shale microscopic model. (**a**,**b**) Original FIB-SEM 3D model: (**a**) is a front view, and (**b**) is a back view. (**c**,**d**) Finite element numerical model, expressed by elastic modulus: (**c**) is a front view, and (**d**) is a back view. (organic matter, carbonate minerals, pyrites).

Figure 6a–f shows the dynamic failure process of microscale organic laminar shale. Referring to the mineral distribution of the sample in Figure 5, it can be seen that at the initial stage of loading (point A), obvious stress concentration occurs at the organic matter lamina and the organic matter around pyrite (Figure 6a). In stages A–B, with the increase in load, the crack first initiates in the organic matter around pyrite (Figure 6b). Before peak point C (including the peak point), there is an obvious displacement discontinuity at the organic lamina (Figure 6c), and a small number of acoustic emission events occur, but there is no obvious macro crack. In stages C–E, with the continuous increase in load, a large number of acoustic emission events are generated, the stress–strain curve decreases in a stepwise pattern, and a small platform appears at point D. The results may be due to the fact that the fracture extension meets the primary fracture, resulting in a sudden increase in strain. In addition, the fracture gradually expands along the organic matter lamina (Figure 6d,e), and the displacement discontinuity gradually increases. Finally, a macro shear plane along the organic matter lamina is formed, and multiple macro cracks are formed along the organic matter lamina. It can be seen that the organic matter lamina plays a decisive role in the final shape of the crack. The macroscopic fracture surface is formed after the peak point, which is also consistent with our general understanding. In E–F, a small number of acoustic emission events occur, the stress–strain curve slowly decreases, and the displacement discontinuity is more pronounced at the main crack (Figure 6f). The strength and modulus of elasticity of the shale were 78.40 MPa and 60,700 MPa, respectively (Figure 6g). Figure 6h depicts the spatial distribution of acoustic emission events in the final damage. A sphere with a specific diameter and color represents an acoustic emission event. The diameter represents the relative magnitude of the released acoustic emission energy. The color represents the type of element failure. Red and blue indicate tensile damage and shear damage, respectively. It can be seen that most of the AE events are caused by tensile damage (red spheres), while only a small number of AE events are caused by sheer damage (blue spheres). Tensile failure is the main cause of rock failure, which is consistent with our general understanding [47]. In addition, through the acoustic emission diagram, we can see that the complex spatial morphology of the main crack and the spatial distribution of the organic matter lamina play decisive roles in the spatial morphology of the crack.

Figure 7 shows the internal failure process of six slices under different strain levels, in which dark blue represents the damage element, and black represents the failure element. The positions of the six slices (i.e., Slice#1, Slice#2, Slice#3, Slice#4, Slice#5, and Slice#6) are located at distances of +Z 7.5, 6.0, 4.5, 3.0, 1.5 and 0 μm. Figure 7a–f correspond to points A–F on the stress–strain curve in Figure 6g, respectively. As can be seen, at point A (Figure 7a), all slices have only damage elements in the organic laminar, and no failure elements appear. Note that the black elements shown on the slice at this point are all primary pores and fissures, rather than new ones under the action of external forces. In stages A–D, with the increase in load, the number of damaged elements increases continuously, and they are distributed in three-phase minerals (Figure 7a–d). More failure elements appear in the vicinity of the pre-existing fracture, and new failure elements gradually conjoin the main fracture and the pre-existing fracture (slice#1). In addition, some of the fractures expand along with the pre-existing fractures and gradually form macroscopic cracks (slice#5). Hard mineral grains (pyrite), on the other hand, are susceptible to global failure, influenced by their own mechanical properties and those of the surrounding minerals (slice#4). In stages D–E, as the load continues to increase, the cracks continue to expand, mainly along the narrower organic laminae, forming shear cracks (Figure 7d,e). Nevertheless, the fracture distribution of each slice differs, primarily due to the heterogeneity of mineral distributions, and especially due to the influence of organic matter lamina distributions. In stages E–F, almost no new cracks are generated, and the work done by the external force is mainly used for crack width expansion (Figure 7e,f). In summary, it can be seen that the differences in mineral distribution lead to significant differences in cracks in different slices at the same strain. The spatial distribution of minerals, especially the spatial distribution of organic laminae, greatly influences the fracture process of cracks. This is consistent with the results of the study [22], which indicate that the heterogeneous mineral distribution controls the fracture occurrence and damage pattern of the samples.

3.2. Mesoscale Simulation

Different scales of study objects can lead to significant differences in the spatial structures of identifiable rock materials [48]. Micro-CT can be widely used to characterize the spatial distribution of microstructures such as pores and fractures in shale reservoirs [49]. Therefore, the three-dimensional X-ray scanning of shale specimens can be used to obtain high-resolution three-dimensional images without loss, and effectively characterize the natural three-dimensional spatial structure of shale. In this section, a micro-CT 3D finite element numerical model characterizing the actual structure of shale was developed by combining the micro-CT scanning technique and the RFPA parallel computing method.

The dynamic splitting process and mechanical properties of mesoscale shale were studied via a Brazilian splitting experiment.

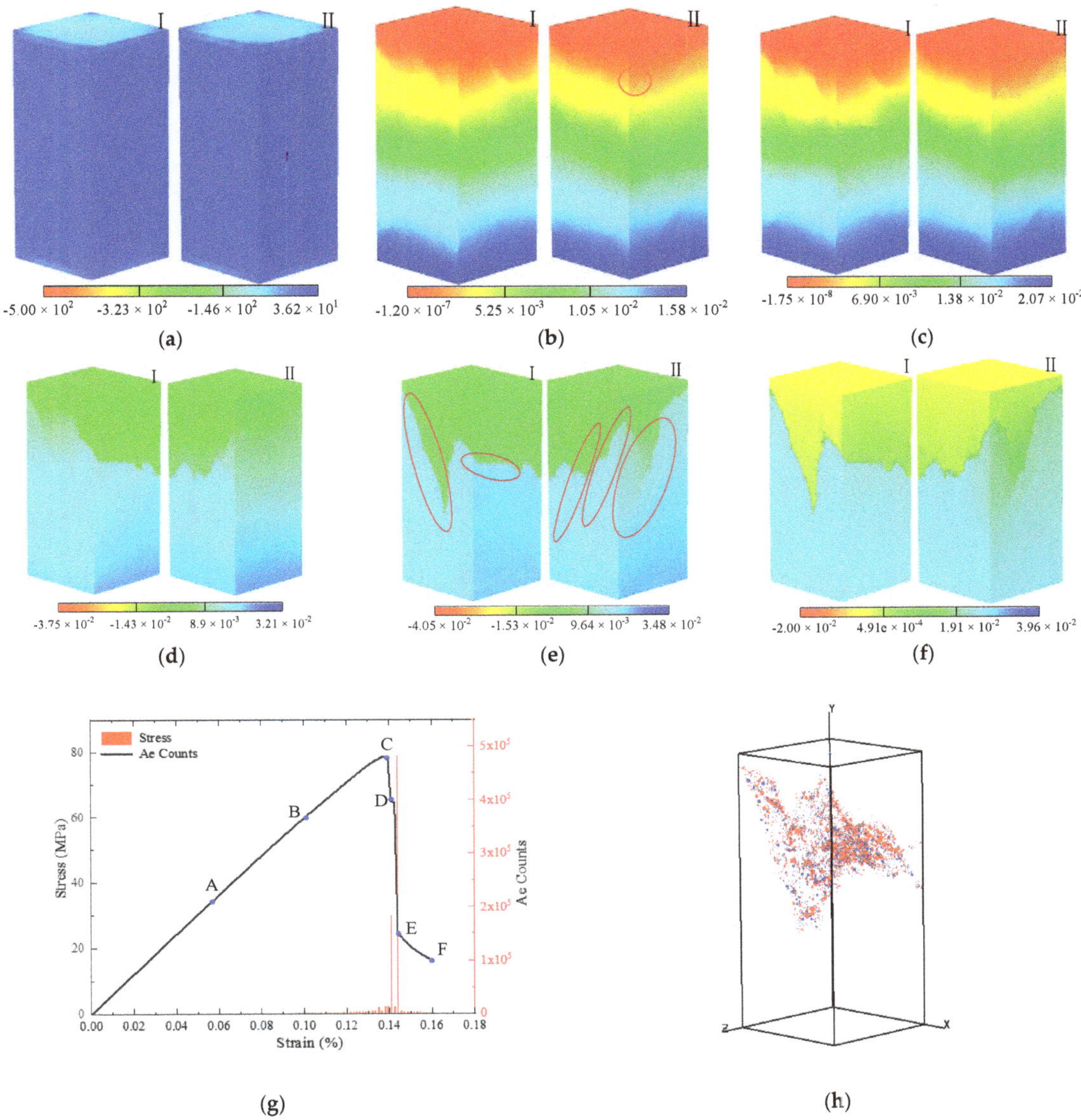

Figure 6. Numerical simulation of the shale sample at the microscale. (**a**) Minimum principal stress. (**b–f**) Vertical displacement contours at different loading stages, (**a–f**) correspond to points A–F in (**g**), respectively. (**I**) front view, and (**II**) back view. (**g**) The spatial distribution of acoustic emission (AE) events. (**h**) The stress–strain curve associated with AE counts for shale. The unit of stress in (**a**) is the MPa, and the unit of displacement in (**b–f**) is the millimeter.

The samples for the experiments in this section were selected from a shale oil and gas reservoir in a block of Qijiagulong Sag, Songliao Basin, with a burial depth of 2350 m. The sample was machined into a cylinder with a diameter of 4 mm and a height of 10 mm along the laminar direction. The CT scan experiment was completed at the Research Institute of Petroleum Exploration and Development, CNPC, Beijing. The experimental equipment was an Xradia 510 Versa high-resolution 3D X-ray microscope from Carl Zeiss Company, Oberkohen, Germany. The voltage was 80 kV, the single exposure time was 1 s, and the scanning resolution was 4 μm.

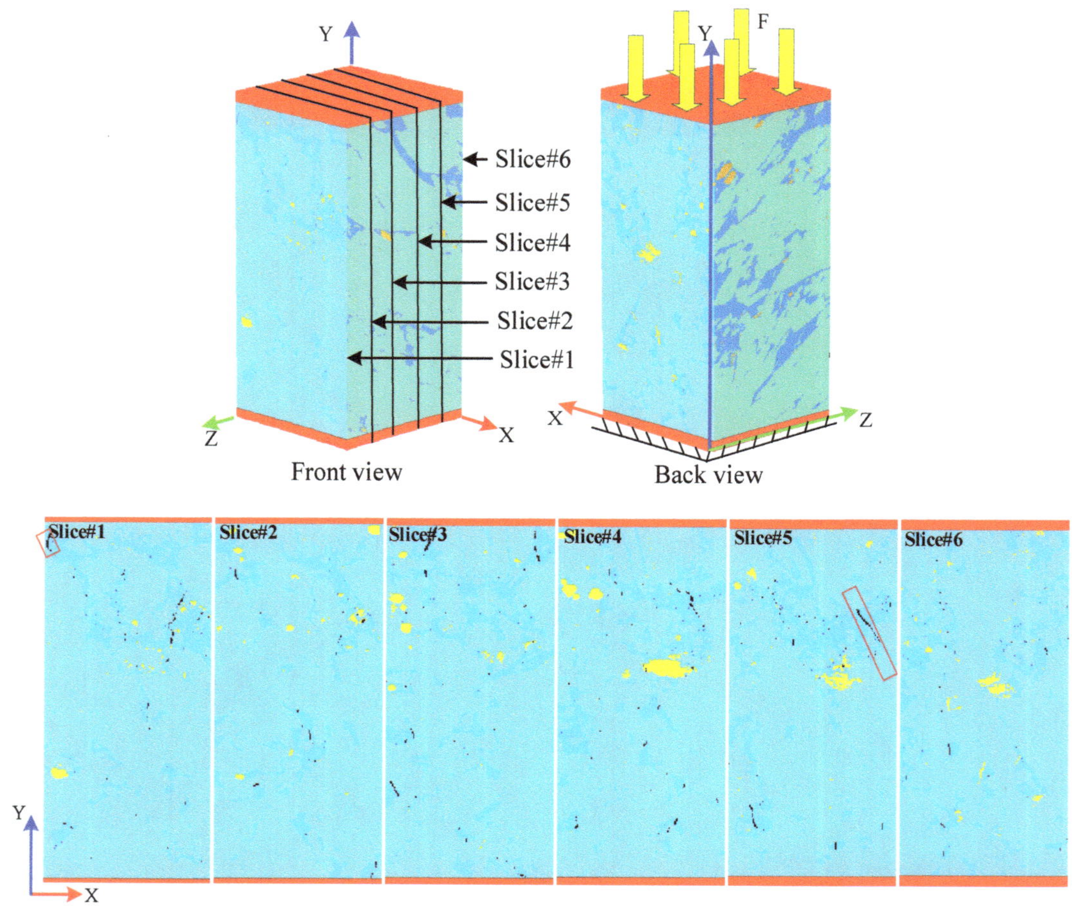

(a)

Figure 7. *Cont.*

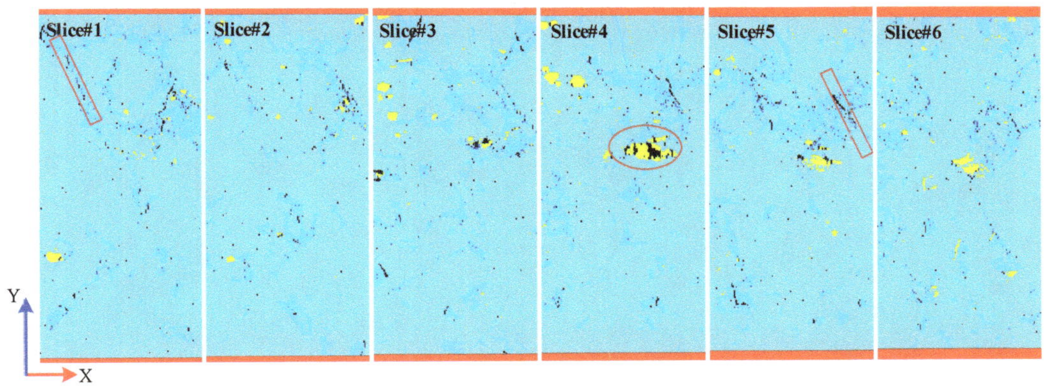

(b)

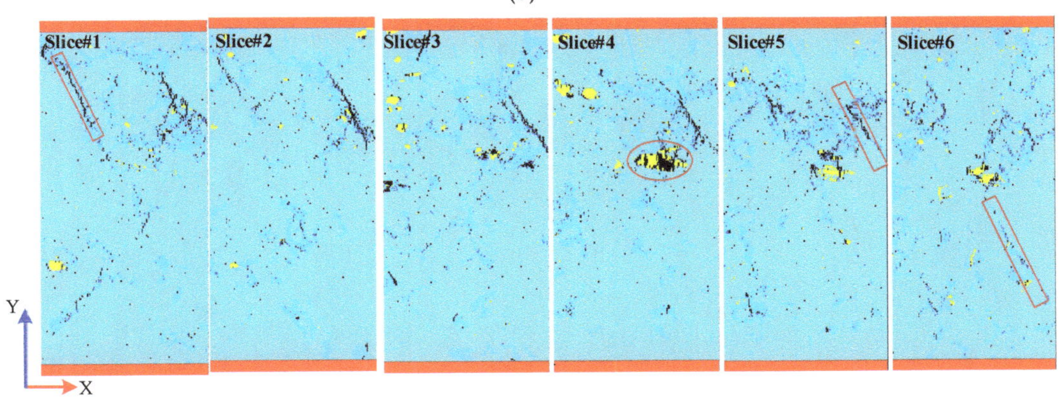

(c)

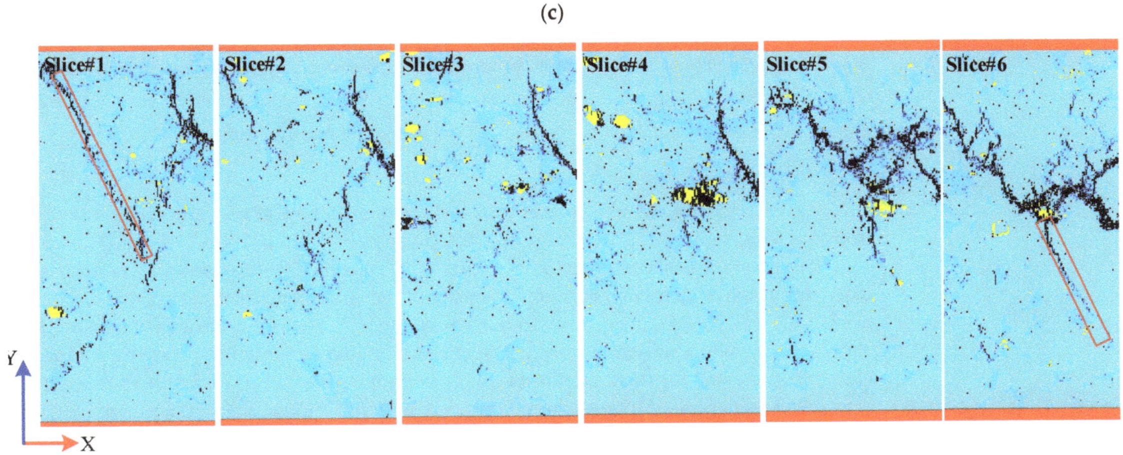

(d)

Figure 7. *Cont.*

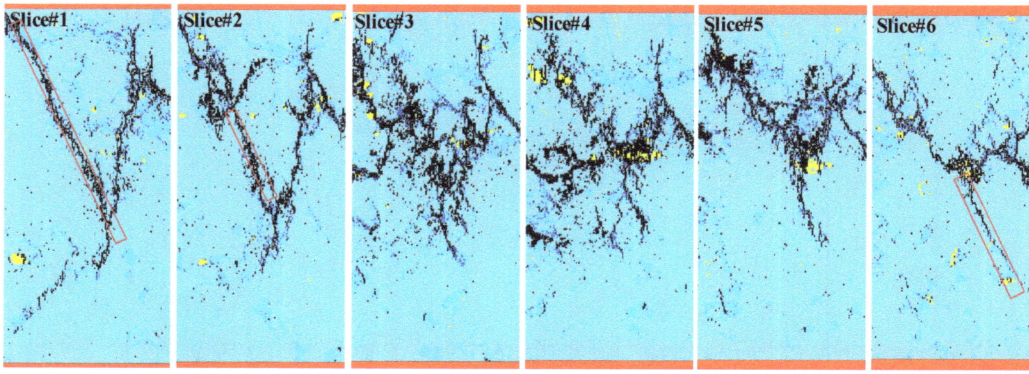

(e)

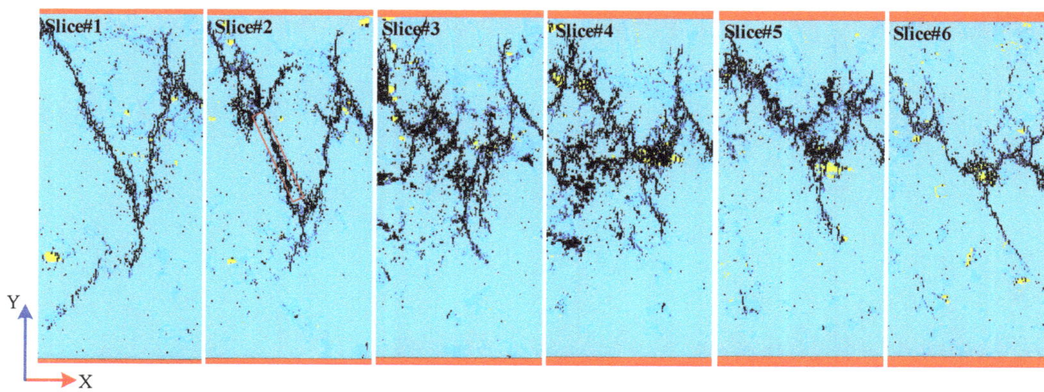

(f)

Figure 7. Internal fracture evolution in micro shale sample sliced along different Z planes. (**a–f**) The slices at different loading stages, corresponding to points A–F in Figure 6g, respectively. (organic matter, carbonate minerals, pyrites, damage elements, fractures).

3.2.1. Mesoscale Numerical Model

Figure 8 shows the CT images of the shale samples. The resolution of a single CT image is 1000 pixels × 1000 pixels, and the pixel size is about 4 μm × 4 μm (Figure 8b). The scan spacing is about 8 μm, and a total of 1250 layers were scanned. Therefore, the diameter of the original 3D micro-CT cylinder of the shale sample obtained was 4 mm and the height was 10 mm (Figure 8a). In our study, the results of the micro-CT image analysis show that the white mineral grains are calcite-dominated carbonates, the gray minerals are clay minerals, and the dark gray minerals are organic matter. The objective of this section is to select the region of interest from Figure 8a, with a diameter of 4 mm and a thickness of 2 mm. As a result of the need to increase computational speed without compromising the computational accuracy of the model, the pixels were reduced 4-fold when constructing the numerical model, and images were taken once at each interval (16 × 16 × 16 μm), for a total of 125 images. Due to the principle of CT imaging, circular artifacts (images that are bright in the middle and dark around) are inevitable in CT images. Therefore, in order to ensure the authenticity of the model and improve threshold segmentation accuracy, this subsection refers to the improved threshold segmentation method proposed by [50] regarding CT images, and the processing results are shown in Figure 8c. Next, the three-dimensional numerical model of Brazilian shale was reconstructed on the mesoscale based on the method introduced in Section 2.3, as shown in Figure 9 ((a) front view, (b) back view), which is represented by the elastic modulus diagram. By comparing Figure 9 with Figure 8a, we can see that the reconstructed numerical model reproduces the mesostructure of the shale, and that the addition of the inhomogeneity coefficient makes the reconstructed numerical model more natural. There are approximately 7.81 million finite elements in the Brazilian disc model. It consists of organic matter, calcite-based carbonate particles, and clay matrix, at 2.98%, 92.88%, and 4.14%, respectively. The Brazilian disc model exhibits a highly inhomogeneous structure, with X-type organic matter and large calcite-dominated carbonate particles distributed on the front side. However, only a few small grains of calcite-dominated carbonates are distributed on the backside of the model. Afterward, the mechanical input parameters from Table 1 are used to establish the meso shale Brazilian disk numerical sample. In the vertical direction, load is applied at a constant rate of 4.8×10^{-5} mm/step.

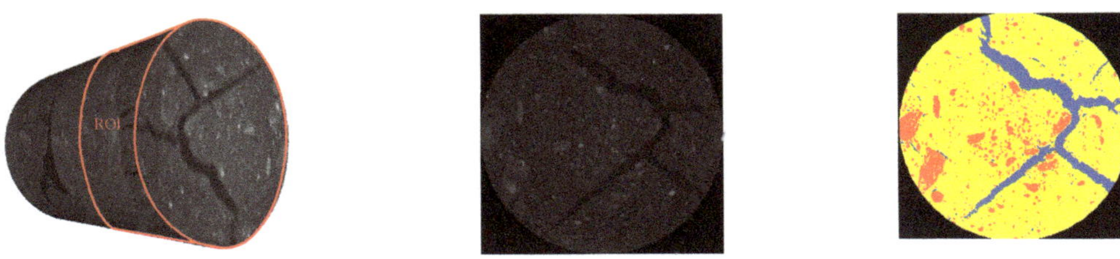

(a) (b) (c)

Figure 8. Mesoscale shale CT image. (**a**) Three-dimensional CT model. (**b**) Single CT image. (**c**) CT image after threshold segmentation (organic matter, clay minerals, carbonate minerals).

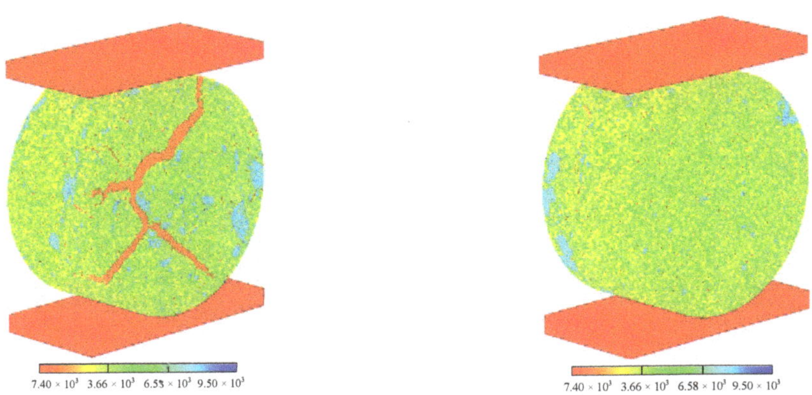

(a) (b)

Figure 9. A three-dimensional numerical model of the Brazilian disc at the mesoscale, expressed by elastic modulus. (**a**) is the front view, and (**b**) is the back view (organic matter, clay minerals, carbonate minerals).

3.2.2. Mesoscale Numerical Modeling Results

The spatial distribution of shale minerals affects the stress state, the location of crack initiation, the expansion path, and the final failure mode of the meso Brazilian disk model. Although the organic matter is only distributed on the front side of the model and the content is only 2.98%, it plays a decisive role in the initiation and expansion of cracks, and the final failure morphology and mechanical behavior.

The dynamic failure process is depicted in Figure 10a–f for the Brazilian disk model of calciteitic shale on the mesoscale. Figure 10a(I) shows that obvious stress concentration occurs in the organic matter at the initial stage of loading; displacement discontinuities are generated as the load increases. Organic matter distribution largely determines the frontal displacement discontinuity of the Brazilian disk, whereas cracks emerge at the frontal displacement discontinuity (Figure 10b). In addition, the back displacement discontinuity is consistent with the splitting experiments of typical Brazilian discs. At point C, the Brazilian disk displacement discontinuity becomes more and more significant. There is a large difference between the displacement discontinuity on the front and the back sides of the Brazilian disk model at point C (Figure 10b). As the load increases in the C–F stages, the cracks continue to grow along with the organic matter of the Brazilian disk, until they eventually form macroscopic cracks. On the back side of the Brazilian disc, no macroscopic cracks are formed. Only a few small cracks form at the intersection of the specimen and the loading plate (Figure 10d–f). Due to the effect of organic matter, the acoustic emission events increase slowly during the numerical experiments, but almost no new acoustic emission events occur after the F point. Likewise, the load–displacement curves show no obvious decrease in brittleness.

Figure 11 shows the internal failure processes of meso Brazilian disk models at different loading stages. Dark blue represents damage elements and black represents failure elements. The six slices (Slice#1, Slice#2, Slice#3, Slice#4, Slice#5, and Slice#6) are located at distances of +Z 2.0, 1.6, 1.2, 0.8, 0.4 and 0 mm. Figure 11a–f correspond to points A–F on the load–displacement curve in Figure 10g, respectively. By observing slice#1 and silce#2, it can be seen that the damage occurs near the center of the Brazilian disc. However, the initial fractures begin on the central axis near the upper and lower loading plates of the Brazilian disk, and continue to expand in the direction of organic matter until they merge into one main fracture. In the lower parts of the Brazilian disc, the fracture no longer propagates along with the organic matter because it is farther away from the central axis, and therefore it turns towards the central axis. The initial stages of loading are not damaged, as can be seen in slices #3–6. In response to the increased external load, some elements are damaged or fail at the central axis, but no macroscopic cracks appear. It is not until after the peak point that cracks are formed near the upper and lower loading plates, respectively. Although the cracks rarely extend, they only remain near the loading plate until the specimen is a mode of completely unstable failure.

In summary, differences in mineral distribution at the mesoscale lead to significant differences in cracking in different slices of the Brazilian disc model under the same strain. The inhomogeneous distribution of minerals leads to large differences in the crack distribution in different slices of the disc model. The spatial distribution of minerals greatly affects the fracture process of cracks, especially for minerals with low mechanical properties or minerals with high mechanical properties, which may not be highly present in the model, but play a decisive role in the mechanical behavior, fracture process, and fracture pattern of the model, which is consistent with the findings of [51].

3.3. Macroscale Simulation

It is usually the cooling process of the lava flow that forms the columnar jointed structures in igneous basalt [52]. The lava flow usually cuts the basalt into regular (quadrilateral or pentagonal) or irregular prismatic prisms. Several hydropower stations around the world have the same geologic structure of columnar joints. Previously, the formation mechanism and controlling factors of CJRM have been investigated [53–55]. In recent years, the construction and application of many large-scale CJRM projects have led to an increase in the interest of engineering geologists in the mechanical properties of CJRMs. The main techniques used to study the structural characteristics of CJRM are in situ tests, laboratory tests, indoor tests, and numerical simulations [52,56,57]. Those resulting from laboratory tests of small-scale rock mass differ significantly from those obtained under field conditions, so there is no advantage to considering rough columnar joint networks on-site. While large-scale rock masses cannot be conveniently sampled on-site, experimental results are often scattered. The numerical simulation method has advantages when it comes to studying the mechanical properties of a rock mass. The model is often simplified in practice, and it is hard to determine whether the joints are distributed accurately. The CJRM reconstructed according to field conditions is essential to studies of the mechanical behavior of this geological structure.

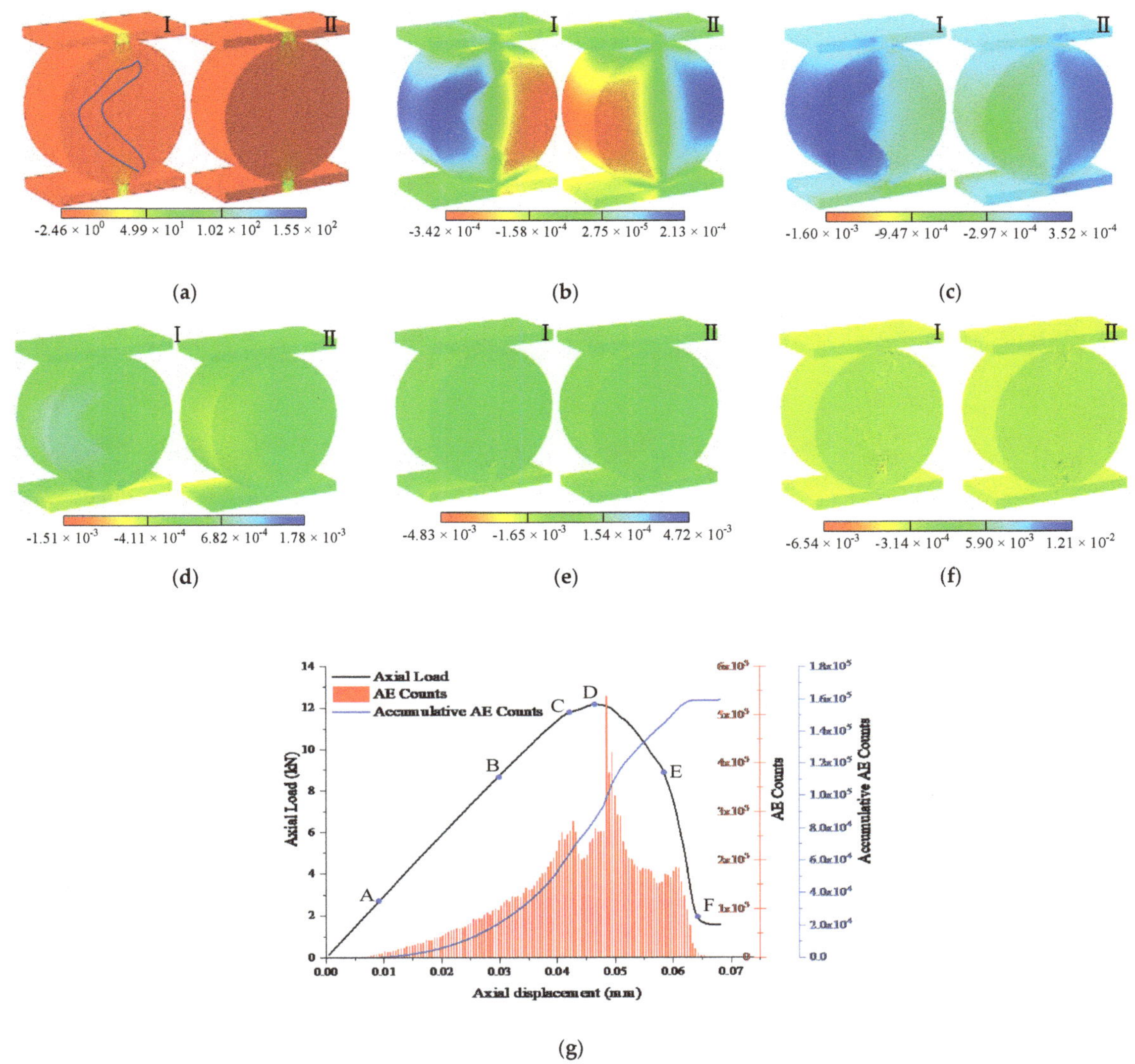

Figure 10. Numerical simulation of the shale sample at the mesoscale. (**a**) Maximum principal stress. (**b–f**) Vertical displacement contours at different loading stages, (**a–f**) correspond to points A–F in (**g**), respectively. (**I**) front view, and (**II**) back view. (**g**) The stress–strain curve associated with AE counts for shale. The unit of stress in (**a**) is the MPa, and the unit of displacement in (**b–f**) is the millimeter.

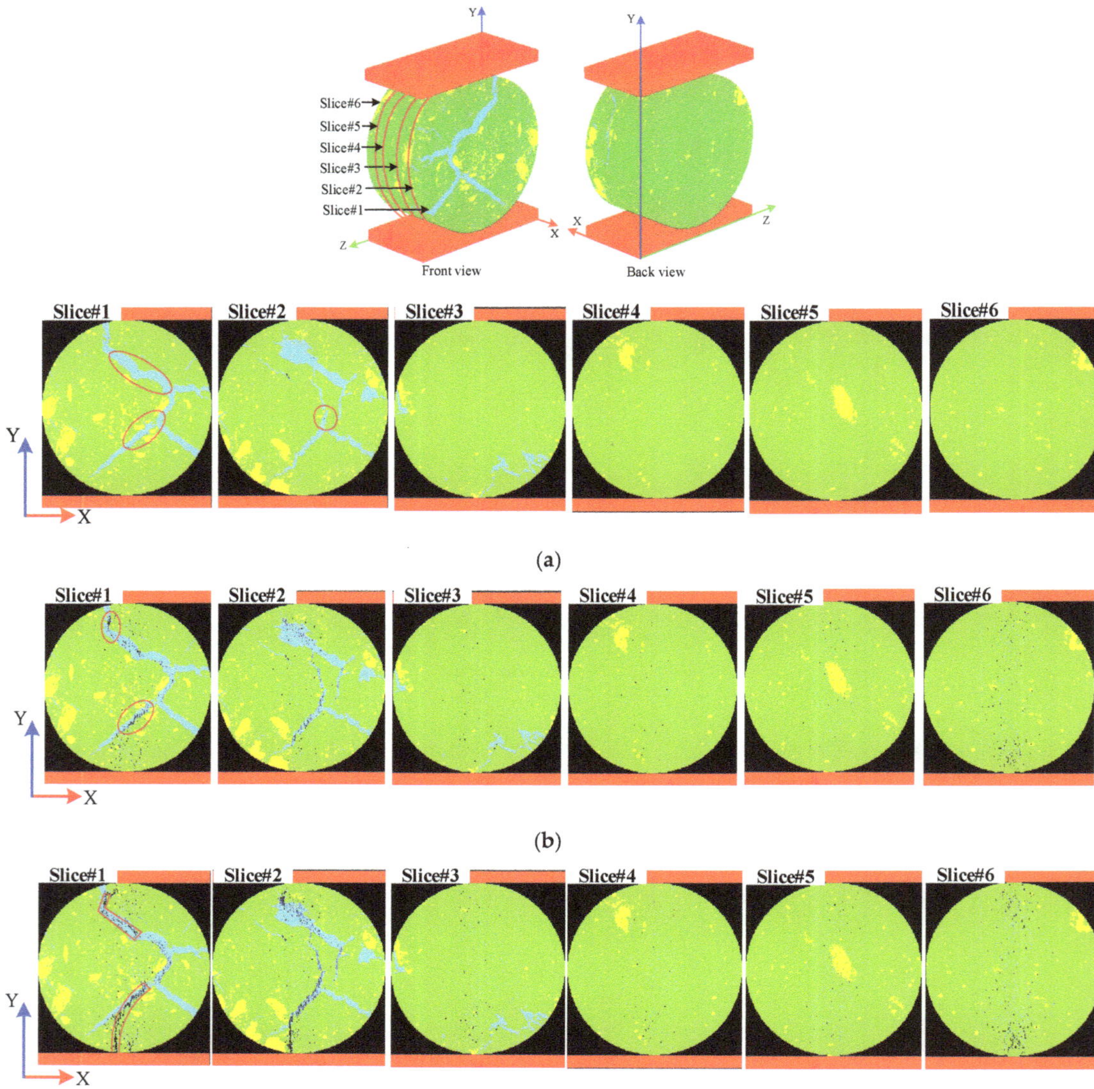

(a)

(b)

(c)

Figure 11. *Cont.*

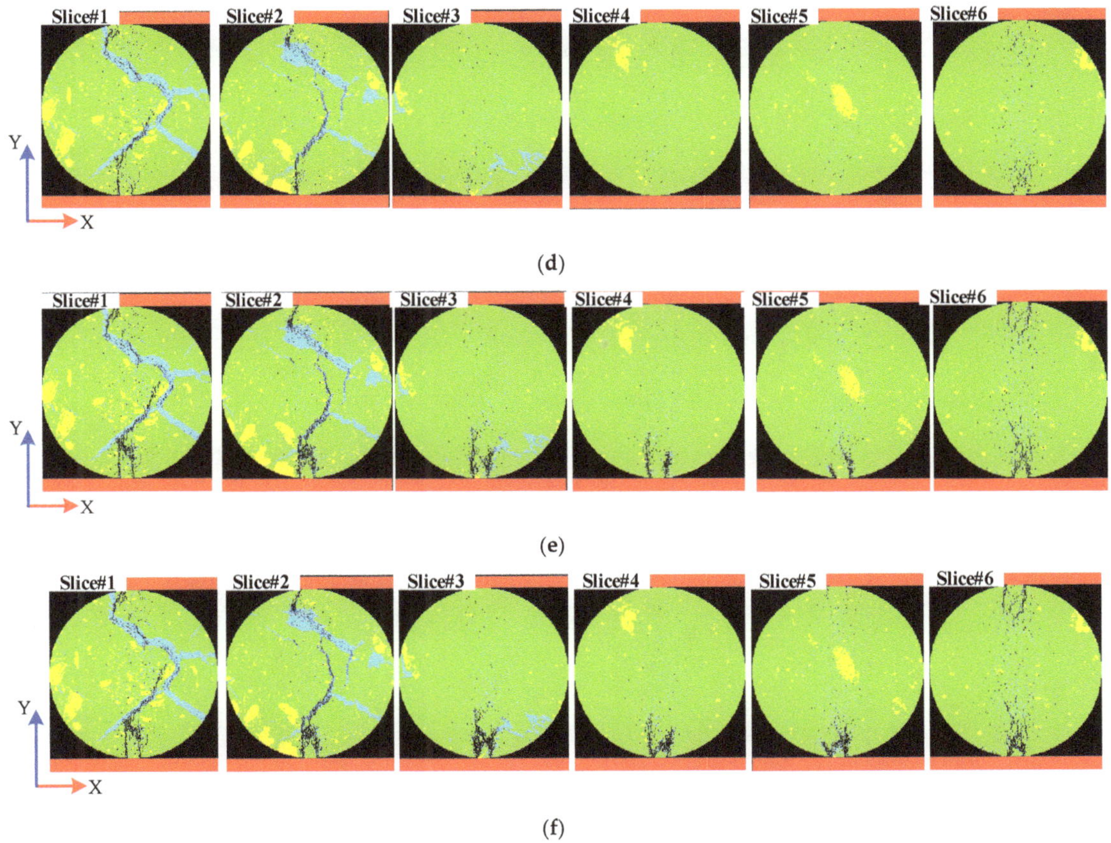

Figure 11. Internal fracture evolution in a shale meso sample sliced along different Z planes. (**a–f**) The slices are at different loading stages, corresponding to points A–F in Figure 10g, respectively. (organic matter, clay minerals, carbonate minerals, damage elements, fractures).

3.3.1. Macroscale Numerical Model of Columnar Jointed Rock Mass

In this section, combined with geological features of the Baihetan hydropower station and based on the CJRM Window images obtained through the sampling window method, a CJRM reflecting the real structure is constructed. Figure 12a,b is the transverse section information image of the CJRM obtained from the selected window on the right slope of the steep-descent basin of the Baihetan hydropower station [35]. Different colors represent different shapes of column-shaped joints, which cut the basalt into columns. In the columnar block cross-sections, the triangular, quadrangular, pentagonal, and hexagonal polygon frequencies are 39.9%, 45.3%, 9.9%, and 4.9%, respectively. Based on the image in Figure 12b, the finite numerical model of CJRM is constructed according to the method and principle introduced in Section 4, as shown in Figure 12c, which is represented by the elastic modulus diagram. Figure 12d depicts the spatial distribution of joints in the model, and the critical information of the numerical model is consistent with that of the digital image. The size of the model is 2.75 m × 2.75 m × 2.75 m; the number of image pixels in each layer is 275 × 275 (the number of single-layer finite element grids is 75,625). A total of 275 layers of images are copied, so the size of the unit is 1 cm × 1 cm × 1 cm, and the total number of model units is about 20 million. The mechanical parameters are derived from [58], as shown in Table 2. The loading method is displacement loading, and the loading rate is 0.1 mm/step.

Table 2. Material properties of the CJRM FE model [58].

Materials	E_L(MPa)	P_L(MPa)	m	ν	ϕ (°)	C/T
Rock matrix	65,100	280	2	0.21	38	10
Joint	1160	10	3	0.31	30	12

3.3.2. Macroscale Numerical Simulation Results

The three-dimensional CJRM reconstructed by digital image processing can actually reflect the spatial distribution of the joints on the engineering site (Figure 12). The spatial distribution of the joints greatly influences the fracture process and the final failure pattern of CJRM. Due to the low strength of the joints, they generate a tensile stress area around the joints in the initial stage of loading (Figure 13), causing the joints to fracture under a relatively small external load. The failure characteristics of the three-dimensional numerical simulation and the in-situ tests are very similar. Most of the failure occurs along the joint planes due to their lower tensile strength. Most longitudinal joints suffer from tensile failure, and the low-angle horizontal joints suffer from compression shear failure (Figure 14). The method proposed provides a useful reference for research into irregular CJRM.

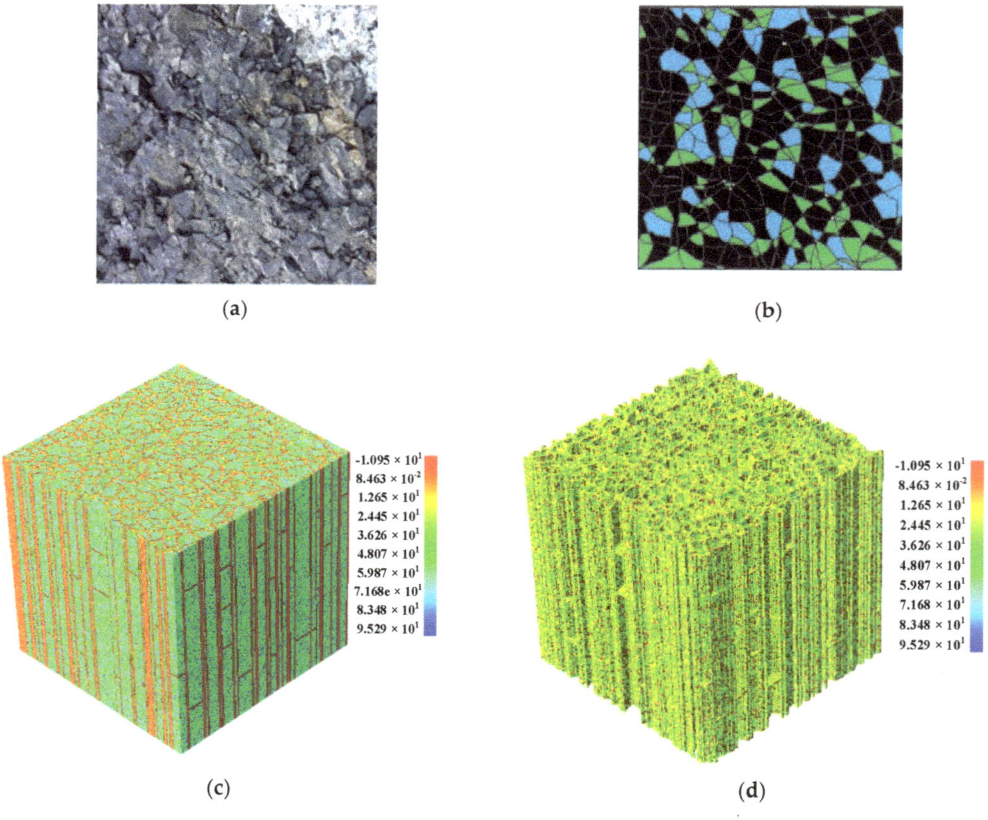

(a) (b) (c) (d)

Figure 12. Configurations of columnar jointed basalt. (**a**,**b**) Structural descriptions of columnar jointed basalt in transverse sections [35]. (**c**) A numerical model for the columnar jointed basalt. (**d**) The spatial shape of the joints is characterized.

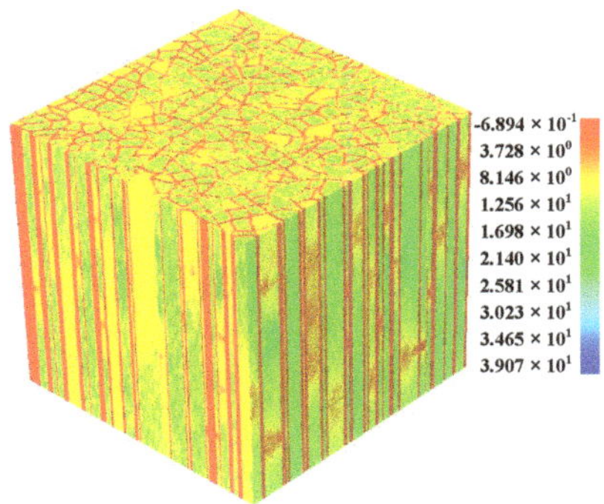

Figure 13. The maximum principal stress distribution; the unit is MFa.

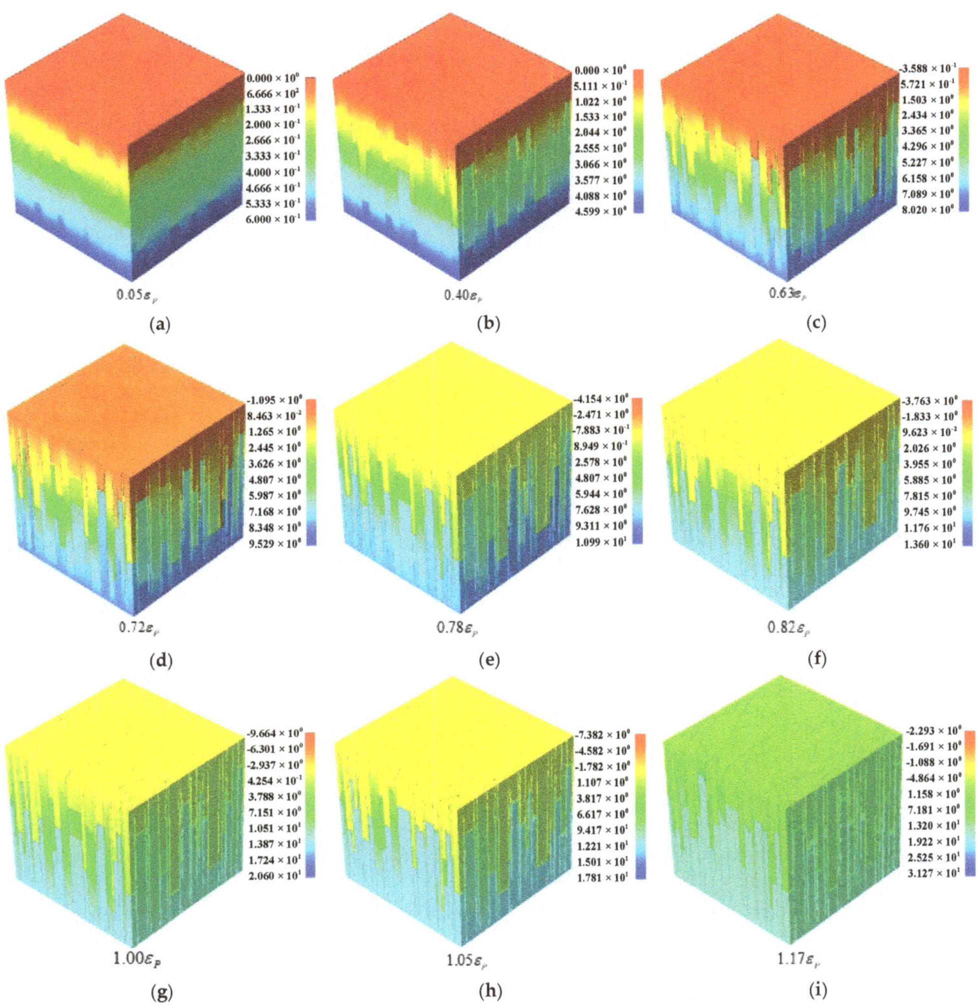

Figure 14. Numerical uniaxial compression test of columnar jointed basalt at the macroscale. (**a–i**) Vertical displacement contours at different loading stages, (**a–i**) correspond to points A–F in Figure 15. The unit of displacement in (**a–i**) is the millimeter.

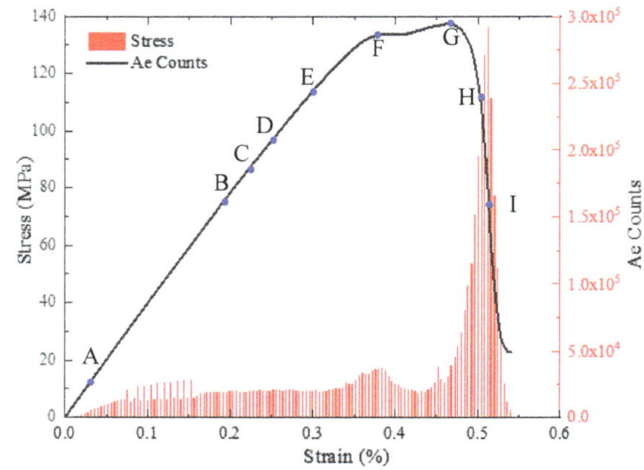

Figure 15. The stress–strain curve associated with AE counts for CJRM.

The stress–strain curves of CJRM are different from those of intact rock samples, and the brittle characteristics of intact rocks are significantly higher than those of CJRM. The stress–strain curve and the AE count for CJRM are shown in Figure 15. There is a stepwise increase when the curve rises to the DE stage before the peak strength. The strain increases from 0.36% to 0.46% at this stage, an increase of 27.8%. The stress increases from 131.2 MPa to 137.5 MPa, a rise of 4.8%. This may be due to the crushed low-angle horizontal joints and the longitudinal blocks being squeezed. After the peak strength, the stress–strain curve of the complete rock sample under uniaxial compression generally decreases sharply. The stress–strain curve of CJRM decreases gradually, without prominent brittle failure characteristics. It can be seen from the changing trend of AE events that AE events occur at low stress levels, and AE events maintain a relatively stable level before peak point E. Due to the low strength of the joints, the joints fracture at low stress levels, and AE events occur. AE events in the A–D interval are mainly caused by longitudinal joint failure, while AE events in the D–E interval are mainly caused by low-angle horizontal joint failure. After the peak point E, the acoustic emission events are mainly concentrated in the E–F interval, and the fracture of the basalt column leads to a large number of AE events. At this stage, the AE events account for about 50% of the total AE events in the fracture process.

The mechanical responses of joints and intact rock vary with the same external displacement (e.g., mismatch stress). Figure 13 shows the maximum principal stress distribution at 0.1 mm displacement. Even under compression, tensile stresses (in red) can be observed in the surroundings of joints. Figure 14 illustrates the vertical displacement field during fracture. Due to the existence of joints when the strain level is 0.40 ε_P (Figure 14b), there are obvious cracks in the longitudinal joints in the middle of the CJRM. With the increase in external displacement, cracks propagate upward and downward along the direction of longitudinal joints simultaneously, and the color of the vertical displacement field is different (Figure 14b–i). The low-level joint is at the junction of the color change, indicating that the displacement is not continuous. At peak point E, macro cracks are apparent along the joint plane on the surface of the CJRM (Figure 14g). Figure 16 describes the damage evolution process of CJRM. The symbols (i.e., A, B, C, D, E, and F) plotted on the stress–strain curve in Figure 8 indicate the strain levels of 0.05, 0.14, 0.40, 0.83, 1.00 and 1.15 ε_P. Figure 16a shows the initial damage pattern, with a strain of 0.05 ε_P. Multiple points of damage co-occur at the longitudinal joint. This shows that the joint strength is low, and the damage occurs at a low-stress level. With the load increase, the number of damage points increases and accumulates in longitudinal joints. Damage points also appear at low and horizontal angles (Figure 16b–e). Finally, multiple macroscopic damage zones are formed.

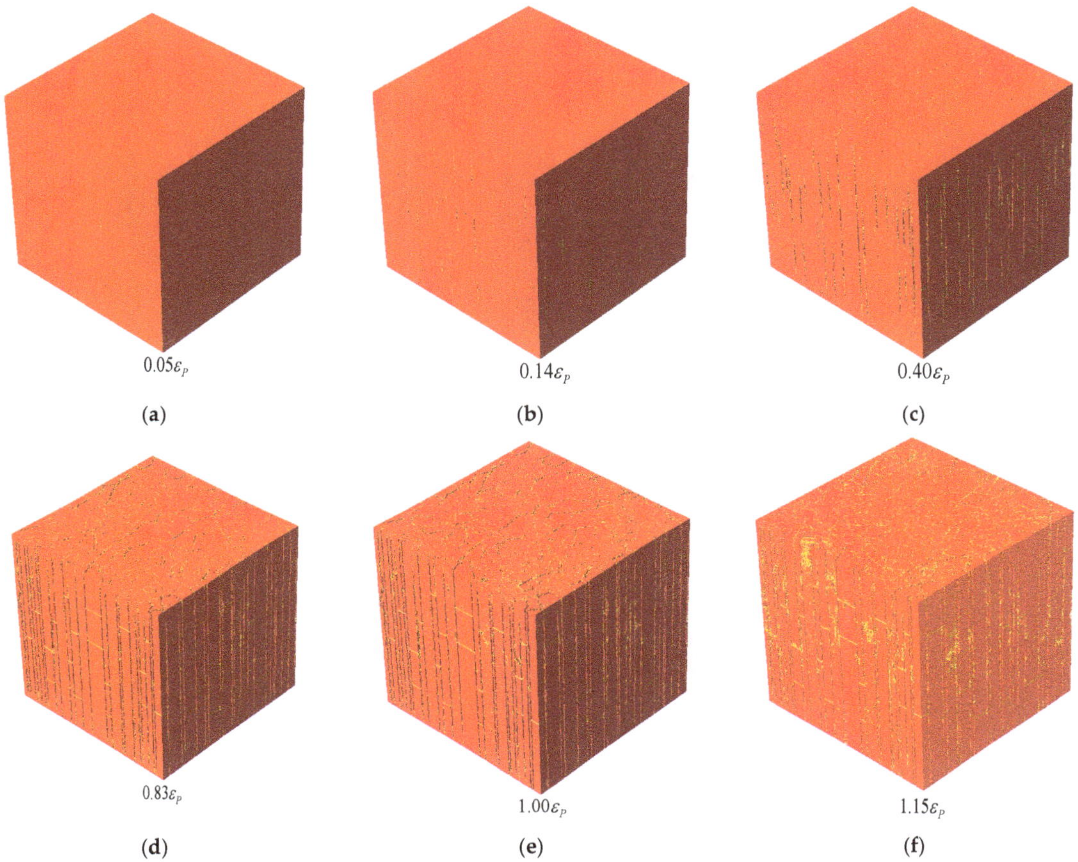

Figure 16. Damage evolution in columnar jointed basalt, where the yellow spots represent the damage occurrence and transparent means that the element is completely damaged, (**a**–**f**) correspond to points A–F in Figure 15.

Figure 17 compares the damage characteristics of CJRM between the numerical test and the in-situ test. It can be seen from Figure 17a–c that the failure characteristics of the three-dimensional numerical sample are very similar to those of the field test, and the failure mode is also consistent. Figure 17d describes the spatial distribution of AE events when the numerical model finally fails. A sphere with a specific diameter and color represents an acoustic emission event. The diameter represents the relative magnitude of the released acoustic emission energy. The color represents the type of element failure. Red and blue indicate tensile damage and shear damage, respectively. It can be seen that most AE events occur at joint locations. Most AE events at low-angle horizontal joints are caused by compression tensile damage (red sphere), while most AE events at longitudinal joints are caused by shear damage (blue sphere).

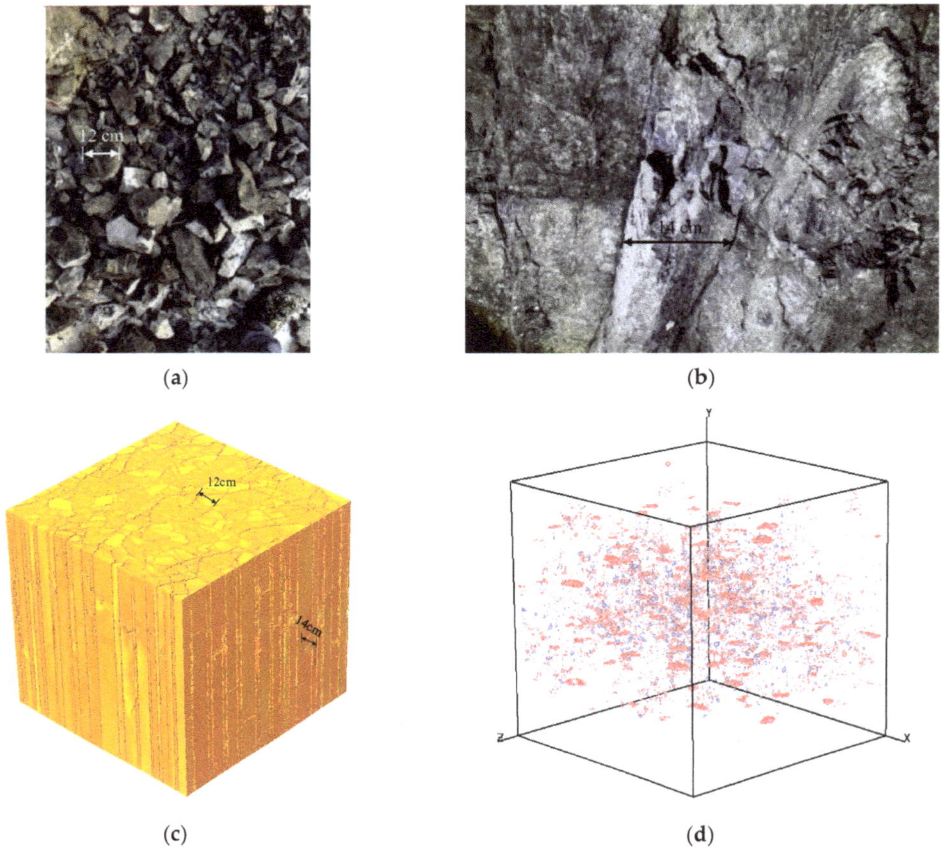

Figure 17. Comparisons of CJRM failure modes between the numerical specimen and the in-situ specimen. (**a**,**b**) The fracture mode of the in-situ specimen; (**a**) is the top view, and (**b**) is the front view [7]. (**c**) The fracture mode of the numerical specimen. (**d**) The spatial distribution of acoustic emission (AE) events.

4. Discussion

The statistical damage mechanical approach has been used to effectively predict rock failure issues [59,60]. Furthermore, it is possible to create realistic numerical rock samples using digital image processing methods. Our study was based on the study of [14,59,60], which considered the non-homogeneity of the structure through digital image techniques, and the non-homogeneity of single minerals through Weibull distributions [50]. That is, the dual inhomogeneity of structure and material is considered simultaneously. Therefore, the multi-scale numerical models we developed were all more realistic in reflecting the spatial rock structure compared with previous studies [34]. Meanwhile, the constructs of the numerical models were integrated into the RFPA3D parallel computing simulator, which simulates with high efficiency the progressive damage process and mechanical property behavior of the high-resolution multi-scale rock models with the help of a cloud computing platform [34]. Consequently, complex, heterogeneous rocks can be modeled successfully with this integration approach.

It is crucial to acquire digital images of rocks at different scales using suitable scanning techniques before constructing numerical models based on digital images of rocks. There are currently a variety of techniques available for obtaining 3D digital images of rocks on the micro-, meso-, and macroscales [41,49,61–63]. Even so, the three-dimensional structural characteristics of engineering-scale rocks are often difficult to determine, particularly for rock reservoirs that are deep underground, such as shale oil and gas reservoirs. In future research, we should integrate multiple technologies, including geological exploration, to construct more realistic engineering-scale numerical models, i.e., digital rock masses. Furthermore, although a modeling and simulation study, from digital cores to digital rock masses, was completed in this paper, the microscopic to macroscopic linkage has not been addressed. Therefore, establishing an effective macroscopic linkage is also the focus of our future research.

5. Conclusions

The statistical meso-damage mechanical model has been applied to study the failure of rocks. However, it is difficult for numerical models to characterize the actual sample structure due to the limitations of modeling methods. Additionally, the bottleneck of element number and calculation efficiency limits numerical modeling. The DIP technology in this paper is used to extract the structural characteristics of samples obtained by micro-CT, FIB-SEM, or sample window methods, reconstruct the 3D models, and then combine the models into a 3D numerical model using a vectorizing method. The 3D finite element numerical model was integrated into the RFPA parallel simulator. The finite element parallel computing system based on the digital image was built, which effectively solved the above problems. By incorporating the intrinsic shale geometric features, i.e., the spatial distribution of mineral composition, the 3D numerical simulations can reasonably predict the mechanical and fracture properties of the shale sample. The calculation system can also help us monitor and analyze the stress, strain, and displacement fields of the sample from any angle. This method is superior to some traditional methods, such as acoustic emission monitoring, in-situ CT scanning monitoring, and the 3DP method, which struggle to accurately and quantitatively monitor the initiation and evolution of micro cracks in real-time. Meanwhile, the results show that the finite element numerical models we constructed from the microscopic digital core to the macroscopic digital rock mass can effectively simulate mechanical behavior. In addition, they can reproduce dynamic fracture processes. That is, the numerical simulation system proposed in this paper is a useful tool for simulating the multi-scale rock model failure process. It has significant potential for use in studying the mechanical properties of rocks at multiple scales and high resolutions. The research basis of this paper is the RFPA numerical code, which has been successfully applied to the study of many brittle rock materials. Therefore, the method proposed in this paper can be used to study brittle rock. However, because the essence of this method is based on digital image modeling, the accuracy of the numerical model constructed in this paper is seriously limited by the accuracy of the digital images.

Author Contributions: Conceptualization and methodology, X.L.; investigation, Z.L., S.M., C.T. and J.T.; formal analysis, writing—original draft preparation, X.L.; writing—review and editing, X.L. and S.M. All authors have read and agreed to the published version of the manuscript.

Funding: This study was supported by the National Natural Science Foundation of China (Grant Nos. 52009016, 51779031), National Key R&D Program of China (2018YFE0196000), and China Postdoctoral Science Foundation (2019M663502).

Conflicts of Interest: The authors declare no conflict of interest.

List of Symbols

Symbol	Property	Unite
D	Damage variable	-
E	Elastic modulus of the damaged elements	MPa
E_m	Mean elastic modulus of element	MPa
E_L	Elastic modulus	MPa
E_0	Elastic modulus of the undamaged elements	MPa
P_m	Mean axial compressive strength of element	MPa
P_L	Axial compressive strength	MPa
T/C	Ratio of tensile to compressive strength	-
$P(u)$	Distribution density	MPa^{-1}
m	Homogeneity index	-
F	Numerical output strength of element	MPa
u	Material parameters of the element (such as Young's modulus and strength) obtained from experiments	MPa
u_0	The mean value of elements put into the numerical simulation program	MPa
ε	Strain tensor	-
$\bar{\varepsilon}$	Equivalent principal strain	-
ε_1	Maximum principal strain	-
ε_2	Middle principal strain	-
ε_3	Minimum principal strain	-
ε_{c0}	Critical strain under peak compressive	-
ε_{cu}	Ultimate compression strain	-
ε_{t0}	Threshold strain	-
ε_{tu}	Ultimate tensile strain	-
ε_p	Strain under peak compressive strength	-
η	Ultimate tensile stress factor	-
ν	Poisson ratio	-
σ	Stress tensor	MPa
σ_1	Maximum principal stress	MPa
σ_3	Minimum principal stress	MPa
σ_c	Fracture closure pressure	MPa
σ_{c0}	Uniaxial compressive strength	MPa
σ_{rt}	Residual tensile stress	MPa
σ_{rc}	Residual compressive stress	MPa
σ_{t0}	Uniaxial tensile strength	MPa
ϕ	Friction angle	°
$f(x,y)$	Initial gray value of the pixel (X, Y)	-
$f\prime(x,y)$	Gray value of the pixel (X, Y) after binarization	-
I	Segmentation threshold	-

References

1. Ju, Y.; Liu, P.; Chen, J.; Yang, Y.; Ranjith, P.G. CDEM-based analysis of the 3D initiation and propagation of hydrofracturing cracks in heterogeneous glutenites. *J. Nat. Gas Sci. Eng.* **2016**, *35*, 614–623. [CrossRef]
2. Zhu, J.B.; Zhou, T.; Liao, Z.Y.; Sun, L.; Li, X.B.; Chen, R. Replication of internal defects and investigation of mechanical and fracture behaviour of rock using 3D printing and 3D numerical methods in combination with X-ray computerized tomography. *Int. J. Rock Mech. Min. Sci.* **2018**, *106*, 198–212. [CrossRef]
3. Ju, Y.; Wang, L.; Xie, H.; Ma, G.; Mao, L.; Zheng, Z.; Lu, J. Visualization of the three-dimensional structure and stress field of aggregated concrete materials through 3D printing and frozen-stress techniques. *Constr. Build. Mater.* **2017**, *143*, 121–137. [CrossRef]
4. Belda, R.; Palomar, M.; Peris-Serra, J.L.; Vercher-Martínez, A.; Giner, E. Compression failure characterization of cancellous bone combining experimental testing, digital image correlation and finite element modeling. *Int. J. Mech. Sci.* **2020**, *165*, 105213. [CrossRef]
5. Amann, F.; Kaiser, P.; Button, E.A. Experimental Study of Brittle Behavior of Clay Shale in Rapid Triaxial Compression. *Rock Mech. Rock Eng.* **2011**, *45*, 21–33. [CrossRef]
6. Daigle, H.; Hayman, N.W.; Kelly, E.D.; Milliken, K.L.; Jiang, H. Fracture capture of organic pores in shales. *Geophys. Res. Lett.* **2017**, *44*, 2167–2176. [CrossRef]

7. Xia, Y.; Zhang, C.; Zhou, H.; Hou, J.; Su, G.; Gao, Y.; Liu, N.; Singh, H.K. Mechanical behavior of structurally reconstructed irregular columnar jointed rock mass using 3D printing. *Eng. Geol.* **2020**, *268*, 105509. [CrossRef]
8. Yang, B.; Xue, L.; Zhang, K. X-ray micro-computed tomography study of the propagation of cracks in shale during uniaxial compression. *Environ. Earth Sci.* **2018**, *77*, 652. [CrossRef]
9. Zhou, J.; Chen, M.; Jin, Y.; Zhang, G.-Q. Analysis of fracture propagation behavior and fracture geometry using a tri-axial fracturing system in naturally fractured reservoirs. *Int. J. Rock Mech. Min. Sci.* **2008**, *45*, 1143–1152. [CrossRef]
10. Guo, T.; Zhang, S.; Qu, Z.; Zhou, T.; Xiao, Y.; Gao, J. Experimental study of hydraulic fracturing for shale by stimulated reservoir volume. *Fuel* **2014**, *128*, 373–380. [CrossRef]
11. Tang, J.; Li, J.; Tang, M.; Du, X.; Yin, J.; Guo, X.; Wu, K.; Xiao, L. Investigation of multiple hydraulic fractures evolution and well performance in lacustrine shale oil reservoirs considering stress heterogeneity. *Eng. Fract. Mech.* **2019**, *218*, 218. [CrossRef]
12. Peng, T.; Yan, J.; Bing, H.; Yingcao, Z.; Ruxin, Z.; Zhi, C.; Meng, F. Laboratory investigation of shale rock to identify fracture propagation in vertical direction to bedding. *J. Geophys. Eng.* **2018**, *15*, 696–706. [CrossRef]
13. Dong, C.Y.; Pater, C.J.D. Numerical implementation of displacement discontinuity method and its application in hydraulic fracturing. *Comput. Methods Appl. Mech. Eng.* **2001**, *191*, 745–760. [CrossRef]
14. Liang, Z.Z.; Xing, H.; Wang, S.Y.; Williams, D.J.; Tang, C.A. A three-dimensional numerical investigation of the fracture of rock specimens containing a pre-existing surface flaw. *Comput. Geotech.* **2012**, *45*, 19–33. [CrossRef]
15. Li, Z.; Li, L.; Huang, B.; Zhang, L.; Li, M.; Zuo, J.; Li, A.; Yu, Q. Numerical investigation on the propagation behavior of hydraulic fractures in shale reservoir based on the DIP technique. *J. Pet. Sci. Eng.* **2017**, *154*, 302–314. [CrossRef]
16. Fang, Z.; Harrison, J. Development of a local degradation approach to the modelling of brittle fracture in heterogeneous rocks. *Int. J. Rock Mech. Min. Sci.* **2002**, *39*, 443–457. [CrossRef]
17. Shimizu, H.; Murata, S.; Ishida, T. The distinct element analysis for hydraulic fracturing in hard rock considering fluid viscosity and particle size distribution. *Int. J. Rock Mech. Min. Sci.* **2011**, *48*, 712–727. [CrossRef]
18. Fatahi, H.; Hossain, M.; Fallahzadeh, S.H.; Mostofi, M. Numerical simulation for the determination of hydraulic fracture initiation and breakdown pressure using distinct element method. *J. Nat. Gas Sci. Eng.* **2016**, *33*, 1219–1232. [CrossRef]
19. Tang, C.A.; Xu, T.; Yang, T.H.; Liang, Z.Z. Numerical investigation of the mechanical behavior of rock under confining pressure and pore pressure. *Int. J. Rock Mech. Min. Sci.* **2004**, *41*, 336–341. [CrossRef]
20. Yue, Z.Q.; Chen, S.; Tham, L.G. Finite element modeling of geomaterials using digital image processing. *Comput. Geotech.* **2003**, *30*, 375–397. [CrossRef]
21. Li, Q.; Xing, H.; Liu, J.; Liu, X. A review on hydraulic fracturing of unconventional reservoir. *Petroleum* **2015**, *1*, 8–15. [CrossRef]
22. Chen, S.; Yue, Z.Q.; Tham, L.G. Digital Image Based Approach for Three-Dimensional Mechanical Analysis of Heterogeneous Rocks. *Rock Mech. Rock Eng.* **2006**, *40*, 145–168. [CrossRef]
23. Zhu, W.C.; Liu, J.H.; Yang, T.C.; Sheng, J.; Elsworth, D. Effects of local rock heterogeneities on the hydromechanics of fractured rocks using a digital-image-based technique. *Int. J. Rock Mech. Min. Sci.* **2006**, *43*, 1182–1199. [CrossRef]
24. Hambli, R. Micro-CT finite element model and experimental validation of trabecular bone damage and fracture. *Bone* **2013**, *56*, 363–374. [CrossRef] [PubMed]
25. Ruggieri, S.; Cardellicchio, A.; Leggieri, V.; Uva, G. Machine-learning based vulnerability analysis of existing buildings. *Autom. Constr.* **2021**, *132*, 103936. [CrossRef]
26. Mahabadi, O.K.; Randall, N.X.; Zong, Z.; Grasselli, G. A novel approach for micro-scale characterization and modeling of geomaterials incorporating actual material heterogeneity. *Geophys. Res. Lett.* **2012**, *39*, L01303. [CrossRef]
27. Wu, M.Y.; Zhang, D.M.; Wang, W.S.; Li, M.H.; Liu, S.M.; Lu, J.; Gao, H. Numerical simulation of hydraulic fracturing based on two-dimensional surface fracture morphology reconstruction and combined finite-discrete element method. *J. Nat. Gas Sci. Eng.* **2020**, *82*, 103479. [CrossRef]
28. Yu, Q.; Yang, S.; Ranjith, P.G.; Zhu, W.; Yang, T. Numerical Modeling of Jointed Rock Under Compressive Loading Using X-ray Computerized Tomography. *Rock Mech. Rock Eng.* **2016**, *49*, 877–891. [CrossRef]
29. Tang, C.A.; Webb, A.A.G.; Moore, W.B.; Wang, Y.Y.; Ma, T.H.; Chen, T.T. Breaking Earth's shell into a global plate network. *Nat. Commun.* **2020**, *11*, 3621. [CrossRef]
30. Wang, S.Y.; Sloan, S.W.; Tang, C.A. Three-Dimensional Numerical Investigations of the Failure Mechanism of a Rock Disc with a Central or Eccentric Hole. *Rock Mech. Rock Eng.* **2013**, *47*, 2117–2137. [CrossRef]
31. Li, T.; Li, L.; Tang, C.; Zhang, Z.; Li, M.; Zhang, L.; Li, A. A coupled hydraulic-mechanical-damage geotechnical model for simulation of fracture propagation in geological media during hydraulic fracturing. *J. Pet. Sci. Eng.* **2019**, *173*, 1390–1416. [CrossRef]
32. Liang, Z.; Wu, N.; Li, Y.; Li, H.; Li, W. Numerical Study on Anisotropy of the Representative Elementary Volume of Strength and Deformability of Jointed Rock Masses. *Rock Mech. Rock Eng.* **2019**, *52*, 4387–4402. [CrossRef]
33. Wu, N.; Liang, Z.-Z.; Li, Y.-C.; Li, H.; Li, W.-R.; Zhang, M.-L. Stress-dependent anisotropy index of strength and deformability of jointed rock mass: Insights from a numerical study. *Bull. Eng. Geol. Environ.* **2019**, *78*, 5905–5917. [CrossRef]
34. Li, G.; Tang, C.-A.; Liang, Z.-Z. Development of a parallel FE simulator for modeling the whole trans-scale failure process of rock from meso- to engineering-scale. *Comput. Geosci.* **2017**, *98*, 73–86. [CrossRef]

35. Xia, Y.; Zhang, C.; Zhou, H.; Chen, J.; Gao, Y.; Liu, N.; Chen, P. Structural characteristics of columnar jointed basalt in drainage tunnel of Baihetan hydropower station and its influence on the behavior of P-wave anisotropy. *Eng. Geol.* **2020**, *264*, 105304. [CrossRef]
36. Wang, G.; Carr, T.R. Methodology of organic-rich shale lithofacies identification and prediction: A case study from Marcellus Shale in the Appalachian basin. *Comput. Geosci.* **2012**, *49*, 151–163. [CrossRef]
37. Chen, S.; Zhu, Y.; Wang, H.; Liu, H.; Wei, W.; Fang, J. Shale gas reservoir characterisation: A typical case in the southern Sichuan Basin of China. *Energy* **2011**, *36*, 6609–6616. [CrossRef]
38. Sone, H.; Zoback, M.D. Mechanical properties of shale-gas reservoir rocks—Part 2: Ductile creep, brittle strength, and their relation to the elastic modulus. *Geophysics* **2013**, *78*, D393–D402. [CrossRef]
39. Ouchi, H.; Agrawal, S.; Foster, J.T.; Sharma, M.M. Effect of Small Scale Heterogeneity on the Growth of Hydraulic Fractures. In Proceedings of the SPE Hydraulic Fracturing Technology Conference and Exhibition, The Woodlands, TX, USA, 24–26 January 2017.
40. Zhao, J.; Zhang, D. Dynamic microscale crack propagation in shale. *Eng. Fract. Mech.* **2020**, *228*, 106906. [CrossRef]
41. Tang, X.; Jiang, Z.; Jiang, S.; Li, Z. Heterogeneous nanoporosity of the Silurian Longmaxi Formation shale gas reservoir in the Sichuan Basin using the QEMSCAN, FIB-SEM, and nano-CT methods. *Mar. Pet. Geol.* **2016**, *78*, 99–109. [CrossRef]
42. Gu, X.; Cole, D.R.; Rother, G.; Mildner, D.F.R.; Brantley, S.L. Pores in Marcellus Shale: A Neutron Scattering and FIB-SEM Study. *Energy Fuels* **2015**, *29*, 1295–1308. [CrossRef]
43. Kumar, V.; Sondergeld, C.; Rai, C.S. Effect of mineralogy and organic matter on mechanical properties of shale. *Interpretation* **2015**, *3*, SV9–SV15. [CrossRef]
44. Veytskin, Y.B.; Tammina, V.K.; Bobko, C.P.; Hartley, P.G.; Clennell, M.B.; Dewhurst, D.N.; Dagastine, R.R. Micromechanical characterization of shales through nanoindentation and energy dispersive x-ray spectrometry. *Géoméch. Energy Environ.* **2017**, *9*, 21–35. [CrossRef]
45. Li, C.; Ostadhassan, M.; Kong, L.; Bubach, B. Multi-scale assessment of mechanical properties of organic-rich shales: A coupled nanoindentation, deconvolution analysis, and homogenization method. *J. Pet. Sci. Eng.* **2019**, *174*, 80–91. [CrossRef]
46. Voltolini, M.; Ajo-Franklin, J. Evolution of propped fractures in shales: The microscale controlling factors as revealed by in situ X-Ray microtomography. *J. Pet. Sci. Eng.* **2020**, *188*, 106861. [CrossRef]
47. Healy, D.; Jones, R.R.; Holdsworth, R.E. Three-dimensional brittle shear fracturing by tensile crack interaction. *Nature* **2006**, *439*, 64–67. [CrossRef]
48. Gou, Q.; Xu, S.; Hao, F.; Yang, F.; Zhang, B.; Shu, Z.; Zhang, A.; Wang, Y.; Lu, Y.; Cheng, X.; et al. Full-scale pores and micro-fractures characterization using FE-SEM, gas adsorption, nano-CT and micro-CT: A case study of the Silurian Longmaxi Formation shale in the Fuling area, Sichuan Basin, China. *Fuel* **2019**, *253*, 167–179. [CrossRef]
49. Ma, Y.; Pan, Z.; Zhong, N.; Connell, L.D.; Down, D.I.; Lin, W.; Zhang, Y. Experimental study of anisotropic gas permeability and its relationship with fracture structure of Longmaxi Shales, Sichuan Basin, China. *Fuel* **2016**, *180*, 106–115. [CrossRef]
50. Yu, Q.; Liu, H.; Yang, T.; Liu, H. 3D numerical study on fracture process of concrete with different ITZ properties using X-ray computerized tomography. *Int. J. Solids Struct.* **2018**, *147*, 204–222. [CrossRef]
51. Xu, H.; Wang, G.; Fan, C.; Liu, X.; Wu, M. Grain-scale reconstruction and simulation of coal mechanical deformation and failure behaviors using combined SEM Digital Rock data and DEM simulator. *Powder Technol.* **2020**, *360*, 1305–1320. [CrossRef]
52. Goehring, L. *On the Scaling and Ordering of Columnar Joints*; University of Toronto: Toronto, ON, Canada, 2008.
53. Goehring, L.; Morris, S.W. Order and disorder in columnar joints. *Eur. Lett.* **2005**, *69*, 739–745. [CrossRef]
54. Chavdarian, G.V.; Sumner, D.Y. Origin and evolution of polygonal cracks in hydrous sulphate sands, White Sands National Monument, New Mexico. *Sedimentology* **2011**, *58*, 407–423. [CrossRef]
55. Hetényi, G.; Taisne, B.; Garel, F.; Médard, É.; Bosshard, S.; Mattsson, H.B. Scales of columnar jointing in igneous rocks: Field measurements and controlling factors. *Bull. Volcanol.* **2011**, *74*, 457–482. [CrossRef]
56. Lin, Z.; Xu, W.; Wang, H.; Zhang, J.; Wei, W.; Wang, R.; Ji, H. Anisotropic characteristic of irregular columnar-jointed rock mass based on physical model test. *KSCE J. Civ. Eng.* **2017**, *21*, 1728–1734. [CrossRef]
57. Jin, C.; Yang, C.; Fang, D.; Xu, S. Study on the Failure Mechanism of Basalts with Columnar Joints in the Unloading Process on the Basis of an Experimental Cavity. *Rock Mech. Rock Eng.* **2015**, *48*, 1275–1288. [CrossRef]
58. Meng, Q.-X.; Wang, H.-L.; Xu, W.-Y.; Chen, Y.-L. Numerical homogenization study on the effects of columnar jointed structure on the mechanical properties of rock mass. *Int. J. Rock Mech. Min. Sci.* **2019**, *124*, 104127. [CrossRef]
59. Tang, C.; Liu, H.; Lee, P.; Tsui, Y.; Tham, L. Numerical studies of the influence of microstructure on rock failure in uniaxial compression—Part I: Effect of heterogeneity. *Int. J. Rock Mech. Min. Sci.* **2000**, *37*, 555–569. [CrossRef]
60. Tang, C.; Tham, L.; Lee, P.; Tsui, Y.; Liu, H. Numerical studies of the influence of microstructure on rock failure in uniaxial compression—Part II: Constraint, slenderness and size effect. *Int. J. Rock Mech. Min. Sci.* **2000**, *37*, 571–583. [CrossRef]
61. Cao, D.; Hou, Z.; Liu, Q.; Fu, F. Reconstruction of three-dimension digital rock guided by prior information with a combination of InfoGAN and style-based GAN. *J. Pet. Sci. Eng.* **2021**, *208*, 109590. [CrossRef]
62. Han, J.; Han, S.; Kang, D.H.; Kim, Y.; Lee, J.; Lee, Y. Application of digital rock physics using X-ray CT for study on alteration of macropore properties by CO2 EOR in a carbonate oil reservoir. *J. Pet. Sci. Eng.* **2020**, *189*, 107009. [CrossRef]
63. Kelly, S.; El-Sobky, H.; Torres-Verdín, C.; Balhoff, M.T. Assessing the utility of FIB-SEM images for shale digital rock physics. *Adv. Water Resour.* **2016**, *95*, 302–316. [CrossRef]

applied sciences

MDPI

Article

Dynamic Mechanical Behavior of the Frozen Red Sandstone under Coupling of Saturation and Impact Loading

Junce Xu [1], Hai Pu [1,2,*] and Ziheng Sha [1]

[1] State Key Laboratory for Geomechanics and Deep Underground Engineering, China University of Mining and Technology, Xuzhou 221116, China; xujunce@cumt.edu.cn (J.X.); zhsha@cumt.edu.cn (Z.S.)
[2] College of Mining Engineering and Geology, Xinjiang Institute of Engineering, Urumqi 830091, China
* Correspondence: haipu@cumt.edu.cn

Abstract: Saturation is one of the critical factors causing frost damage to rock masses in alpine regions, and dynamic stress perturbations further complicate the damage process. Therefore, the effects of water content and loadings should be considered in the construction and maintenance of rock structures during winter in cold regions. In this study, the effects of saturation and impact loading on the dynamic mechanical behavior of frozen red sandstone were investigated using a low-temperature split Hopkinson pressure bar system (LT-SHPB). By combining low-field nuclear magnetic resonance (LF-NMR) and scanning electron microscopy (SEM), the dynamic evolution of the microstructure of the frozen sandstone due to changes in saturation was investigated. The results indicated that the increase of saturation reshapes the pore structure of the frozen sandstone and promotes the expansion of pores of different sizes during freezing, while at complete saturation the frozen samples are mainly developed with meso- and macropores. The dynamic strength, elastic modulus, and brittleness index of the frozen sandstone under impact loading, which are limited by the critical saturation Src, tend to increase and then decrease with saturation. For the four impact loads, the dynamic strength of the samples increased by 21.2%, 27.1%, 32.5%, and 34.3% when the saturation was increased from 0 to 50%, corresponding to 1.38, 1.43, 1.51, and 1.56 times the dynamic strength of the fully saturated samples, respectively. In contrast, the ultimate deformation capacity of the frozen sandstone showed an opposite trend with saturation. As the impact load increases, the dynamic strength, elastic modulus, and peak strain of the frozen sandstone show a significant strengthening effect due to the increase in strain rate, while its brittleness index gradually decreases, dropping by 11.2% at full saturation. In addition, the energy dissipation capacity of the frozen sample first increases and then decreases with increasing saturation, with the enhancement effect of saturation on energy dissipation smaller than the weakening effect.

Keywords: saturation; frozen sandstone; SHPB; dynamic mechanics; brittleness

Citation: Xu, J.; Pu, H.; Sha, Z. Dynamic Mechanical Behavior of the Frozen Red Sandstone under Coupling of Saturation and Impact Loading. *Appl. Sci.* **2022**, *12*, 7767. https://doi.org/10.3390/app12157767

Academic Editors: Zhengzhao Liang, Bei Jiang and Nuwen Xu

Received: 6 July 2022
Accepted: 1 August 2022
Published: 2 August 2022

1. Introduction

Under the influence of surface runoff and groundwater migration or seepage, water is not evenly distributed in the rock mass, resulting in significant variations in saturation. At low temperatures (<0 °C), the water in the pores turns into ice, resulting in a volume increase of approximately 9% and reshaping the pore structure [1,2]. As a result, there are differences in the mechanical responses of rocks with different saturations after freezing [3]. In addition, rock structures in cold regions often suffer from dynamic loading [4]. For example, rock slopes in open-pit mines are affected by frost damage and are also susceptible to blasting, leading to geological hazards on the slopes [5]. Therefore, a comprehensive understanding of the dynamic mechanical behavior of frozen rock with different saturation is essential for the safe operation of rock structures in cold regions.

The mechanical properties of frozen rock under static loading have been extensively studied. At low temperatures, the static mechanical properties of rocks, such as strength, modulus, cohesion, and internal friction angle, are significantly improved [6,7]. The presence of pore ice is a key factor in improving the properties of rocks at low temperatures [8,9]. For example, dry rocks show little change in mechanical properties at low temperatures, while the properties of saturated rock (e.g., strength or modulus) show an increasing tendency, with the effect being exacerbated in colder environments [10,11]. Unlike other factors such as temperature, freezing time, etc., the mechanical response of rocks at low temperatures is closely related to saturation, and their mechanical properties are controlled by the coupling of water weakening and ice solidification. However, when saturation exceeds a critical value, the volume growth of pore ice leads to damage, which is manifested by a rapid increase in local strain and the number of internal microcracks [12]. By performing strain and acoustic emission tests on frozen sandstone with different satu-rations, Huang et al. [7] found that significant strain and acoustic emission signals were present during freezing only when the saturation exceeded 80%. However, current studies on the mechanical behavior of rock at low temperatures have mainly focused on saturated rock, especially under quasi-static loading, and may not fully reflect the mechanical behavior under dynamic loading.

The dynamic mechanical characteristics of rocks at low temperatures have attracted more attention with the increase in rock structures in cold regions [13,14]. The split Hopkinson pressure bar (SHPB) system is the main equipment for dynamic testing [15,16]. The SHPB test system has been used to investigate the dynamic mechanical properties and deformation behavior of rock under freeze–thaw action [17–19]. It was found that the dynamic strength and deformation resistance decreased with freeze–thaw cycles [20]. However, for some rock projects built in cold regions, the study of rock dynamics at low temperatures is of more practical importance than the cyclic freeze–thaw action [21]. For example, wet rocks in seismic- or blast-disturbed zones show softening effects under the coupling of low temperature and dynamic loading, which is the main source of geological hazards [22]. To this end, Chen et al. [5] conducted a series of SHPB tests on frozen saturated sandstones and concluded that the dynamic mechanical properties of saturated frozen rocks are significantly different from those of dry rocks due to pore ice. As noted by Weng et al. [23], saturation is a key factor affecting the dynamic strength of frozen rock. However, research on dynamic mechanical properties of the rock at low temperatures is limited and focuses mainly on dry or saturated conditions.

Therefore, samples with different saturations were prepared using red sandstone from Xinjiang, China, and impact tests were conducted using a LT-SHPB system. Based on the results, the effects of saturation on dynamic strength, deformation, brittleness, and energy conversion of the frozen sample were investigated. In addition, LT-NMR and SEM were used to investigate the changes in the microstructure of the samples. Finally, the effects of saturation on the microstructure and mechanical response of the sample were discussed.

2. Material and Methods

2.1. Material and Samples Preparation

The distribution of cold regions in China is shown in Figure 1. Xinjiang Province belongs to the shallow and medium-depth seasonal permafrost regions, which makes rock work in this region vulnerable to frost damage [4]. At the same time, the area is a base for open-pit coal mining in China, and winter workings are prone to landslides and other geological hazards [5]. Therefore, the red sandstone used for testing was obtained from an open-pit mine in Urumqi (Figure 1), which is structurally intact, uniformly granular, and brownish red in color. The primary mineral composition of the rock was determined by X-ray diffraction (XRD) (Figure 2), revealing that the sandstone is composed mainly of quartz (47.3%), feldspar (31.5%), and illite (9%), with minor amounts of calcite and chlorite.

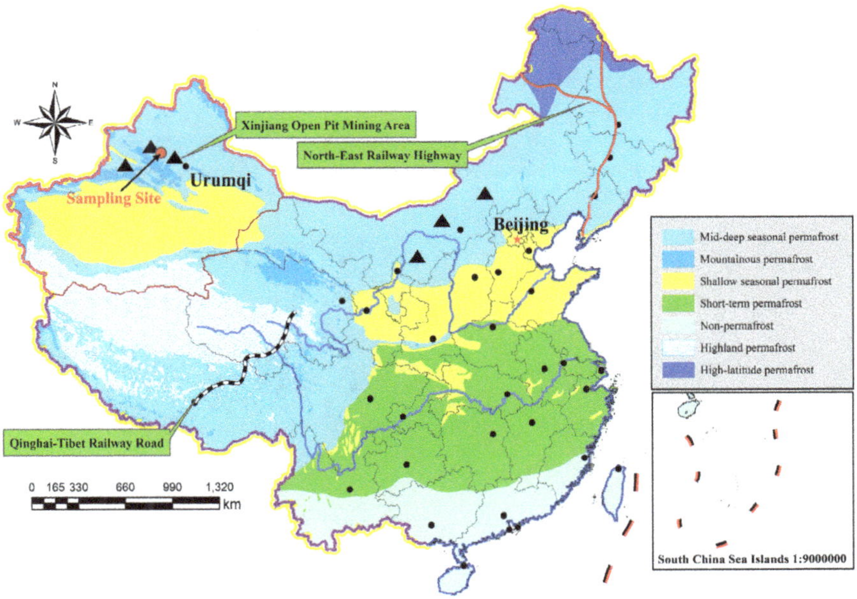

Figure 1. Distribution of cold regions in China and sampling points for the red sandstone sample; the red circles refer to the approximate locations of the sandstone sampling.

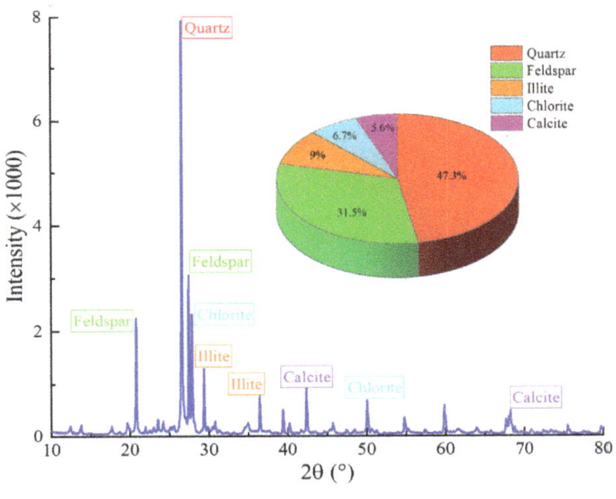

Figure 2. XRD result for the tested sandstone.

As shown in Figure 3, fresh red sandstone block was processed into standard samples with a diameter of 50 mm and a length-to-diameter ratio of 1 [24]. To reduce the influence of rock structure, all samples were taken from the same block and in the same direction. Moreover, both ends of the sample had to be ground to a tolerance of 0.02 mm, with no more than 0.1% rad from the perpendicular to the axis. Once the samples were machined, an ultrasonic detector was used to measure the P-wave velocity and select samples with similar P-wave velocity. In addition, five samples with a length and diameter of 50 × 100 mm were selected for physical and mechanical tests, with the results listed in Table 1.

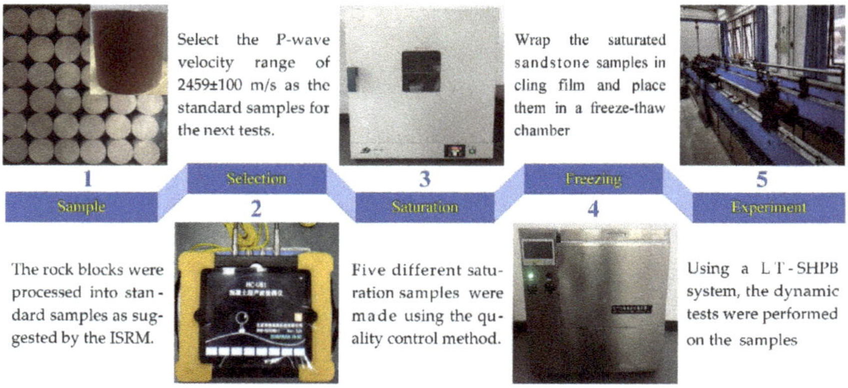

Figure 3. Experimental procedure and equipment.

Table 1. The physical and mechanical parameters of selected samples.

V_P(m/s)	ρ_d/(g·cm^{-3})	ρ_{sat}/(g·cm^{-3})	n/(%)	σ_p/(MPa)	η/(%)
2459	2.14	2.31	16.57	13.84	7.94

Notes: V_P, P-wave velocity; ρ_d, dry density; ρ_{sat}, saturation density; n, porosity; σ_p, uniaxial compressive strength; η, water content.

To obtain the samples with different saturations (Figure 3), the selected samples were first dried in a drying oven at 105 °C for 48 h until their mass remained constant. The samples were then saturated with pure water for 24 h using a vacuum saturator. Thereafter, the samples were evaporated in a drying oven at 50 °C and weighed at intervals of 0.5 to 1 h to obtain the specified saturation [25]. The saturations for the test were set to dry (0%), 30%, 50%, 80%, and fully saturated (100%). Finally, considering the winter temperatures at the sampling site, the prepared samples were placed in a freeze–thaw chamber to freeze at −25 °C for 12 h [4]. Following the methods described above, frozen samples with different saturations were obtained for SHPB tests at low temperatures.

2.2. Experimental System

2.2.1. LT-NMR System

In this study, a LT-NMR analysis system developed by Suzhou Niumag Ltd. is used. The resonance frequency of the device is 12.8 MHz, while the main magnetic field strength is approximately 0.3 T. The transverse relaxation time T_2 can be tested with the CPMG pulse sequence. The samples were frozen at 0%, 30%, 50%, 80%, and 100% saturation in a freeze–thaw chamber at −25 °C for 24 h. After freezing, samples were placed in water at 20 °C for 24 h before LT-NMR testing [6].

2.2.2. LT-SHPB System

As shown in Figure 4, a LT-SHPB system developed by China University of Mining and Technology was adopted to conduct the impact tests. The system consists of five main components: an impact loading system, a compression bar system, a data acquisition system, a damping system, and a low-temperature freezing system. In the experimental system, the bars and the striker are made of high-strength Cr40 alloy steel with diameters, P-wave velocities, yield strengths, and elastic modulus of 50 mm, 5400 m/s, 800 MPa, and 208 GPa, respectively. The low-temperature freezing system consists of a temperature controller, a liquid nitrogen tank, an electric heating wire sensor, and a cold chamber. The cold chamber, as the key structure of the system, has a good insulating effect. Its inner insulation layer is mainly composed of aluminum silicate acupuncture blanket and insulating brick.

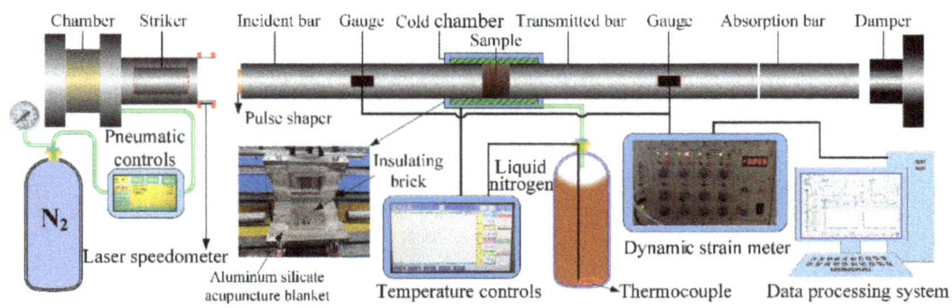

Figure 4. The LT-SHPB system.

Before the test, the chamber temperature was set to −25 °C. The frozen sandstone sample was then quickly placed in the cold chamber. Molybdenum disulfide was applied at the ends as a lubricant to reduce the friction effect between the bars and the sample. It is worth mentioning that the sample should remain in the chamber for at least 0.5 h [5]. To obtain the desired waveform, a circular rubber pad with a diameter of 15 mm and a thickness of 2 mm was used as a pulse shaper. A compressive wave was generated when the high-pressure gas propelled the striker against the incident bar. Part of the compressive wave propagated through the sample to the transmission bar. As a result, the strain signal generated on the bar was measured by the strain gauges, as shown in Figure 5. Based on the one-dimensional stress wave propagation theory, the three-wave analysis method was used to calculate the strain and stress [26], as shown in Equation (1).

$$\begin{cases} \sigma = \frac{A_r E_0}{2}[\varepsilon_i(t) + \varepsilon_r(\mathrm{t}) + \varepsilon_t(t)] \\ \varepsilon = \frac{C_s}{L_s}\int_0^t [\varepsilon_i(t) - \varepsilon_r(t) - \varepsilon_t(t)]dt \\ \dot{\varepsilon} = \frac{C_s}{L_s}[\varepsilon_i(t) - \varepsilon_r(t) - \varepsilon_t(t)] \end{cases} \quad (1)$$

where σ, ε, and $\dot{\varepsilon}$ refer to dynamic stress, dynamic stress, and strain rate, respectively; A_r is the ratio of the cross-sectional area of the bar to the sample; E_0, C_s, and L_s are the elastic modulus, the wave velocity of the bar, and the length of the sample, respectively; $\varepsilon_i(t)$, $\varepsilon_r(t)$, and $\varepsilon_t(t)$ refer to the incident, reflected, and transmitted strain, respectively. In addition, the striker impact velocity is measured by a laser beam measurement system.

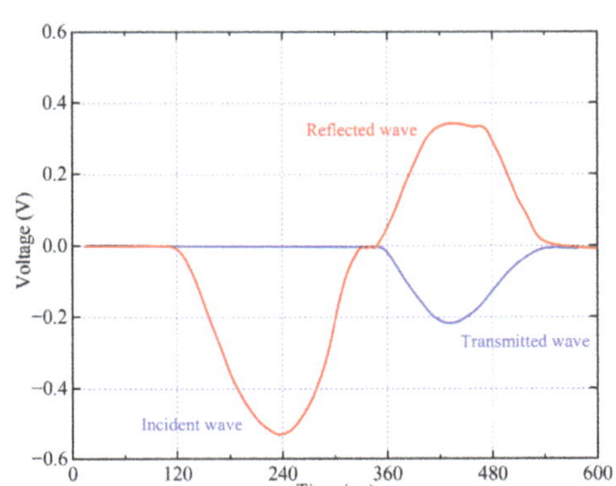

Figure 5. Typical strain gauge signals.

2.3. Dynamical Testing Scheme

As mentioned earlier, samples with five different saturations were selected for the impact tests. Four test groups were designed using impact velocity as the loading condition:

4 m/s, 5 m/s, 6 m/s, and 7 m/s. Each group contains three samples. As a result, 60 samples were required to carry out the tests, with five additional samples prepared as spares. For simplicity, the samples were numbered, e.g., D-80-7 denotes the frozen samples at 80% saturation with an impact velocity of 7 m/s at low temperatures. Moreover, the uniformity of the stress at both ends of the samples must be verified before performing the dynamic tests. As shown in Figure 6, the superposition of incident and reflected waves largely coincides with the transmitted waves, which means that the stresses at both ends of the sample are in equilibrium during loading [24].

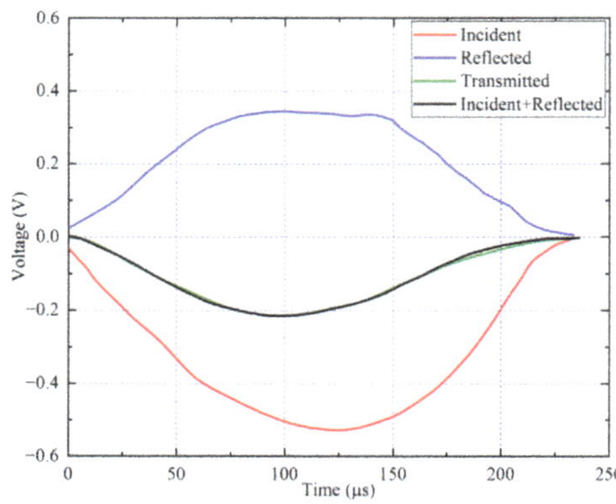

Figure 6. Verification of test stress uniformity hypothesis.

3. Experimental Results and Analysis

3.1. LT-NMR Results

Low-field NMR tests were performed on the samples to investigate the effect of water content on the microstructure at low temperatures. In the NMR study, the intensity of the magnetic resonance signal (MR) represents the amount of hydrogen-containing liquid in the sample, while the change rate of the MR signal over time (relaxation time T_2) is related to the state of the molecules in the liquid. Therefore, analyzing the relaxation time of the hydrogen-containing liquid could provide a comprehensive view of the pore space distribution in the rock [27]. For a hydrogen-containing fluid (water) in the rock, the transverse relaxation time T_2 is:

$$\frac{1}{T_2} = \rho(\frac{S}{V})_{\text{pore}} \tag{2}$$

where T_2 is the transverse relaxation time (ms); ρ indicates the surface relaxation strength, which is related to the surface of the rock particles and the nature of the cement [13]; and S/V refers to the specific surface of the pores in the rock, which is related to the shape of the pore, e.g., if the pore is spherical, it is taken as 3/R. Therefore, observing the characteristics of the T_2 spectrum of the sample can indirectly obtain the distribution pattern of different pore sizes.

The distribution curves of the T_2 spectrum of the sandstone samples for 0%, 30%, 50%, 80%, and 100% are shown in Figure 7. The T_2 spectrum distribution mainly has three peaks, indicating that three sizes of pore structures were mainly formed in the sample. Moreover, the area under the peaks is mainly between 0.1~10 ms. According to Yao et al. [28], there is a correlation between T_2, pore diameter, and water type. For example, when T_2 is 10~100 ms, the pore size is 0.1~1 μm and is called mesopore, which is mainly composed of capillary water [27]. As shown in Figure 7, for the dried sample, the area of the T_2 spectrum per size decreases compared to other saturated samples. As saturation increases, the area occupied by the second peak gradually increases from 2.19% (dry) to 9.32% (100%), showing that the increase in saturation has a significant effect on the development of mesopores in the sample. When the saturation exceeds 80%, the area of the third peak increases by 3.57%, indicating that the macropores of the sample have developed further. Thus, saturation has a strong influence on the pore development of the sample at low temperatures. When the saturation is less than 80%, the micropores and mesopores in the sample spread predominantly, while the macropores of the sample gradually develop when the saturation is greater than 80%. In other words, the higher the water content, the more significant the increase in porosity of the sample after freezing. This indicates that the increasing moisture accelerates the expansion of the pores inside the sample and increases the sensitivity of the sample to the freeze heave forces.

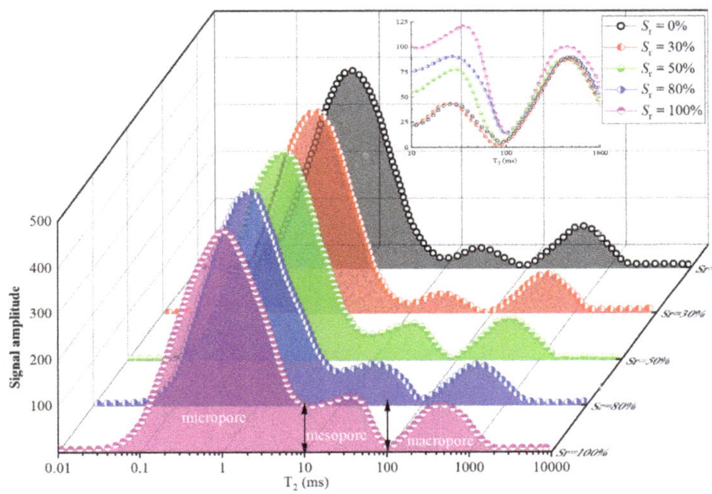

Figure 7. Distribution of the T_2 spectrum of the samples with different saturations.

3.2. Dynamic Impact Test Results

3.2.1. Dynamic Stress–Strain Behavior

In this study, four impact velocities were determined by adjusting N_2 pressure and the position of the striker [15]. The dynamic stress–strain curves of the sample with different saturations are shown in Figure 8a–e, while Figure 8f presents a typical stress–strain curve of the sample. The stress–strain behavior of frozen sandstone at low temperature has four stages, as shown in Figure 8f: (I) the initial compression stage, in which defects such as pores or microcracks in the sandstone gradually close under the action of compressive stress waves; (II) the linear elastic stage, in which elastic energy is rapidly accumulated; (III) the plastic deformation stage, in which defects such as microcracks in the sandstone begin to develop rapidly, characterized by a rapid increase in strain and a slow increase in stress; the initial and final points of the plastic stage are the yield point and the peak point of the sample with coordinates (σ_e, ε_e) and (σ_p, ε_p); and (IV) the post-peak stage, which is rarely mentioned because the samples are detached from the compression bars during this stage [5].

As can be seen in Figure 8, the stress–strain curves of the dried specimen show a distinct compression stage and decrease with the increasing impact velocity. Since the pore water freezes into ice with some bearing capacity at low temperatures, stage I gradually decreases with increasing saturation. For example, the curves show no compaction process when the saturation reaches 50%. On the other hand, thermal shrinkage of some mineral particles reduces the defect volume, thus reducing the initial damage. Moreover, the plastic stage III is closely related to saturation and impact velocity, which first decreases and then increases with increasing saturation. According to Equation (1), the dynamic stress versus time is shown in Figure 9. As the impact velocity increases, the plastic deformation occurs earlier, which is consistent with the results of Xing et al. [29]. In addition, the time of plastic deformation first increases and then decreases with increasing saturation. Moreover, the time to reach the dynamic stress peak is negatively correlated with the impact velocity, while it is positively correlated with the impact strength. To clearly describe the changes in dynamic behavior, several parameters, namely peak strength, elastic modulus, and brittleness index (*BI*), were used to assess the effects of saturation and impact loading on the dynamic mechanical properties of the frozen sample. The results are listed in Table 2.

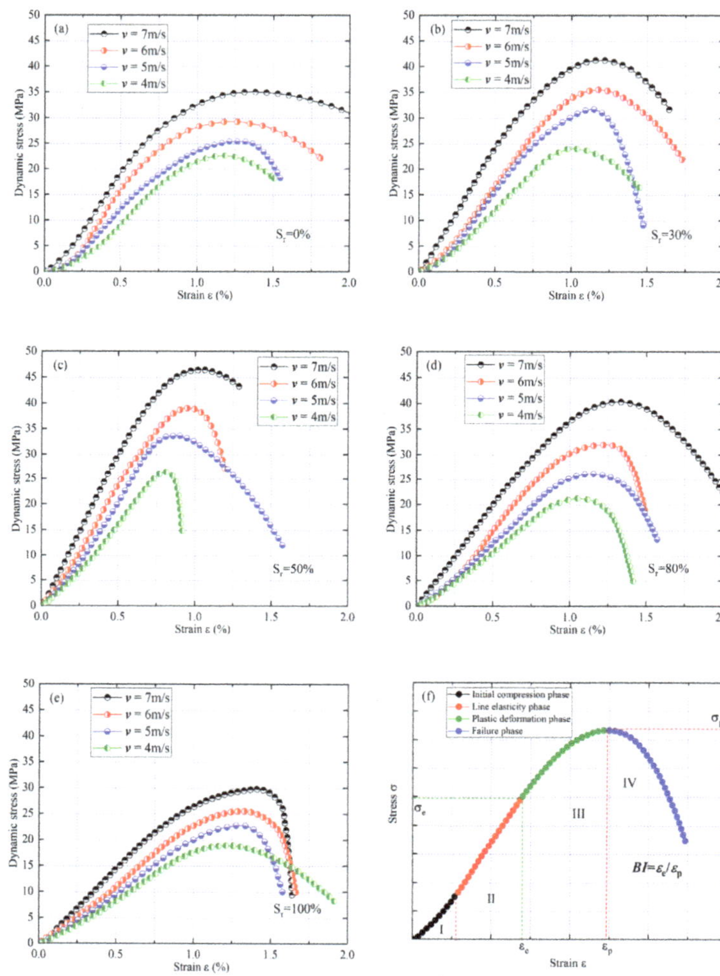

Figure 8. Dynamic stress–strain curves of the sample at low temperatures: (**a**–**e**) denote samples with saturation of 0%, 30%, 50%, 80%, and 100%, respectively; (**f**) shows the deformation phase divisions of typical stress–strain curves.

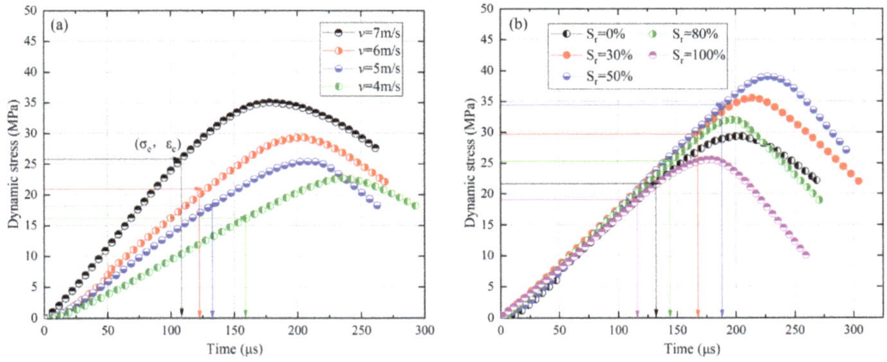

Figure 9. Dynamic stress versus time for frozen samples: (**a**) S_r = 0%; (**b**) v= 6 m/s.

Table 2. Dynamic mechanical parameters of sandstone under different conditions.

Samples	Impact Velocity (m/s)	Dynamic Strength (MPa)	Dynamic Modulus (GPa)	Peak Strain (%)	Brittleness Index
D-0-4	4	22.1	1.95	1.167	0.728
D-0-5	5	25.7	2.04	1.247	0.694
D-0-6	6	29.3	2.28	1.286	0.657
D-0-7	7	34.6	2.59	1.353	0.602
D-30-4	4	24.6	2.39	1.015	0.753
D-30-5	5	31.3	2.83	1.094	0.721
D-30-6	6	35.5	3.03	1.176	0.697
D-30-7	7	42.8	3.42	1.218	0.665
D-50-4	4	26.4	3.26	0.812	0.862
D-50-5	5	33.8	3.8	0.891	0.819
D-50-6	6	39.7	4.07	0.961	0.760
D-50-7	7	46.5	4.47	1.041	0.682
D-80-4	4	21.3	2.02	1.049	0.788
D-80-5	5	26.2	2.27	1.154	0.745
D-80-6	6	32.6	2.58	1.238	0.703
D-80-7	7	40.3	3.06	1.318	0.675
D-100-4	4	19.4	1.71	1.134	0.744
D-100-5	5	23.6	1.83	1.292	0.712
D-100-6	6	25.5	1.92	1.331	0.684
D-100-7	7	28.8	2.11	1.411	0.659

Dynamic Strength

According to the data obtained from Table 2, the relationship between dynamic strength as a function of impact velocity and saturation of the frozen sandstone is shown in Figure 10. The dynamic strength of the sample at low temperatures increases linearly with increasing impact velocity. For example, when the impact velocity increased from 4 m/s to 7 m/s, the dynamic strengths of the sample with different saturations increased by 1.53 (0%), 1.77 (30%), 1.74 (50%), 1.89 (80%), and 1.48 (100%) times, respectively. However, it can be seen in Figure 10b that the dynamic strength of the sample does not increase uniformly with saturation. The dynamic strength (v = 5 m/s) increased by 24.9%, 32.7%, 2.86%, and −7.31% when the saturation increased from 0 to 100%. This indicates that the enhancement effect of pore ice is gradually suppressed by frost damage when the saturation of red sandstone exceeds 80%. At the same time, the enhancement effect of the frozen samples is less than the weakening effect at low temperatures when the saturation increases from 0 to 100% (Figure 10b). In general, the dynamic strength of frozen rock at low temperatures still has the strain rate effect, but it increases and then decreases with saturation.

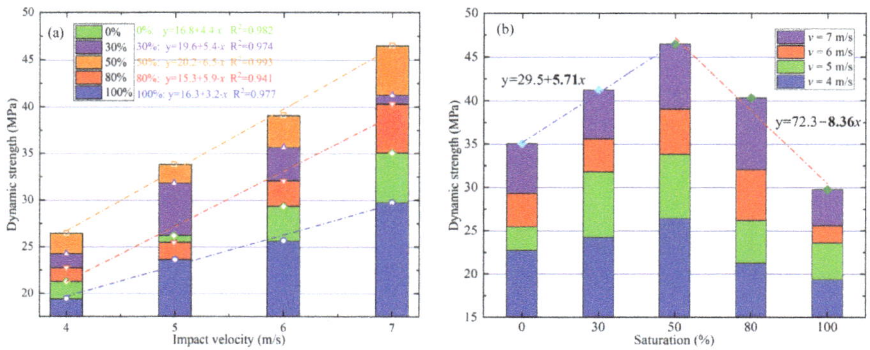

(**a**) (**b**)

Figure 10. Dynamic peak strength versus impact velocity (**a**) and saturation (**b**) of the sample.

In addition, as shown in Figure 11, a dynamic strength increase factor (DIF) is introduced to evaluate the strength increase rate for the frozen samples. The DIF is determined using Equation (3), where $f_v(d)$ is the dynamic strength at the impact velocity v and $f(s)$ is the static strength [30]. From this, it can be seen that the DIF of the sample increases with increasing saturation and then decreases. For an impact velocity of 6 m/s, DIF increases from 2.11 to 2.87 as the saturation increases from 0 to 50%, while it decreases to 2.31 as saturation increases to 80%. As a result, it is reasonable to infer a critical saturation (S_{rc}) between 50% and 80% for the samples at low temperatures. When the saturation is greater than the critical saturation, the strength enhancement caused by the pore ice is gradually inhibited by the damage caused by the volume expansion. As the samples approach full saturation, the strength increase is further suppressed by the cracking and expansion of the pores under the action of the frost heave force.

$$DIF = \frac{f_v(d) - f(s)}{f(s)} \tag{3}$$

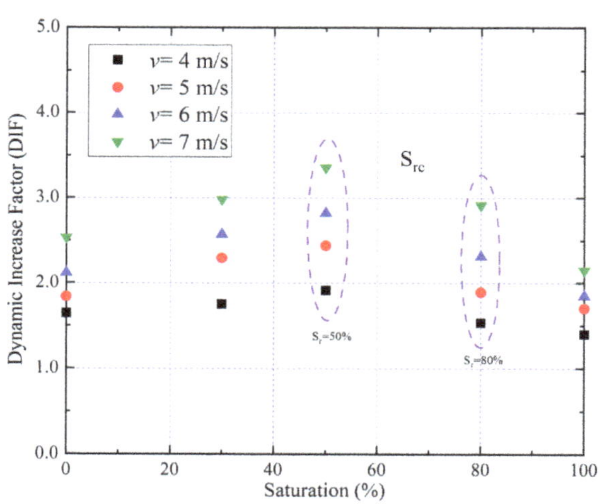

Figure 11. DIF versus impact velocity and saturation of the sample.

Indeed, the volume expansion (9%) caused by the pore ice exerts a frost heave force on the rock skeleton, thereby affecting cracking and expansion in the rock. Huang et al. [7]

found that the sample saturation is a crucial factor for controlling the frost heave force. The relationship between frost heave force P and saturation S_r is expressed in Equation (4) [31].

$$\begin{cases} P = \frac{[\beta n S_r u - n(1-S_r)]K_r K_i}{K_r S_r + K_i} H[u-\chi] \\ H[u-\chi] = \begin{cases} 0 & (u \leq \chi) \\ 1 & (u > \chi) \end{cases} \\ \chi = \frac{1-S_r}{\beta S_r} \end{cases} \quad (4)$$

where β is the volume expansion coefficient of ice; n represents the porosity; u denotes the freezing rate; K_r and K_i are the bulk modulus of the rock skeleton and the ice, respectively; and $H\,[u-\chi]$ is the step function. From Equation (4), the frost heave force is affected by the porosity, saturation, freezing rate, and bulk modulus of ice and rock skeleton. Here, β, K_r, and K_i are set to 0.09, 20.83 GPa, and 5 GPa, respectively [32]. The step function defaults to 1 and S_r values range from 0.8 to 1. Substituting the parameters into Equation (4) yields the relationship between frost heave stress, saturation, and freezing rate, which is shown in Figure 11.

As shown in Figure 12, the frost heave force is positively related to saturation; yet, even at 100% saturation, the effective heave force can still occur at lower freezing rates. As the freezing rate increases, the intensity factor of the frozen stress in the pore space exceeds the critical intensity factor K_{IC}, leading to the initiation and propagation of cracks and thus internal damage to the sample [33]. As a result, the dynamic strength of the sample increases and decreases with increasing saturation. However, due to the weak cementation of the rock in Xinjiang [34], the critical saturation of the red sandstone in the actual tests is between 50% and 80%, which is less than the predicted threshold of 91% for frost deformation failure.

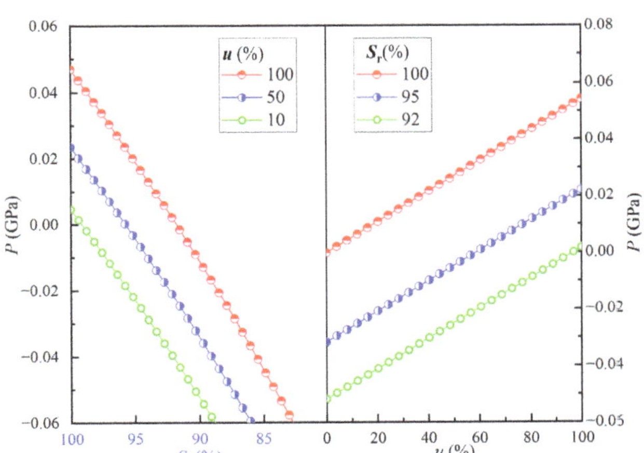

Figure 12. Frost heaving force versus saturation and freezing rates.

Dynamic Elastic Modulus

Following Figure 8f, the tangential modulus was adopted as the dynamic elastic modulus (E_d) in this study. The variation of E_d with impact velocity and saturation is shown in Figure 13. In Figure 13a, the E_d of the sample at low temperatures initially shows an increasing trend with saturation. For example, when the saturation of the sample increases from 0 to 50%, the E_d increases by 72.6% from 2.59 GPa to 4.47 GPa at an impact velocity of 7 m/s. However, as saturation increases (>50%), the increasing trend is suppressed and the E_d of the sample reduces by 18.5% to 2.11 GPa at complete saturation. The increase in deformation resistance is due to the pore ice acting as a support effect for the pore wall and enhancing the cementation between the wall and the ice, facilitating the propagation of stress waves. As a result, the pore ice reduces the stress concentration at the pore tip [2], leading to an increase in dynamic modulus. However, as saturation increases, new microcracks appear in the sample under the influence of freeze heave forces and the enhancement of pore ice gradually decreases, leading to a decrease in dynamic elastic modulus. Moreover, as shown in Figure 13b, the E_d of the sample increases linearly with the impact velocity at low temperatures. Under the five saturation conditions, the E_d increases rapidly by 32.8 %, 43.1 %, 37.1 %, 51.5 %, and 23.4%, respectively, when the impact velocity increases from 4 m/s to 7 m/s, indicating that the stiffness property of the frozen sample has a significant impact on the strengthening effect.

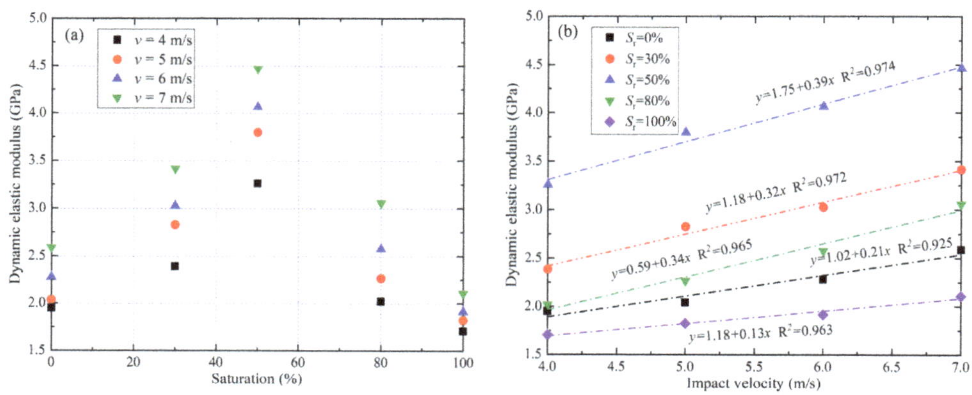

(a) (b)

Figure 13. Dynamic elastic modulus versus saturation (**a**) and impact velocity (**b**) of the sample.

Dynamic Brittleness Index

Brittleness is an important parameter to describe the deformation and failure process of the rock [35]. Various external factors such as water content and temperature inevitably change the brittle state and failure pattern of rock [24]. Therefore, a brittleness index (*BI*) is used to characterize the effect of saturation on the pre-peak deformation characteristics of the frozen sample, as follows:

$$BI = \frac{\varepsilon_e}{\varepsilon_p} \quad (5)$$

where ε_e and ε_p refer to the yield point strain and peak point strain in the stress–strain curve, respectively. If *BI* has the value of 1, the sample is considered to be completely brittle, i.e., the larger the value of *BI*, the more brittle the sample is. From Table 2, the minimum value of *BI* is 0.602 and the maximum value is 0.862. The surface of *BI* for frozen sandstone as a function of initial saturation and impact velocity is shown in Figure 14.

As shown in Figure 14, the *BI* values of the sample decrease with increasing impact velocity, suggesting that impact velocity promotes the transition from very brittle to ductile frozen sandstone. Thus, *BI* values decrease by 17.4 %, 11.6 %, 20.9 %, 14.3 %, and 11.3 % at five saturation states, respectively, when the impact velocity is increased from 4 m/s to 7 m/s. In terms of energy, an increase in *BI* value means an increase in the elastic energy accumulation stage (II) and a decrease in the energy release stage (III) within the sample. Therefore, microcracks are less likely to propagate in the frozen sample at low impact velocities, consistent with the experimental results of Chen et al. [5]. In contrast, the *BI* value of the frozen sample tends to rise with increasing saturation due to pore ice. As a result, the stress–strain curves show a shortening in the stage (III) and an increase in the stage (II). However, when saturation exceeds critical saturation, pore ice plays a role in eliminating internal defects and drives crack propagation by frost heave forces. Therefore, two effects play a critical role in controlling the deformation behavior of the frozen rock. For the tested rock, when the saturation is close to 80 %, the crack development within the sample under frost heave forces plays an important role. For example, when the saturation increases from 50 % to 80 %, the *BI* values decrease by 8.54 %, 9.04 %, 7.49 %, and 1.02 % at four impact velocities, respectively. Therefore, it can be inferred that the saturation distribution of the rock masses in the cold regions could affect the deformation characteristics under the impact loading.

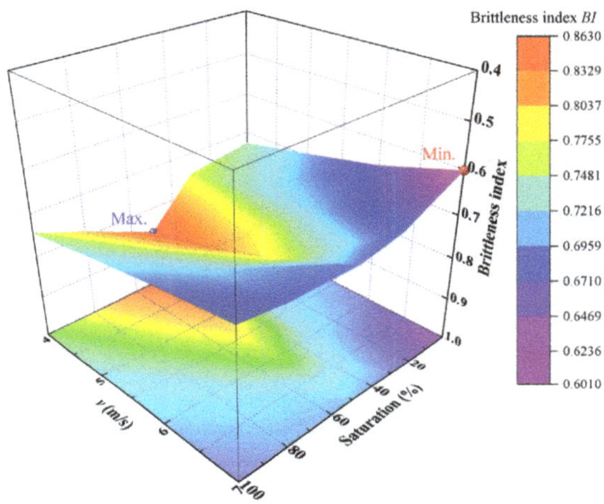

Figure 14. *BI* values versus impact velocity and saturation of the sample.

3.3. Energy Evolution

The failure of rock under impact loading can be considered as the result of the mutual transformation of the different forms of energy. During impact testing, the main forms of energy include the incident energy E_{in}, reflected energy E_{re}, transmitted energy E_{ti}, dissipated energy E_{di}, and other magnetic heat energy, with dissipated energy reflecting impact energy converted into fracture energy and other energy. From a microscopic point of view, energy dissipation reflects the continuous development of microdefects in the rock, while macroscopically it is represented as the gradual loss of rock strength. The energy evolution of the frozen sample under the impact load is calculated by Equation (6) [2], where A_s is the cross-sectional area of the incident bar.

$$\begin{cases} E_{in} = E_0 C_s A_s \int_0^t \varepsilon_i{}^2(t)dt \\ E_{re} = E_0 C_s A_s \int_0^t \varepsilon_r{}^2(t)dt \\ E_{tr} = E_0 C_s A_s \int_0^t \varepsilon_t^2(t)dt \\ E_{di} = E_{in} - (E_{re} + E_{tr}) \end{cases} \tag{6}$$

The energy conversion for frozen samples with different saturations is shown in Figure 15. As the impact velocity increases, all forms of energy show an increasing trend, with a quadratic polynomial accurately reflecting the changes. Moreover, the dissipated energy gradually overtakes the reflected energy as the saturation increases (Figure 15a). However, when the saturation increases above 50%, the reflected energy gradually increases faster than the dissipated energy (Figure 15b). This is due to the continuous increase in pore ice, which leads to a decrease in the wave impedance of the samples. To better assess the effect of saturation on the dissipated energy of the frozen sample, Equation (7) was used to calculate the dissipated energy per unit volume, η_d. The evolution of η_d with saturation for different impact velocities is shown in Figure 16. Taking 5 m/s as an example, η_d of the dried sample is 0.67 J/cm^3 at low temperatures. As saturation increases, η_d begins to increase and reaches 0.76 J/cm^3 and 0.82 J/cm^3 at 30% and 50% saturation, corresponding to an increase of 13.4 % and 22.4 %, respectively, over the dried sample. However, at 80 % and 100 % saturation, there is a significant decrease in η_d to 0.56 J/cm^3 and 0.52 J/cm^3, respectively. Therefore, it can be inferred that saturation has a considerable effect on the energy dissipation capacity of the frozen sample, where the range of 0~50% increases the energy dissipation capacity, while 50~100 % shows the effect of weakening. The effect of increasing the dissipated energy is less than the effect of weakening, indicating that the rock sample is sensitive to the work caused by the internal frost heave forces. In other words, when the saturation increases above 50 %, the frost heave forces promote the formation of more microcracks in the sample, which decreases the dissipated energy demand.

$$\eta_d = \frac{E_{di}}{V_s} \tag{7}$$

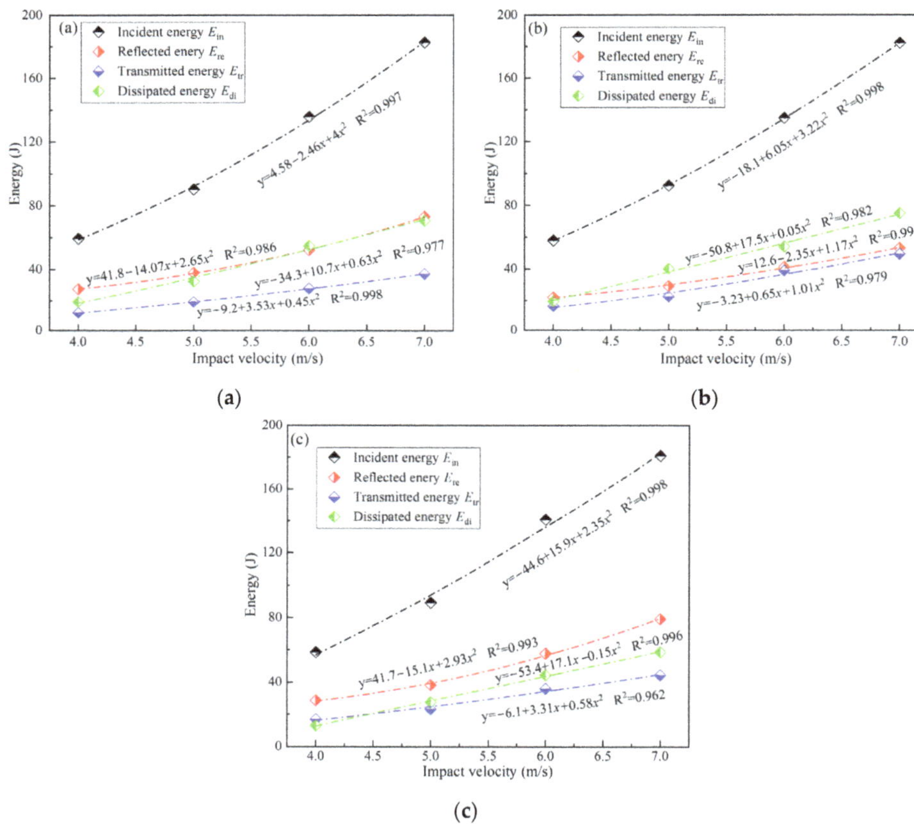

Figure 15. Evolution of rock energy distribution with saturation of (**a**) S_r = 0%, (**b**) S_r = 50%, and (**c**) S_r = 100%.

3.4. Macro- and Microscopic Failure Features

The fracture behavior of rock is closely related to the impact velocity and microstructural changes during impact testing [36]. Hence, studying the fracture state during impact can show how saturation affects the mechanical properties of rock at low temperatures. After the tests, frozen sandstone fragments with different saturations were collected and dried. The results are presented in Table 3. It is obvious that the sandstone samples are completely broken at different impact velocities due to their relatively low strength. At low impact velocity (4 m/s), there were large broken blocks in the samples with different saturations, especially blocks approximately 5.9 cm in length in the sample with 50% saturation.

At this time, the pore ice within the sample played a role in supporting the pore wall and cementation, so that some of the crack development at the pore tip was suppressed under the action of the stress waves, reflecting the increase in the volume of macroscopically fractured blocks. Meanwhile, the plastic deformation before the peak during the impact test implies the development and extension of the internal microcracks, so the increase of plastic deformation indicates the increase of fragmentation degree. As can be seen in Figure 8, the plastic deformation increases with saturation and then decreases, which corresponds to the same trend in the fragmentation degree. Moreover, plastic deformation increases with the impact velocity, while the powder particles of the broken samples gradually increase in size. At an impact velocity of 7 m/s, the samples were broken into powder, indicating that the frozen sample mainly changed from tensile failure to complex fracture. Therefore, the final failure morphology of the samples can be used to infer the role of pore ice.

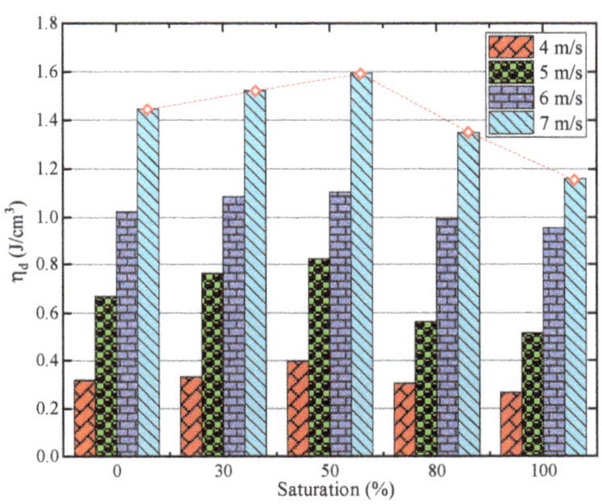

Figure 16. Dissipated energy per unit volume (η_d) of samples at low temperatures.

In addition, as shown in Figure 17, SEM images are used to infer changes in the microscopic morphology of the red sandstone after freezing. The images demonstrate that saturation has a significant effect on the surface morphology of the sample, and that many microcracks have formed after the freezing treatment. In the original state, the internal structure of the sandstone is relatively loose (Figure 17a) and has many small primary cracks, pores, and other defects, some of which can be up to 4 μm in width. After freezing, the surface structure becomes dense due to thermal contraction of the mineral particles, and some of the original cracks close, with a maximum crack width as small as 2 μm (Figure 17b). In Figure 17c, some of the previously closed cracks reactivate and propagate at low saturation as water freezes to ice at low temperatures. When the saturation increases to 50% (Figure 17d), some cracks develop further and connect, but most cracks are less than 1 μm in width, and the damaging effect of pore ice on the matrix is still less than its strengthening effect. However, when the sample saturation reaches 80%, the increasing pore water facilitates the continued expansion of the pore ice, leading to the development of larger cracks, some of which are up to 6 μm in width (Figure 17e). When the sample is fully saturated, a block appears on the fracture surface, indicating that the volume of ice expands even after the cracks are penetrated, causing the block to be extruded; meanwhile, the cracks on the fracture surface reach approximately 9 μm in width (Figure 17f). In summary, as the saturation increases, microcracks begin to develop inside the sample, while the strengthening of the pore ice prevents the manifestation of the internal damage to a certain extent. However, when the saturation exceeds 50%, the pores inside the sample cannot provide sufficient expansion space for the ice, resulting in a fracture network inside the sample. As a result, the strength and deformation resistance tend to decrease as the sample approaches full saturation.

Table 3. Impact failure modes of the frozen sample.

S_r	S_r = 0%	S_r = 30%	S_r = 50%	S_r = 80%	S_r = 100%
Impact velocity 4 (m/s)	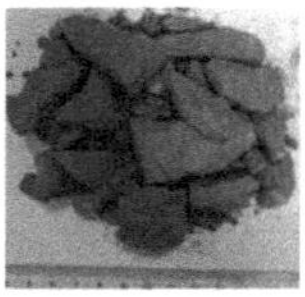	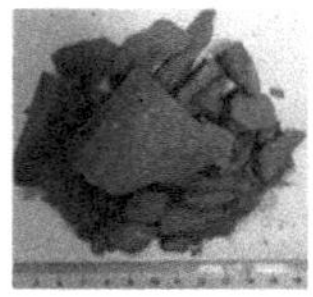		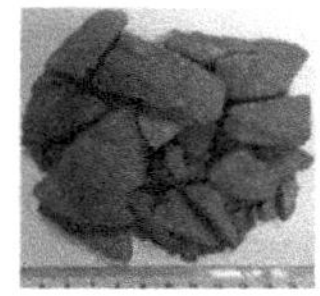	
Impact velocity 5 (m/s)	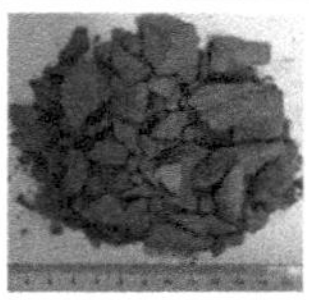				
Impact velocity 6 (m/s)			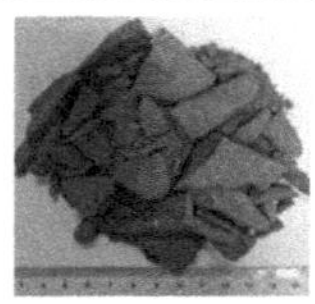		
Impact velocity 7 (m/s)		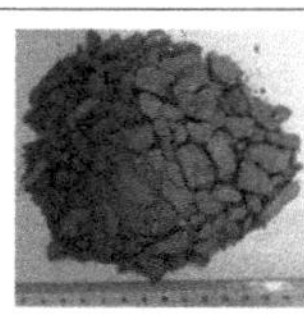			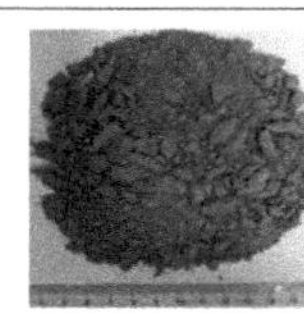

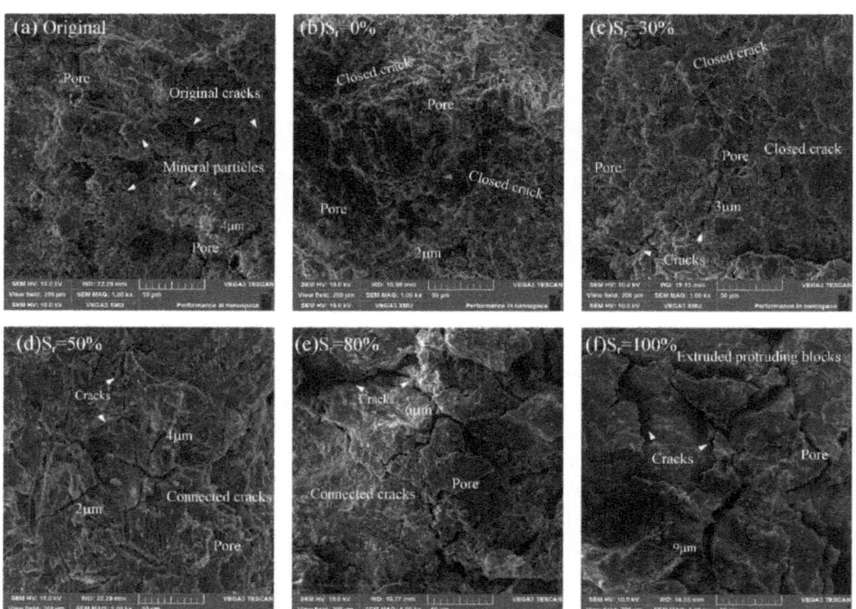

Figure 17. Microscopic characteristics of frozen sandstone sample with different saturation from SEM images (1000×).

4. Discussion

Due to groundwater migration and seepage, pore water in natural rock is not uniformly distributed, resulting in significant variations in rock saturation [3]. When the temperature drops below 0 °C, the pore water in the rock begins to freeze into ice. The pore ice remolds the pore structure, leading to dramatic differences in mechanical properties (e.g., strength, deformation). Under impact testing, the pore ice reduces the stress concentration in the crack tips, preventing the formation of some cracks caused by compressive stress waves. For example, as saturation increases, the plastic deformation of the sample first decreases and then increases. Moreover, the enhanced phenomenon is observed in static tests, such that the strength, elastic modulus, and brittleness are usually greatly increased at low temperatures [7,32]. However, as saturation increases, the rock interior is damaged by the expansion of pore ice, weakening the strengthening effect [9]. When the two effects coexist, the saturation of the rock determines which process dominates by affecting the ice content in the pore. In other words, saturation is crucial for enhancing or damaging pore ice at low temperatures. Consequently, the influence of saturation on the mechanical properties of the samples needs further discussion.

In the dry or low-water-content frozen samples, where the internal pores are mainly occupied by air, the changes in dynamic properties during freezing are related only to the shrinkage properties of the mineral particles. As the saturation increases, the presence of pore ice becomes the main reason for the change in mechanical properties [6]. When the saturation of the sample is low, the growth of pore ice volume is limited due to the lack of pore water [37]. At this time, the pore ice in the rock has a strengthening effect, which mainly consists of three aspects: the supporting effect increases the dynamic strength or deformation resistance of the rock; the crack-filling effect reduces the stress concentration at the crack tips; and the adhesion forces between the unfrozen film water with the rock matrix and the ice improve the mechanical properties of the pore in terms of tensile and shear strength (Figure 18). When the pore water freezes into ice, the dynamic mechanical properties of the sample tend to strengthen, while the degree of fracture decreases. Moreover, the plastic deformation of the sample under dynamic compressive stress decreases with increasing saturation, which is due to the limitation of microcracks development inside the sample. However, when the saturation gradually increases (>50%), the dynamic strength of the sample decreases significantly, especially at 100% saturation. The continuous expansion of the pore ice gradually leads to the germination of microcracks in the specimens. Therefore, microcracks are more likely to develop earlier during impact loading, leading to an increase in plastic deformation (Figure 8) [1]. When saturation exceeds 50% in this work, there is sufficient water in the rock to provide a basis for the growth of frost heave during freezing. Therefore, the role of pore water in the strengthening or weakening effect needs further investigation.

As shown in Figure 18c, the pore distribution characteristics of the sample mainly include main pores, secondary pores, branch pores, and isolated pores. Since the freezing point of water in pores is inversely proportional to the pore radius, the water in the main pores freezes first and then grows into the secondary pores [27]. When the water content of the sample is low (S_r = 30%), most of the unfrozen water migrates through the water film into the main pores and ice growth stops after a certain volume because there is insufficient water [25]. Consequently, the T_2 spectrum of the red sandstone in Figure 7 shows no significant change. At this time, the pore ice mainly has a strengthening effect. As saturation increases, the ice in the main pores continues to grow due to sufficient unfrozen water and drives some of the water into the secondary pores. If the pore is closed, a water pressure P_w is created in the pore. When the P_w exceeds the tensile strength of the rock skeleton, triggering the spreading of the secondary pore is the primary reason for the change in the second peak of the T_2 spectrum. When the pore is connected, the hydrostatic pressure P_i generated by the pore ice drives the unfrozen water seepage to form damage and exerts a dynamic water pressure on the rock skeleton, weakening the deformation resistance of the rock matrix [38]. The pressure Pi also provides a supporting force for the pore wall. At the same time, the mechanical properties of ice are similar to those of rock, which reduces the wave impedance under the compressive stress wave and effectively reduces the dissipated energy. Therefore, cracks are less likely to develop in samples under impact loading due to the effect of pore ice, indicated macroscopically by an increase in the elastic phase and a decrease in the plastic deformation phase (Figure 8).

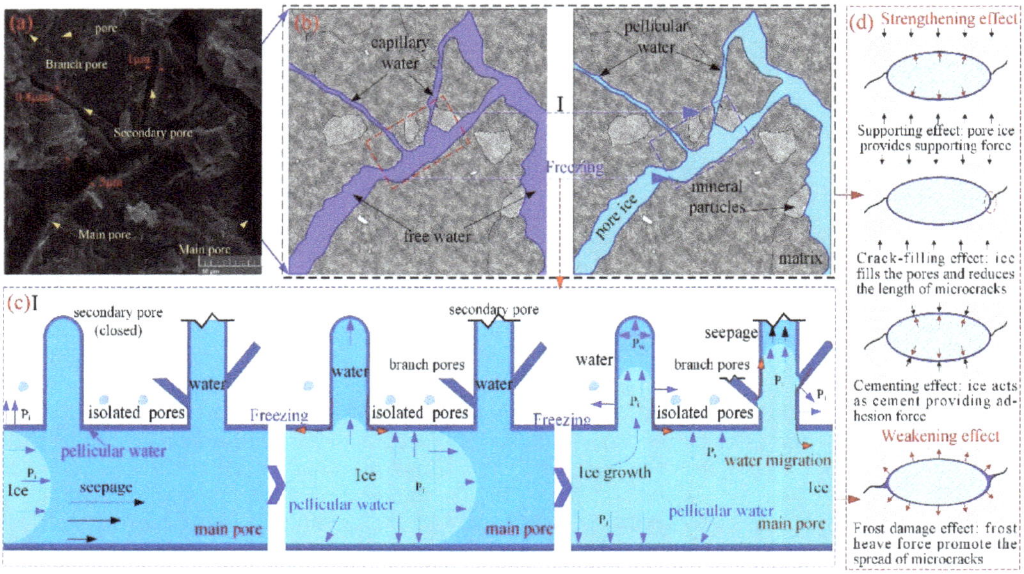

Figure 18. Expansion behavior and influence mechanism of pore ice during sandstone freezing: (**a**) SEM image; (**b**) abstracted pore structure model; (**c**) freezing process; (**d**) pore ice effects.

In addition, the ice in the main pore gradually shifts to the secondary pores during the freezing process, with the capillary mechanism playing an important role in this process [25]. Due to the relatively low chemical potential of the ice in the main pore, the supercooled water in the small pores migrates along the thin water film to the larger pores [6], resulting in a continuous growth of ice in the main pores and a gradual increase in pressure P_i. When the difference between the ice pressure P_i and the water pressure P_w is $(2\gamma_{sl}/R) \cdot \cos\theta$ (γ_{sl} is the interfacial tension between ice and water; R is the equivalent capillary radius; and θ is the contact angle [27]), the pore ice gradually develops toward the capillary pores and generates frozen expansion, which promotes the pore development. In general, the dynamic mechanical properties of frozen rock are not always enhanced with increasing saturation, with the strengthening effect suppressed when the saturation exceeds a critical value. The strengthening effect is less than the weakening effect when the saturation of the red sandstone exceeds 80%. Therefore, from the test results, the dynamic strength, modulus, and dissipated energy of the frozen sandstone gradually decrease when the saturation exceeds 80%, while the ultimate deformation capacity and broken degree progressively increase. In summary, as saturation increases, pore ice gradually transitions from filling, supporting, and bonding effects to frost damage.

5. Conclusions

The effects of saturation and impact loading on the dynamic mechanical behavior of the frozen red sandstone in the Urumqi area were investigated using a LT-SHPB system. Moreover, LT-NMR and SEM methods were employed to analyze the microstructural evolution of the frozen sandstone with different saturations. The main conclusions are as follows:

1. During freezing at low temperatures, when the saturation was below the critical saturation, the sample was dominated by the development of micropores, while meso- and macropores mainly propagated within the saturated sample, as evidenced by an increase in the second and third peaks in the T_2 spectrum of 9.32% and 3.57%, respectively. Under low temperature conditions, meso- and macropores were more sensitive to saturation and more susceptible to crystallization pressure.
2. As saturation increased, the dynamic strength, elastic modulus, and brittleness of the frozen sandstone tended to increase and then decrease. At four impact loads, the strength of the samples increased by 21.2%, 27.1%, 32.5%, and 34.3% when the saturation increased from 0% to 50%, which corresponded to 1.38, 1.43, 1.51, and 1.56 times the strength of the fully saturated samples, respectively. In addition, the dynamic mechanical properties of the frozen sandstone showed a significant effect of impact strengthening with increasing impact velocity.
3. The energy dissipation capacity of the frozen red sandstone was significantly affected by its saturation at low temperatures. When the saturation increased from dry to 50%, the energy dissipation capacity during impact gradually increased, while the energy dissipation capacity gradually decreased as saturation increased from 50% to 100%. The strengthening effect reached its maximum when the saturation of the sample was 50%. In addition, the enhancement effect of saturation on energy dissipation was smaller than the weakening effect.

Author Contributions: Methodology, J.X. and H.P.; investigation, J.X. and Z.S.; resources, J.X.; data curation, Z.S. and J.X.; writing—original draft preparation, J.X.; writing—review and editing, J.X. and H.P.; visualization, Z.S. and J.X.; supervision, H.P. All authors have read and agreed to the published version of the manuscript.

Funding: This research was funded by the National Natural Science Foundation of China (No. 51974296, No. 52061135111).

Institutional Review Board Statement: Not applicable.

Informed Consent Statement: Not applicable.

Data Availability Statement: Data available on request due to privacy restrictions.

Conflicts of Interest: The authors declare no conflict of interest.

References

1. Wang, T.; Sun, Q.; Jia, H.; Shen, Y.; Li, G. Fracture Mechanical Properties of Frozen Sandstone at Different Initial Saturation Degrees. *Rock Mech. Rock Eng.* **2022**, *55*, 3235–3252. [CrossRef]
2. Zhou, Z.; Yude, E.; Cai, X.; Zhang, J. Coupled Effects of Water and Low Temperature on Quasistatic and Dynamic Mechanical Behavior of Sandstone. *Geofluids* **2021**, *2021*, 9926063. [CrossRef]
3. Song, Y.; Yang, H.; Tan, H.; Ren, J.; Guo, X. Study on damage evolution characteristics of sandstone with different saturations in freeze-thaw environment. *Chin. J. Rock Mech. Eng.* **2021**, *40*, 1513–1524.
4. Xu, J.; Pu, H.; Sha, Z. Mechanical behavior and decay model of the sandstone in Urumqi under coupling of freeze–thaw and dynamic loading. *Bull. Eng. Geol. Environ.* **2021**, *80*, 2963–2978. [CrossRef]
5. Chen, Y.; Huidong, C.; Ming, L.; Pu, H. Study on dynamic mechanical properties and failure mechanism of saturated coal-measure sandstone in open pit mine with damage under real-time low-temperature conditions. *J. China Coal Soc.* **2022**, *47*, 1168–1179.
6. Liu, B.; Sun, Y.; Yuan, Y.; Liu, X.; Bai, X.; Fang, T. Strength characteristics of frozen sandstone with different water content and its strengthening mechanism. *J. China Univ. Min. Technol.* **2020**, *49*, 1085–1093.
7. Huang, S.; Cai, Y.; Liu, Y.; Liu, G. Experimental and Theoretical Study on Frost Deformation and Damage of Red Sandstones with Different Water Contents. *Rock Mech. Rock Eng.* **2021**, *54*, 4163–4181. [CrossRef]
8. Jia, H.; Zi, F.; Yang, G.; Li, G.; Shen, Y.; Sun, Q.; Yang, P. Influence of Pore Water (Ice) Content on the Strength and Deformability of Frozen Argillaceous Siltstone. *Rock Mech. Rock Eng.* **2020**, *53*, 967–974. [CrossRef]
9. Wang, T.; Sun, Q.; Jia, H.; Ren, J.; Luo, T. Linking the mechanical properties of frozen sandstone to phase composition of pore water measured by LF-NMR at subzero temperatures. *Bull. Eng. Geol. Environ.* **2021**, *80*, 4501–4513. [CrossRef]
10. Yamabe, T.; Neaupane, K.M. Determination of some thermo-mechanical properties of Sirahama sandstone under subzero temperature condition. *Int. J. Rock Mech. Min. Sci.* **2001**, *38*, 1029–1034. [CrossRef]

11. Kodama, J.; Goto, T.; Fujii, Y.; Hagan, P. The effects of water content, temperature and loading rate on strength and failure process of frozen rocks. *Int. J. Rock Mech. Min. Sci.* **2013**, *62*, 1–13. [CrossRef]
12. Chen, T.C.; Yeung, M.R.; Mori, N. Effect of water saturation on deterioration of welded tuff due to freeze-thaw action. *Cold Reg. Sci. Technol.* **2004**, *38*, 127–136. [CrossRef]
13. Li, J.; Kaunda, R.B.; Zhou, K. Experimental investigations on the effects of ambient freeze-thaw cycling on dynamic properties and rock pore structure deterioration of sandstone. *Cold Reg. Sci. Technol.* **2018**, *154*, 133–141. [CrossRef]
14. Zhang, J.; Deng, H.; Deng, J.; Ke, B. Development of Energy-Based Brittleness Index for Sandstone Subjected to Freeze-Thaw Cycles and Impact Loads. *IEEE Access* **2018**, *6*, 48522–48530. [CrossRef]
15. Gong, F.; Yan, J.; Luo, S.; Li, X. Investigation on the Linear Energy Storage and Dissipation Laws of Rock Materials Under Uniaxial Compression. *Rock Mech. Rock Eng.* **2019**, *52*, 4237–4255. [CrossRef]
16. Jacob, B.J.; Misra, S.; Parameswaran, V.; Mandal, N. Control of planar fabrics on the development of tensile damage zones under high-speed deformation: An experimental study with granite and gneiss. *J. Struct. Geol.* **2020**, *140*, 104418. [CrossRef]
17. Liu, X.; Liu, Y.; Dai, F.; Yan, Z. Tensile mechanical behavior and fracture characteristics of sandstone exposed to freeze-thaw treatment and dynamic loading. *Int. J. Mech. Sci.* **2022**, *226*, 107405. [CrossRef]
18. Cai, X.; Zhou, Z.; Zang, H.; Song, Z. Water saturation effects on dynamic behavior and microstructure damage of sandstone: Phenomena and mechanisms. *Eng. Geol.* **2020**, *276*, 105760. [CrossRef]
19. Song, Z.; Wang, Y.; Konietzky, H.; Cai, X. Mechanical behavior of marble exposed to freeze-thaw-fatigue loading. *Int. J. Rock Mech. Min. Sci.* **2021**, *138*, 104648. [CrossRef]
20. Yang, R.; Fang, S.; Guo, D.; Li, W.; Mi, Z. Study on Dynamic Tensile Strength of Red Sandstone Under Impact Loading and Negative Temperature. *Geotech. Geol. Eng.* **2019**, *37*, 4527–4537. [CrossRef]
21. Yang, R.; Fang, S.; Li, W.; Yang, Y.; Yue, Z. Experimental Study on the Dynamic Properties of Three Types of Rock at Negative Temperature. Geotech. *Geol. Eng.* **2019**, *37*, 455–464.
22. Zakharov, E.V. Effects of negative temperatures on crushing rocks of various deposits in Yakutia. *Obogashchenie Rud* **2021**, *2021*, 3–9.
23. Weng, L.; Wu, Z.; Liu, Q.; Wang, Z. Energy dissipation and dynamic fragmentation of dry and water-saturated siltstones under sub-zero temperatures. *Eng. Fract. Mech.* **2019**, *220*, 106659. [CrossRef]
24. Xu, J.; Pu, H.; Sha, Z. Experimental Study on the Effect of Brittleness on the Dynamic Mechanical Behaviors of the Coal Measures Sandstone. *Adv. Civ. Eng.* **2021**, *2021*, 6679333. [CrossRef]
25. Wan, Y.; Chen, G.Q.; Sun, X.; Zhang, G.Z. Triaxial creep characteristics and damage model for red sandstone subjected to freeze-thaw cycles under different water contents. *Chin. J. Geotech. Eng.* **2021**, *43*, 1463–1472.
26. Li, X. (Ed.) *Rock Dynamic Fundamentals and Applications*; Science Press: Beijing, China, 2014; ISBN 987-7-03-040425-1.
27. Cheng, H.; Chen, H.; Cao, G.; Rong, C.; Yao, Z.; Cai, H. Damage mechanism of porous rock caused by moisture migration during freeze-thaw process and experimental verification. *Chin. J. Rock Mech. Eng.* **2020**, *39*, 1739–1749.
28. Yao, Y.; Liu, D.; Che, Y.; Tang, D.; Tang, S.; Huang, W. Petrophysical characterization of coals by low-field nuclear magnetic resonance (NMR). *Fuel* **2010**, *89*, 1371–1380. [CrossRef]
29. Xing, H.Z.; Zhang, Q.B.; Ruan, D.; Dehkhoda, S.; Lu, G.X.; Zhao, J. Full-field measurement and fracture characterisations of rocks under dynamic loads using high-speed three-dimensional digital image correlation. *Int. J. Impact Eng.* **2018**, *113*, 61–72. [CrossRef]
30. Dai, F.; Huang, S.; Xia, K.; Tan, Z. Some fundamental issues in dynamic compression and tension tests of rocks using split Hopkinson pressure bar. *Rock Mech. Rock Eng.* **2010**, *43*, 657–666. [CrossRef]
31. Kang, Y.; Liu, Q.; Zhao, J.; Zhang, F. Research on frost deformation characteristics of rock and simulation of tunnel frost deformatlon in cold region. *Chin. J. Rock Mech. Eng.* **2012**, *31*, 2518–2526.
32. Liu, B.; Sun, Y.; Han, Y.; Liu, N.; Li, T. Laboratory investigation on mechanical and hydraulic properties of sandstone under freeze–thaw cycle. *Environ. Earth Sci.* **2022**, *81*, 146. [CrossRef]
33. Liu, S.; Xu, J.; Liu, S.; Wang, P. Fractal study on the dynamic fracture of red sandstone after F-T cycles. *Environ. Earth Sci.* **2022**, *81*, 152. [CrossRef]
34. Sun, L. *Structural Evolution and Rock Pressure Activity Regularity of Weakly Cemented Strata of the Large Mining Height Work Face in Western China*; University of Science and Technology Beijing: Beijing, China, 2016; Volume 3.
35. Wu, N.; Liang, Z.Z.; Li, Y.C.; Li, H.; Li, W.R.; Zhang, M.L. Stress-dependent anisotropy index of strength and deformability of jointed rock mass: Insights from a numerical study. *Bull. Eng. Geol. Environ.* **2019**, *78*, 5905–5917. [CrossRef]
36. Zheng, G.H.; Xu, J.Y.; Wang, P.; Fang, X.Y.; Wang, P.X.; Wen, M. Physical characteristics and degradation model of stratified sandstone under freeze-thaw cycling. *Rock Soil Mech.* **2019**, *40*, 632–641.
37. Wang, B.; Sun, P.; Luo, T.; Zhang, T.; Yang, W. Freezing Pressurized Water into a Standard Cylindrical Ice Sample in a Triaxial Cell. *Geofluids* **2021**, *2021*, 6678966. [CrossRef]
38. Wu, J.; Wang, H.; Zong, Q.; Xu, Y. Experimental Investigation of Dynamic Compression Mechanical Properties of Frozen Fine Sandstone. *Adv. Civ. Eng.* **2020**, *2020*, 8824914. [CrossRef]

applied sciences

MDPI

Article

Tensile Properties and Tensile Failure Criteria of Layered Rocks

Min Gao [1,2,3], Zhengzhao Liang [2,*], Shanpo Jia [3] and Jiuqun Zou [1]

1 State Key Laboratory of Mining Response and Disaster Prevention and Control in Deep Coal Mines, Anhui University of Science and Technology, Huainan 232001, China; gaomin@aust.edu.cn (M.G.); 2020136@aust.edu.cn (J.Z.)

2 State Key Laboratory of Coastal and Offshore Engineering, Dalian University of Technology, Dalian 116024, China

3 Institute of Unconventional Oil & Gas, Northeast Petroleum University, Daqing 163318, China; jiashanporsm@163.com

* Correspondence: liangzz@dlut.edu.cn

Abstract: Rocks are less resistant to tension than to compression or shear. Tension cracks commonly initiate compression or shear failure. The mechanical behavior of layered rocks under compression has been studied extensively, whereas the tensile behavior still remains uncertain. In this paper, we study the effect of layer orientation on the strength and failure patterns of layered rocks under direct and indirect tension through experimental and numerical testing (RFPA^{2D}: numerical software of Rock Failure Process Analysis). The results suggest that the dip angle of the bedding planes significantly affects the tensile strength, failure patterns, and progressive deformation of layered rocks. The failure modes of the layered specimens indicate that the tensile strength obtained by the Brazilian disc test is not as accurate as that obtained by the direct tension test. Therefore, the modified Single Plane of Weakness (MSPW) failure criterion is proposed to predict the tensile strength of the layered rocks based on the failure modes of direct tension. The analytical predictions of the MSPW failure criterion agrees closely with the experimental and numerical results. In rock engineering, the MSPW failure criterion can conveniently predict the tensile strength and reflect the failure modes of layered rocks (such as shale, slate, and layered sandstone) with satisfactory accuracy.

Keywords: layered rocks; anisotropy; failure mode; tension test; tensile failure criterion

Citation: Gao, M.; Liang, Z.; Jia, S.; Zou, J. Tensile Properties and Tensile Failure Criteria of Layered Rocks. *Appl. Sci.* **2022**, *12*, 6063. https://doi.org/10.3390/app12126063

Academic Editor: Ricardo Castedo

Received: 22 May 2022
Accepted: 14 June 2022
Published: 15 June 2022

1. Introduction

Many types of rock mass, such as shale, slate, and sandstone, exhibit inherent anisotropic properties [1–4]. The uniaxial compressive and tensile strength of layered rocks are crucial parameters to consider at the design and construction stage of geotechnical structures such as tunnels, slopes, and other underground engineering projects. They are also vitally important for the development of unconventional oil and gas, such as shale gas [5,6]. However, only the anisotropic compressive (or shear) strength has been extensively studied from experimental, theoretical, and numerical points of view [7–11], but the anisotropic tensile strength still needs to be further investigated because tensile failure is extremely important in rock engineering. In fact, the tensile strength of layered rocks was previously assumed to be nil in many engineering applications. There are situations, however, where such an assumption is not safe. As a result, the tensile failure of rock and the resulting instability can lead to serious disasters, such as rock burst in hydropower caverns, landslide, dam break, and so on [12]. Therefore, understanding the tensile properties of layered rocks is extremely important to assess the stability of geotechnical and petroleum engineering projects.

There are two main methods to assess the tensile strength of layered rocks, including the direct uniaxial tension test and the indirect tension test, such as the Brazilian disc test (BDT) and other indirect testing methods [13]. The Brazilian disc test was officially proposed by the International Society for Rock Mechanics [14] as a suggested method for determining the tensile strength of rock materials. The Brazilian disc test is the most commonly adopted method to determine the indirect tensile strength of rock materials [15–23]. However, during the Brazilian test, a small eccentricity of the axial load may cause large non-uniformities of stress in specimens [24], or stress concentrations may be induced by caps. For transversely isotropic rock masses, failure under the Brazilian test is not necessarily a pure tensile failure; in some cases, it can even be pure shear failure, but it is often a combination of both tensile and shear failure [25–27]. Although these disadvantages are obviously present in the Brazilian test, it is still widely used in laboratory experimental and numerical studies to investigate the tensile properties of transversely isotropic rocks [28–34].

The direct tensile test, which is theoretically the simplest and most effective method for tensile strength determination, is in fact difficult to carry out in the laboratory [35–38]. Over the past few decades, the anisotropic tensile behavior of layered rocks has been studied by a few researchers through direct tension [39,40]. Some anisotropic tensile criteria were investigated to predict the tensile strength of layered rocks based on the direct tension test and analytical methods [41–43]. Ma et al. analyzed the anisotropic tension strength criteria of layered rocks and selected the best one to predict the fracture pressure of the inclined wells in layered formations [44,45]. Shang et al. investigated the anisotropic direct tensile behavior of laminated and transversely isotropic Midgley Grit sandstone using a particle-based discrete element method [46]. However, the tensile failure modes and failure mechanisms of layered rocks still need to be further studied. Furthermore, the failure criterion of the layered rocks under direct tension should be evaluated based on the failure modes.

In this work, numerical tests are undertaken to investigate the progressive failure and mechanical behavior of layered rocks under direct and indirect tension using a numerical code RFPA^{2D}, which was developed based on the finite element method, statistical theory, and damage mechanics [47]. The influence of layer orientation on tensile behavior is discussed. The main failure modes and failure mechanisms of layered rocks under direct and indirect tension are analyzed. The Single Plane of Weakness failure criterion is then modified based on the failure modes and failure mechanisms of direct tension. The modified SPW (MSPW) failure criterion has been verified by experimental and numerical results. The analytical predictions of the MSPW failure criterion agrees closely with the experimental and numerical results.

2. Experimental Study on Layered Rocks by the Brazilian Test

2.1. Test Equipment and Samples

Layered rock samples (shale in Figure 1) with different dip angles, θ (the angle between the bedding plane and the horizontal plane), as shown in Figure 2, were selected for the experiments. According to the method suggested by the International Society for Rock Mechanics (ISRM), the samples were carefully drilled from a large fresh shale block. They were machined into Brazilian disc cylinders (25 mm in thickness and 50 mm in diameter) with seven different dip angels ($0°, 15°, 30°, 45°, 60°, 75°, 90°$). These samples are displayed in Figure 1. The experiments were performed using a Rock Mechanics Testing System (RMT-150C). The axial displacement loading mode was selected, with a controlled rate of 0.0005 mm/s until the failure of the tested shale sample. A photograph and a schematic diagram of the experimental setup is presented in Figure 2.

Figure 1. Layered shale samples with different dip angles under the Brazilian test.

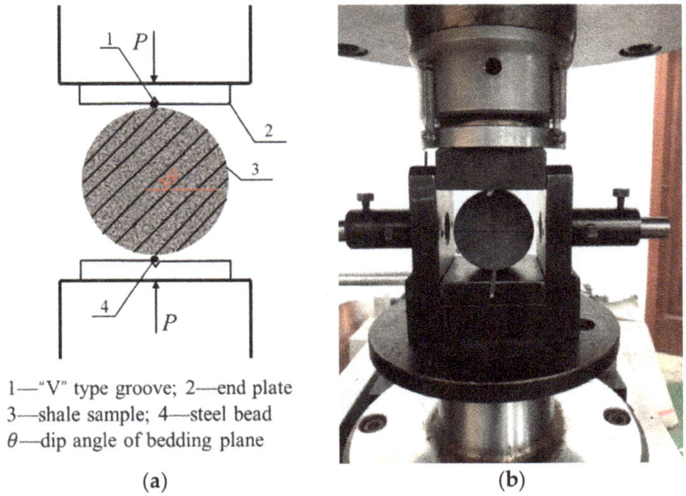

Figure 2. Loading schematic of the layered shale Brazilian test. (**a**) Loading schematic diagram; (**b**) Photograph of experimental setup.

According to the theory of the Brazilian disc test, the tensile strength of rock samples can be expressed as (negative sign represents tensile stress):

$$\sigma_t = -\frac{2P_t}{\pi D t} \tag{1}$$

where P_t is the failure load of the rock sample, D is the diameter of the rock sample, and t is the thickness of the rock sample.

2.2. Failure Mode of the Shale Samples

Figure 3 shows the typical failure modes of layered shale samples with different dip angles under the Brazilian test. When the dip angle of the sample was 0°, the shale sample almost cracked along the center line of the disc. When the dip angle of the shale sample increases to 30°, the shale sample first fractured near the loading end, and then cracked along the bedding plane. With the increasing dip angle of the samples, the shale samples almost presented shear failure along the bedding plane. The failure mode of the shale sample with a dip angle of 90° was similar to the sample at 0°. The failure modes of shale samples with different dip angles can be summarized as follows: splitting failure along the center line of the disc, composite failure with tensile and shear cracks, and shear failure along the bedding plane. The failure modes of the shale samples show anisotropic characteristics, and also indicate that only the shale samples with dip angles of 0° and 90°

experience splitting failure along the center line of the disc, which satisfies the assumption of Equation (1). In another words, the tensile strength of the shale samples with different dip angles obtained by the Brazilian test are not quite accurate, except for the shale samples with dip angles of 0° and 90°.

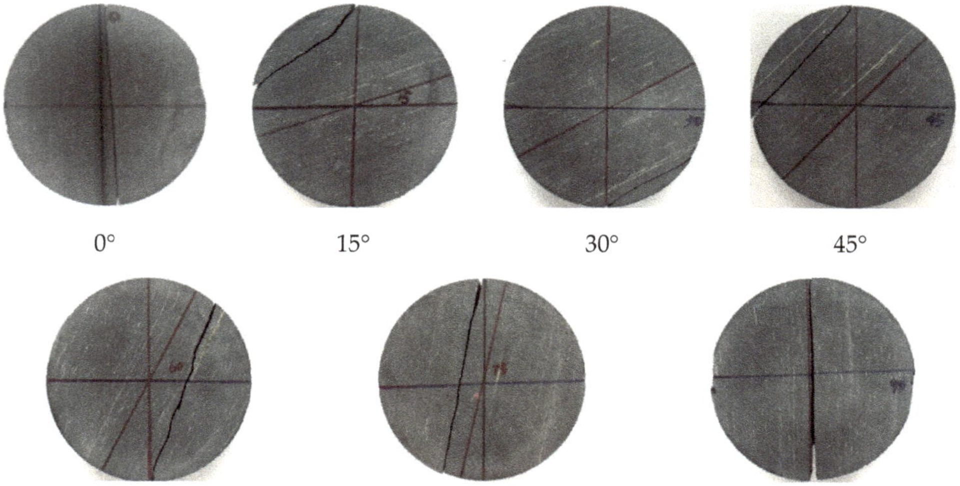

0° 15° 30° 45°

60° 75° 90°

Figure 3. Typical failure modes of layered shale under the Brazilian test.

2.3. *Tensile Strength of the Shale Samples*

Figure 4 shows the variable relationship between the tensile strength of shale samples and the dip angles of the bedding plane. Figure 4 demonstrated that the tensile strength of the shale samples obviously represents anisotropic characteristics. By increasing the dip angle of the bedding plane, the tensile strength of the shale sample decreases. The tensile strength of the shale sample with a dip angle of 0° is the largest. This tensile strength can be considered as the tensile strength of the rock matrix, because the sample is cracked along the center line of the disc. Meanwhile, the tensile strength of the shale sample with a dip angle of 90° is the smallest. The failure mode of the sample is closer to that of the sample with a dip angle of 0°. However, the tensile strength of the sample with a dip angle of 0° is regarded as the tensile strength of the bedding plane. In Figure 4, it can be clearly seen that the tensile strength of the shale samples changes significantly when the dip angle of the bedding plane increases from 0° to 15°. Relatively, the tensile strength of the shale samples changes gradually when the dip angle of the bedding plane increases from 15° to 75°. The variation of the tensile strength of the shale samples is closely related to the failure modes, as analyzed above.

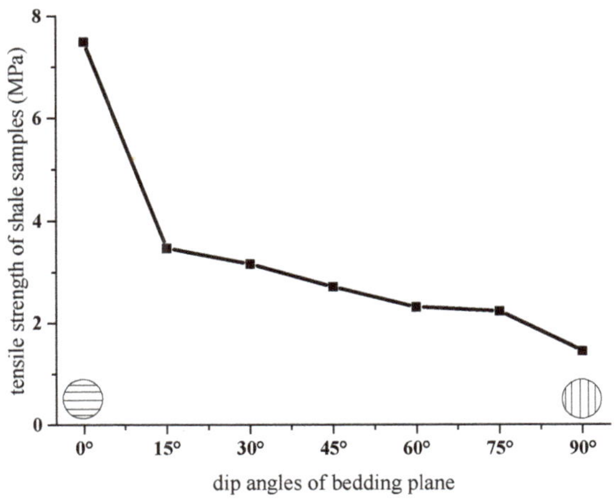

Figure 4. Average tensile strength of layered shale versus dip angles of the bedding plane.

3. Numerical Study of Layered Rocks

3.1. Description of the Numerical Code

The $RFPA^{2D}$ code was developed by considering the deformation of an elastic material containing an initial random distribution of micro-features to simulate the progressive failure, including the simulation of the failure process, failure-induced seismic events, and failure-induced stress redistribution. There are two main features of $RFPA^{2D}$ that are different from the traditional finite element method: (a) The heterogeneity of rock properties is introduced into the model; the $RFPA^{2D}$ code can simulate the non-linear deformability of a quasi-brittle behavior with an ideal brittle constitutive law for the local material. (b) A simple elastic–brittle–plastic model is implemented in the $RFPA^{2D}$ code. The element maintains linear deformation before the elemental stresses satisfy the failure criterion. The element has a residual strength after failure, and the stiffness of the element will decrease gradually. In this manner, the code can simulate strain-softening and discontinuous mechanical problems in a continuum mechanics mode. Both maximal tensile stress criterion and maximal tensile strain criterion are adopted in the model. The maximal tensile stress criterion is used to predict the brittle failure from the elastic stage to the residual deformation stage, and the maximal tensile strain criterion is applied to predict when the element loses its carrying capacity completely.

For heterogeneity, the material properties for elements are randomly distributed throughout the specimen by following a Weibull distribution:

$$\varphi = \frac{m}{\sigma_0}\left(\frac{\sigma}{\sigma_0}\right)^{m-1} exp\left[-\left(\frac{\sigma}{\sigma_0}\right)^{m}\right] \tag{2}$$

where, φ represents the strength of the elements distribution rule, σ is the element strength, and σ_0 is the mean strength of all the elements for the specimen. For an elastic modulus, the same distribution is used. We define m as the homogeneity index of the rock. A user-friendly pre- and post-processor is integrated to generate the finite element mesh and prepare the input data. The $RFPA^{2D}$ code can not only simulate the initiation of cracks, but can also model the crack propagation, coalescence, and interaction between multi-cracks. Details of the $RFPA^{2D}$ code can be found in the published literature [47].

3.2. Brazilian Disc Test of Layered Rock Specimens by $RFPA^{2D}$

3.2.1. Numerical Layered Specimens of Brazilian Test

Numerical rock specimens with seven different dip angles of 0° 15°, 30°, 45°, 60°, 75°, and 90° were prepared, and are shown in Figure 5. The mechanical parameters of the rock material and the joint material are listed in Table 1. The loading rate was 0.0008 mm/step in the axial direction.

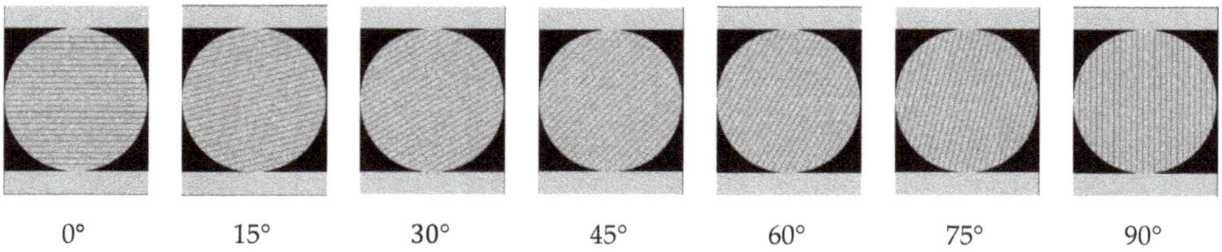

Figure 5. Numerical layered rock specimens under the Brazilian disc test.

Table 1. Elemental parameters of the rock and joint material.

Materials	Parameters	Elastic Modulus	Tensile Strength
Rock	Homogeneity index	4	4
	Mean value	50,000 MPa	45 MPa
Joint	Homogeneity index	3	3
	Mean value	1000 MPa	4 MPa

The homogeneity indices for the rock material and the joint material were 4.0 and 3.0, respectively. Because the tensile strength of the layered rocks depends on the combination of the rock material and the joint material, the strength of the transversely isotropic rock specimen with a layer dip of 90° in Liao's experiments is regarded as the tensile strength of the rock material in the present simulations, and the strength of the transversely isotropic rock specimen with a layer dip of 0° is regarded as the tensile strength of the joint material. A series of numerical tests was carried out to obtain the macro-tensile strength and the elastic modulus parameters according to Liao's experimental study [40]. The mechanical parameters of these two materials are listed in Table 1.

To obtain the mechanical parameters of these two materials, two numerical specimens made of the rock material or the joint material only were also prepared to conduct indirect tensile tests. The elastic modulus and the tensile strength of them are listed in Table 2.

Table 2. Macro-mechanical parameters of the rock and joint material.

Materials	Elastic Modulus (MPa)	Tensile Strength (MPa)
Rock	40,712.5	15.3
Joint	415.8	1.1

3.2.2. Results and Analysis

Figure 6 shows the tensile strength of the layered rock specimens under the numerical Brazilian test, which is calculated by Equation (1). The tensile strength of the layered rock specimens decreases with the increasing dip angle; the maximum tensile strength is 11.49 MPa at a dip angle of 0°, which corresponds to the direct tension strength at a dip angle of 90°. Figure 7 shows the horizontal stress at the center of the specimen vs. the vertical loading displacement. The pre-peak behavior of the specimens with different dip angles is quite similar, showing a linear elastic behavior. Figures 6 and 7 indicate that the tensile strength and loading curves of the layered rock specimens under the Brazilian test are very sensitive to the dip angle.

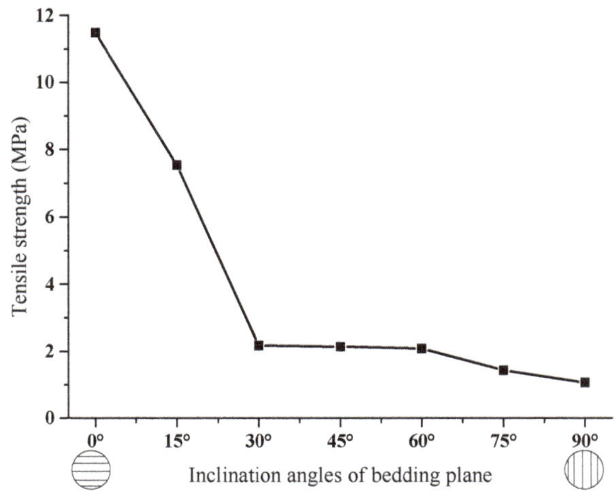

Figure 6. Tensile strength of layered rock specimens under the Brazilian test.

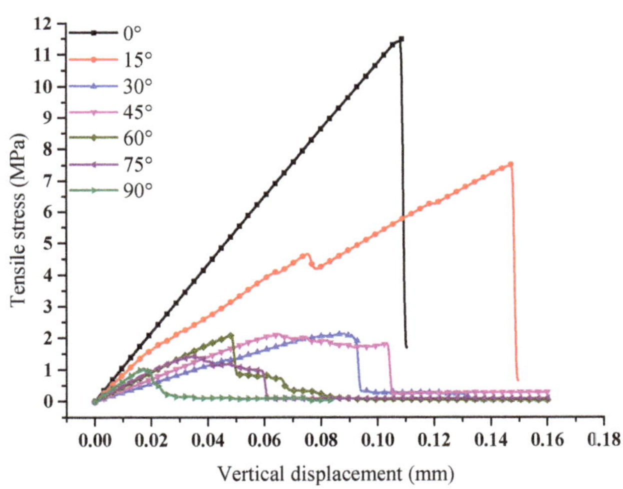

Figure 7. Horizontal stress (tensile stress) at the center of specimen vs. vertical displacement.

Figure 8 presents the failure modes of the layered rock specimens under the Brazilian test. It can be clearly seen that the failure mode of the specimen at 0° is similar to the isotropic rocks. The cracks generate at the center of the specimen, then propagate in a direction towards to the loading ends, and finally penetrate the whole specimen. After that, there are some small shear cracks near the loading ends. The main failure mechanism in this case is rock material tensile splitting. The tensile strength of the specimen at 0° is 11.49 MPa, which is similar to the tensile strength in the direct tension test (11.42 MPa) and the laboratory test (12.50 MPa). The failure modes of the specimens at 15°, 30°, and 45°. are quite complicated, and they show a mixed failure mode combining tensile splitting and shear failure. With the increasing dip angle, the shear cracks along the bedding planes increase. Shear cracks generate at the contacts of the rock material and the joint material, but they do not propagate any farther along the bedding planes. After that, the tensile cracks generate in the rock material, then propagate to shear cracks. Finally, the shear cracks and tensile cracks penetrate the specimen, which results in the specimen failure. The failure modes of the specimens at 60° and 75° are relatively simple, showing a shear failure along the bedding planes. Shear cracks generate at the contacts of the rock material and the joint material, and the cracks propagate along the bedding plane until they penetrate the whole specimen, which finally result in the specimen shear failure. The main crack does not cross the center of the specimen, and no tensile cracks are generated in the center of the disc. The failure mode of the specimen at 90° is similar to that of the specimen at 0°; the cracks generate almost in the center of the disc and propagate towards the loading ends until the cracks penetrate the whole specimen. The tensile strength of the specimen at 90° is 1.06 MPa, which is much lower than the tensile strength of the specimen at 0°. The main failure mechanism in this case is joint material tensile splitting. In conclusion, there are three main failure modes of the layered rock specimens under the Brazilian test, including tensile splitting failure, mixed tensile splitting and shear failure, and shear failure.

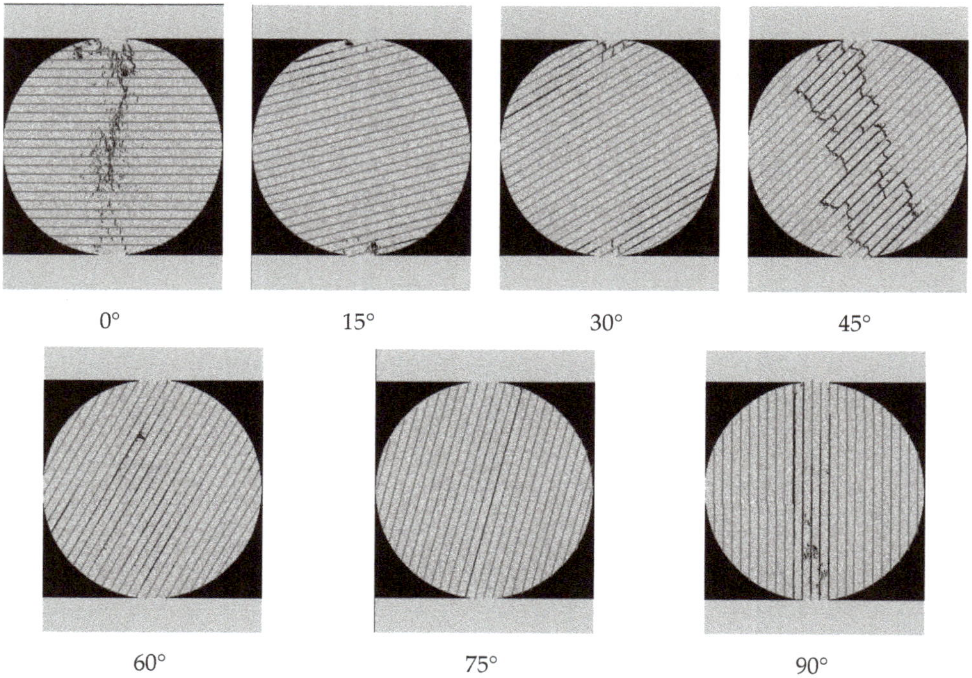

Figure 8. Failure modes of the layered rock specimens under the Brazilian test.

3.3. *Direct Tension Test on Layered Rock Samples by RFPA^{2D}*

3.3.1. Preparation of the Numerical Layered Specimens

Tien et al. used three model materials to prepare artificial transversely isotropic rock blocks [8]. In this study, the layered rock specimens are assumed to be made up of rock material and joint material, which determines the mechanical properties of the layered rocks. In this investigation, seven rock specimens with a dip angle, θ, of 0°, 15°, 30°, 45°, 60°, 75°, and 90°, with respect to the loading direction, are made up of the two different materials. Furthermore, some extra specimens with a layer dip angle of 80° and 85° were prepared to verify the numerical results compared with Liao's experimental study [40]. The specimen size was 200 mm × 100 mm, and it was discretized into 600 × 300 (180,000 elements) meshes.

It is much more difficult to obtain efficient satisfactory transversely isotropic rock samples in laboratory experiments. With RFPA^{2D}, it is more convenient to prepare layered rock samples with different rock materials. As shown in Figure 9, the light gray rock layers and the dark gray joints constitute the numerical specimens, respectively.

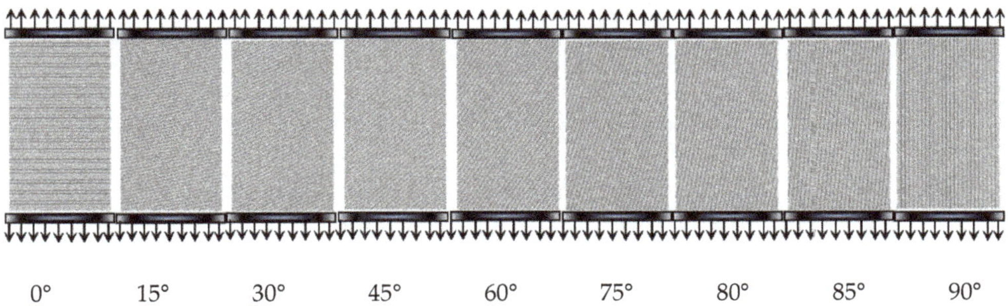

Figure 9. Layered rock specimens with different dip angles subjected to direct tension.

In all cases, the specimens underwent plane strain compression, imposed by the relative motion of the upper and lower loading plates by applying an external displacement at a constant rate of −0.0008 mm/step in the axial direction. The stress, as well as the deformation and the energy released, in each element were computed in each step during loading. The external displacement in the axial direction was slowly increased step by step.

3.3.2. Results and Analysis

The complete stress–strain relations of the layered rock specimens with dip angles, θ, of 0°, 15°, 30°, 45°, 60°, 75°, 85°, and 90° are shown in Figure 10. All stress–strain curves exhibit strong linear elasticity before the rock samples reach their peak tensile strength. The relationship between the stress and strain shows that the tensile strength increases as the dip angle increases from 0° to 90°. This means that the tensile strength is anisotropic due to the existence of weak joints. The critical strain, at the point where the specimens reach the peak tensile load, also increases when the dip angle increases from 0° to 85°. However, the critical strain of the specimen with a dip angle of 90° decreases compared with the specimen at 85°. We can consider that the critical strain may be influenced by the failure modes of the layered rock specimens.

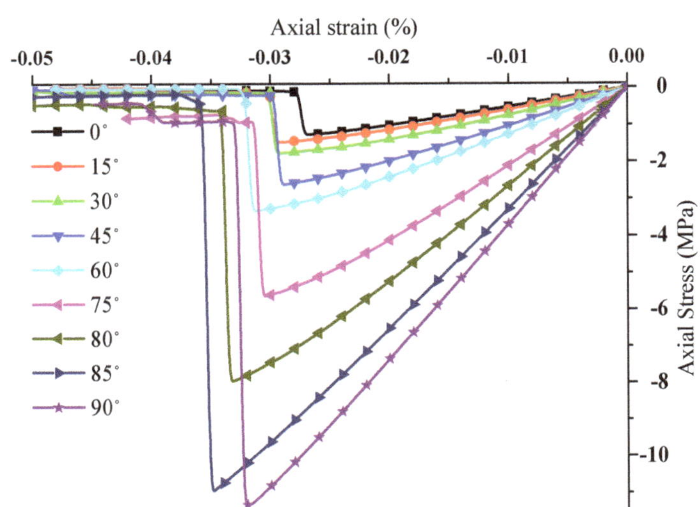

Figure 10. Complete stress–strain curves for the layered rock specimens.

The relationships between tensile strength and the tensile modulus with different dip angles are shown in Figure 11. The tensile strength of the specimen with a dip angle of 0° is very close to the tensile strength of the specimen with a dip angle of 15°. We can also see that, by varying the dip angle, the tensile modulus of the specimens follows the same trend as the tensile strength. As the dip angle increases from 0° to 90°, the tensile modulus increases gradually.

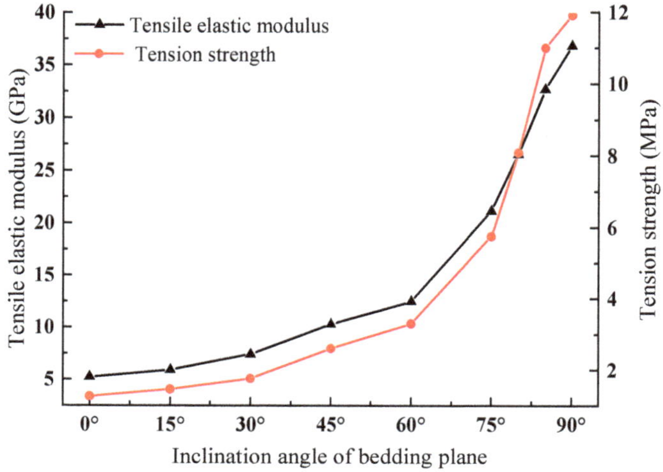

Figure 11. The tensile strength and tensile modulus of specimens with different dip angles.

It is obvious that the strength and tensile modulus of the specimens with the lower dip angles, from 0° to 15°, are closer to those of the joint material, whereas the strength and tensile modulus of the specimens with higher dip angles, from 75° to 90°, are closer to the tensile modulus of the rock material. We can conclude that the deformation behavior and tensile strength of layered rocks with lower dip angles depend on the existence of joints. However, the influence of joints on the layered rocks with higher dip angles is slight. Figure 12 shows the simulated tensile strength under direct tension, which has a good agreement with experimental results.

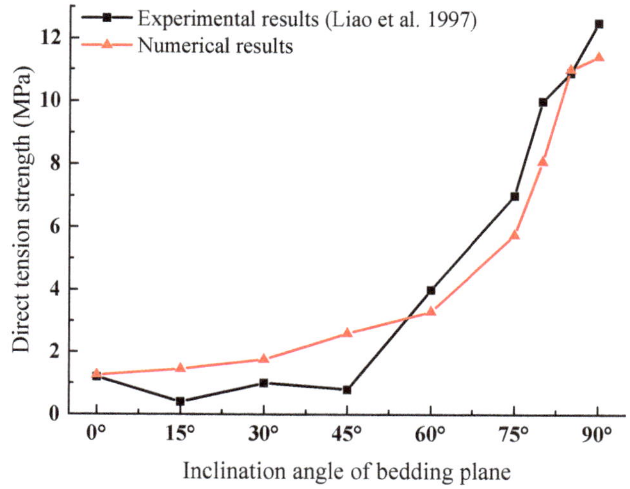

Figure 12. Tensile strength of the layered rock mass with different dip angles [40].

Figure 13 shows typical failure patterns of specimens with different dip angles. The fracture process of the specimen with a dip angle of 85° is displayed in Figure 14. The colors in Figures 13 and 14 represent the value of minimal principal stress. We can observe that the fracture of specimens with a lower dip angle, from 0° to 60°, is caused by the propagation of joints, and the fractures occur along joints from one side to the other. More than two fractures are observed in the specimens with a dip angle from 0° to 45°, whereas only one fracture can be seen along the joint from the upper-right to the lower-left corner in the specimen with a dip angle of 60°.

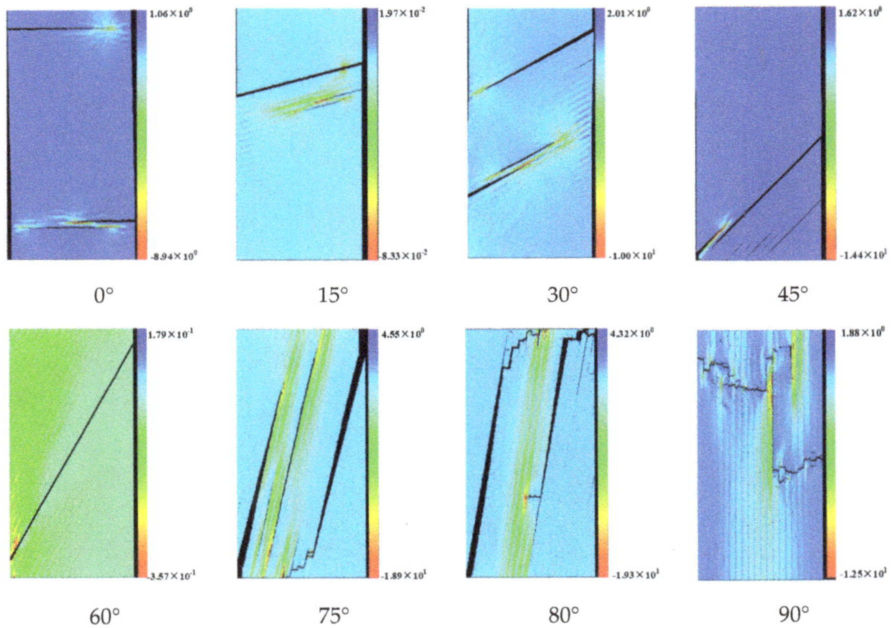

Figure 13. Fracture patterns of specimens with a layer dip angle from 0° to 80° and 90°.

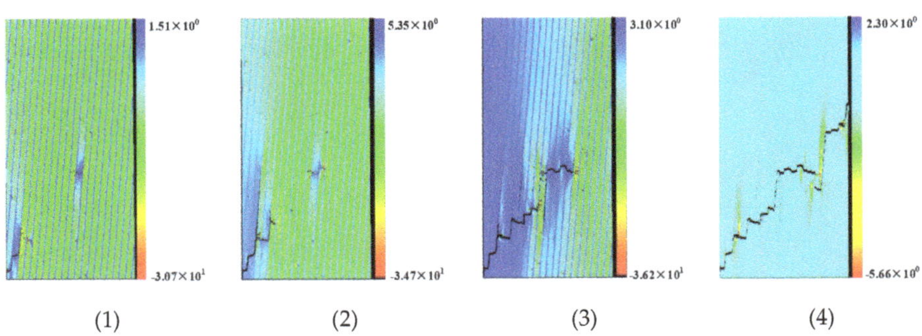

Figure 14. The failure process of the specimen with a layer dip angle of 85°.

However, for the specimens with dip angles of 75°, 80°, and 85°, fractures propagate partly along and partly through the joints. The failure plane exhibits a saw-toothed shape, and the same phenomenon was also observed in Youash's tests [39] and Liao's experiments [40]. For the specimens with dip angles of 75° and 80°, the cracks initiate along the joints from the left or right boundary, then propagate along the joints. Before they reach the other end, the new cracks initiate and propagate in the rock material perpendicular to the joints. The cracks perpendicular to the loading direction only occur near the top or the bottom end of the specimens.

For specimens with layers parallel to the tensile direction with angles of 85° and 90°, the fracture plane also exhibits a clear saw-toothed shape, but only a few parts of the failure plane occur along the foliation. The phenomenon of the saw-toothed failure plane reflects the fact that progressive failure occurs during the direct tensile tests. The progressive failure may include primary tensile cracks perpendicular to the tensile loading direction and secondary cracks parallel to the joints.

Figure 14 shows the progressive failure process of the specimen with a dip angle of 85°. It is interesting to see that the cracks do not initiate perpendicular to the loading direction in the joint material but in the rock material. They often run across many layers before beginning to propagate along the joints. The saw-toothed fracture pattern is caused by alternate propagation perpendicular to the loading direction and along the joints.

Figure 15 shows the failure patterns of sandstone of Lyons, tested in direct tension with different inclined angles by Youash in 1969 [39]. In the direct uniaxial tension tests, we can see that, when the inclined angle is less than 45°, the sandstone of Lyons mainly undergoes tensile failure along the bedding plane. The sandstone of Lyons also shows the saw-toothed fracture pattern when the inclined angle of bedding plane is greater than 45° and less than 90°. When the inclined angle of the bedding plane equals 90°, the sandstone of Lyons exhibits failure of the rock material, and the failure plane is perpendicular to the loading direction.

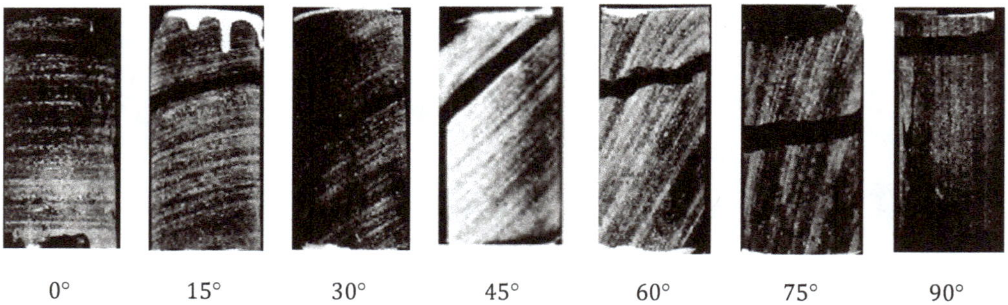

Figure 15. The failure patterns of sandstone of Lyons formation tested in direct tension [39].

The failure modes of the layered rock specimens obtained by numerical simulations under direct tension are similar to the experimental results [39,40], and the uniaxial tensile failure modes of the layered rocks can be divided into three types: (1) tensile failure along the bedding plane; (2) progressive saw-toothed failure; (3) tensile failure of rock material. The diagram of failure modes is shown in Figure 12 [40,44].

3.3.3. Failure Mechanism

Figure 16 shows the failure mechanisms of the layered rock specimens with different dip angles under direct tension, and these correspond to the three main failure modes of layered rock specimens. The layered rocks in Figure 16 is mainly composed of two materials, including rock material and joint material (bedding plane). Figure 16a illustrates that when the dip angle of the bedding plane is equal to 0°, with the increasing axial tensile load, the bedding plane will be fractured by tension while the axial tensile stress is greater than the tensile strength of the bedding plane (T_0). In this configuration, the layered rock specimen will be fractured along the bedding plane, and the tensile strength of the layered rock specimen can be expressed as $\sigma_t = \sigma_1 = T_0$.

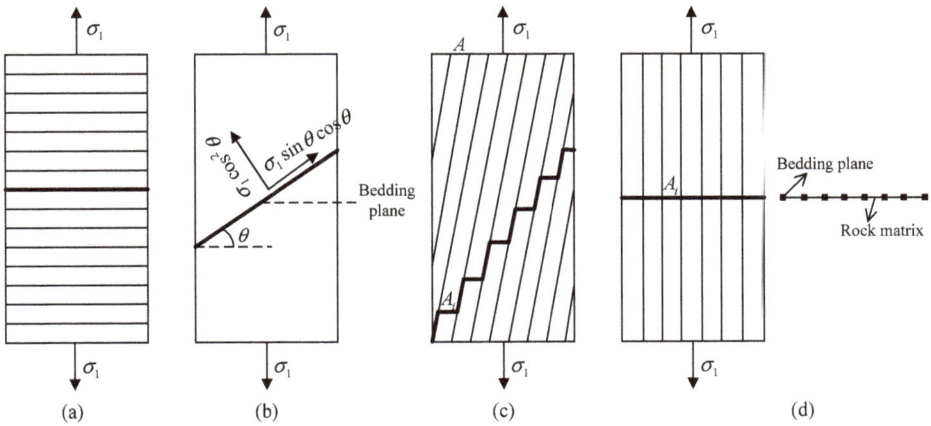

Figure 16. The tensile failure mechanism analysis of the layered rocks.

Figure 16d shows that, when the dip angle of the bedding plane is equal to 90°, with the increasing axial tensile load, the bedding plane first reaches its tensile strength, and then the bedding plane is cracked. However, in contrast to Figure 16a, the rock material is subjected to the axial tensile load at the cross section of the layered rock specimen. At the cross section, the bedding plane can be simplified as a single point when the layer thickness is small; therefore, it has little influence on the bearing capacity of the specimen while the bedding plane is first cracked. Then, as the axial tensile load increases to the tensile strength of the rock material (T_{90}), the layered rock specimen will be fractured along the cross section, resulting in tensile failure, and the failure plane is generally perpendicular to the loading direction. In this configuration, the tensile strength of the layered rock mass can be expressed as $\sigma_t = \sigma_1 = T_{90}$.

Figure 16b shows that when the dip angle of the bedding plane is greater than 0° and less than a specified angle, the layered rock mass is also fractured along the bedding plane. It can be seen in Figure 16b that the normal stress at the bedding plane is $\sigma_n = \sigma_1 \cos^2\theta$. In order to make the bedding plane fracture by tension, the normal stress at the bedding plane should be equal to the tensile strength of the bedding plane, which is expressed as $\sigma_n = T_0$. We can then obtain the tensile strength of the specimen as $\sigma_t = \sigma_1 = T_0/\cos^2\theta$. This demonstrates that the tensile strength of layered rock mass increases with the increasing dip angle.

With the increasing dip angle, the failure patterns of the layered rock specimens change. This will result in a progressive saw-toothed failure, as shown in Figure 16c. It is assumed that the area of the cross section is A, and the area of each small matrix exhibited the saw-toothed shape is A_i, so $\sigma_1 = F/A$ and $\sigma_i = F_i/A_i$. In order to make each small matrix exhibit the saw-toothed shaped cracks, the tensile stress, σ_i, needs to satisfy $\sigma_i = T_{90}$. It can then be surmised that $F_i = T_{90}A_i$. Therefore, $F = \Sigma F_i = T_{90}\Sigma A_i$, $\sigma_t = \sigma_1 = F/A = T_{90}\Sigma A_i/A$. It can be seen that, when the dip angle of the bedding plane is less than 90°, $\Sigma A_i < A$, that is to say, $\sigma_t = \sigma_1 < T_{90}$. We can see that as the layered rock mass generates the progressive saw-toothed failure, the tensile strength of the layered rock mass at this specific angle is less than the tensile strength of the layered rock specimens at a dip angle of 90°. From the point of view of geometry and elastic mechanics, this can explain the variation of tensile strength of the layered rock specimen with different dip angles of the bedding plane.

3.4. Discussion

3.4.1. Failure Modes of Layered Rock Specimens under Direct Tension Test

The main failure modes of layered rocks under the direct tension test are analyzed in Section 3.2; however, the failure modes of layered rocks highly depends on the anisotropic coefficient ($k = \frac{T_{90}}{T_0}$). If the anisotropic coefficient is less than 1 ($k < 1$), that is to say, the tensile strength of the bedding plane is stronger than the tensile strength of the rock material, it can be hypothesized that the bedding plane is the strong binding material, therefore, as the layered rock specimen is subjected to the uniaxial tensile load, the rock material will be fractured first, whatever the layer orientation is. If the anisotropic coefficient is equal to 1 ($k = 1$), which means the tensile strength of the bedding plane is the same as the tensile strength of the rock material, the tensile strength of the layered rocks may present isotropic characteristics. However, layered rocks may not only fracture along the bedding plane or the rock material due to the different elastic parameters between the bedding plane and the rock material. If the anisotropic coefficient is greater than 1 ($k > 1$), the failure modes may present as the three main patterns. In conclusion, the failure modes of layered rock specimens subjected to a direct tensile load may be greatly influenced by the anisotropic coefficient.

3.4.2. Tensile Strength of Layered Rock Specimens

For layered rocks, many Brazilian tests have been conducted to investigate the tensile behavior [48–52]. The numerical results in Section 3.2.2 shows that there are three main failure modes of layered rock specimens under the Brazilian test. Figure 17 also displays the fracture patterns of Mosel slate discs with different dip angles in laboratory tests [53]. Figure 17 shows that the layered rock specimens with different dip angles are not always fractured along the center diameter line under Brazilian tests. The failure modes are different because of the influence of the bedding planes. The typical failure patterns can be divided into three types: layer activation, central fracture, and non-central fracture. However, the precondition of Equation (1) to calculate the tensile strength of rocks should be satisfied so that Brazilian disc specimens are fractured along the center diameter line. Therefore, the tensile strength of the layered rocks calculated by Equation (1) is not reasonable, except for the specimens with dip angles of 0° and 90°, which are fractured along the center diameter line. Similar results can be found in other papers [54–57]. This demonstrates that the tensile strength of the layered rocks obtained by the Brazilian test is erroneous because of the influence of bedding planes. Specially, the tensile strength of the layered rock specimens with dip angles of 0° and 90° can be predicted by the Brazilian test because their failure modes meet the precondition of Equation (1). However, the tensile strength of the layered rock specimens obtained by the direct tension test is relatively accurate. Therefore, we suggest that the direct tension test is more suitable than the Brazilian disc test to measure the tensile strength of layered rock specimens.

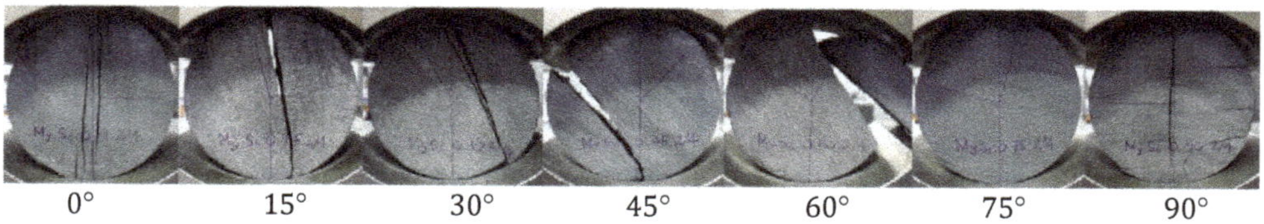

Figure 17. Fracture patterns of Mosel slate discs under different foliation-loading angles [53].

4. Modified Anisotropic Tensile Failure Criterion

4.1. Single Plane of Weakness (SPW) Criterion

Lee and Pietruszczak [43] proposed the SPW criterion of layered rocks based on Jaeger's SPW theory [9], and the SPW criterion can be given as:

$$T(\theta) = \begin{cases} \frac{T_0}{\cos^2\theta} \; 0^\circ \le \theta \le \theta^* \\ T_{90} \; \theta^* \le \theta \le 90^\circ \end{cases} \tag{3}$$

where θ^* is defined as:

$$\theta^* = \cos^{-1}\sqrt{\frac{T_0}{T_{90}}} \tag{4}$$

From Equation (3), it can be obtained that when the dip angle of the bedding plane is less than the critical angle, θ^*, the layered rock specimens in the direct tension tests are fractured along the bedding plane, which is in close compliance with the experimental results. However, when the dip angle of the bedding planes is greater than the critical angle, θ^*, the failure mode of the layered rock specimens is considered to be rock material tensile fracture under direct tension. In this situation, the tensile strength of the layered rock specimen is equal to the tensile strength of the rock material, which is inconsistent with the actual situation and the experimental results.

4.2. Modified SPW Criterion (MSPW Criterion)

To some degree, the SPW criterion can reflect the failure modes of layered rocks under direct tension. However, the failure modes may be different to the actual failure modes under the direct tension test when the dip angle is greater than the critical angle, θ^*. Therefore, based on the failure modes of the layered rocks under the direct tension test, the SPW criterion is modified in this paper.

The saw-toothed failure mode due to direct tension can be seen when the dip angle is higher, as shown in Figure 16c. The tension-fractured rock material is taken for the analysis. The rock material is not only subjected to the force generated by the normal stress, σ_1, but is also subjected to the force generated by the shear stress of the bedding plane. The rock material will be tension fractured as long as the total stress on the rock material reaches its tensile strength, T_{90}; meanwhile, the tensile strength, σ_1, of the layered rock specimen can be obtained:

$$T_{90} = \sigma_1 + 2\sigma_1 \sin^2\theta\cos\theta \tag{5}$$

$$\sigma_1 = \sigma_t = \frac{T_{90}}{1 + 2\sin^2\theta\cos\theta} \tag{6}$$

Therefore, the modified SPW (MSPW) criterion can be written as:

$$T(\theta) = \begin{cases} \frac{T_0}{\cos^2\theta} & 0^\circ \le \theta \le \theta^* \\ \frac{T_{90}}{1+2\sin^2\theta\cos\theta} & \theta^* < \theta \le 90^\circ \end{cases} . \tag{7}$$

where the critical angle, θ^*, is given by:

$$\theta^* \cong \arctan\left(\sqrt{\frac{T_{90}}{T_0}} - 1\right) \tag{8}$$

4.3. Verification of the MSPW Criterion

The numerical tests are based on the experiments of layered rocks carried out by Liao et al. [40]. Figure 18 shows the result comparisons between the laboratory test, numerical simulation, the Liao criterion, the SPW criterion, and the MSPW criterion.

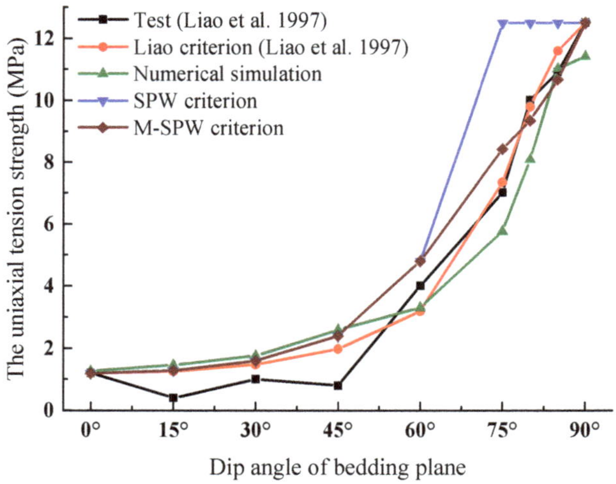

Figure 18. The result comparisons of laboratory test, simulations, and theoretical analysis [40].

Figure 18 illustrates that the results predicted by the MSPW criterion are closer to the experimental results. However, the results predicted by the SPW criterion greatly deviate from the test results at a higher dip angle. The tensile strength predicted by the Liao criterion has a small error compared with the test results as a whole, but the material constant in the Liao criterion is difficult to determine. We can also see that the numerical results simulated by RFPA^{2D} can well reflect the test results, and it is a good numerical method to simulate the failure process of rock samples under the direct tension test. By comparison, it is not difficult to find that the MSPW criterion is relatively accurate whether it reflects the uniaxial tensile failure modes of the layered rocks or predicts the uniaxial tensile strength of the layered rocks.

4.4. Determination of the Tensile Strength of Layered Rocks

Figures 3 and 17 present the failure modes of the layered rock specimens under the Brazilian test; it can be seen that the two specimens with 0° and 90° are almost fractured along the center diameter line. The experimental and numerical results clearly show that the tensile strength of the two specimens obtained by the Brazilian test are similar to the results by the direct tension test and laboratory test. This illustrates that although the tensile strength of the layered rocks obtained by the Brazilian test is controversial, the tensile strength of the two layered rock specimens of 0° and 90° obtained by the Brazilian test is reasonable.

From the expression of the MSPW criterion (Equation (7)), we can see that if the two parameters (T_0 and T_{90}) are determined, the tensile strength of the layered rocks with other dip angles can be predicted. However, the two parameters (T_0 and T_{90}) are difficult to obtain by the direct tension test, which is harder to carry out in the laboratory. Nevertheless, as we analyzed previously, the tensile strength of the layered rock specimens with dip angles of 0° and 90° (T_0 and T_{90}) can be obtained by the Brazilian test, whose results are reasonable. Thus, based on the MSPW criterion, we can simply predict the tensile strength of the layered rocks with different dip angles.

5. Conclusions

This study focuses on the tensile behavior of a transversely isotropic rock with varied layer dip angles subjected to direct and indirect tension tests. The following conclusions can be drawn:

(1) The layered rock specimens display an anisotropic mechanical behavior when subjected to direct tension load. The tensile stress–strain behavior of the layered rocks depends on the direction of the bedding planes with respect to the tensile load. However, the direct tension test of layered rocks in the laboratory needs to be studied in more depth in the future.
(2) The numerical results show that the dip angle has a significant influence on the fracture characteristics during the progressive failure, such as peak strength, failure patterns, and deformational behavior. The failure modes of the layered rock specimens are characterized by tensile failure along the bedding plane, progressive saw-toothed failure, and tensile failure of the rock material under direct tension.
(3) Based on the failure modes of the layered rocks, the SPW failure criterion is modified. The theoretical results of the modified SPW (MSPW) failure criterion show a good agreement with the experimental and the numerical results. The MSPW failure criterion can accurately describe the tensile strength when the dip angle of the bedding plane is close to 45°.
(4) Based on the MSPW criterion, a method to determine the tensile strength of the layered rocks is proposed, which can simply predict the tensile strength of the layered rocks.

Author Contributions: Conceptualization, S.J.; Data curation, J.Z.; Formal analysis, M.G.; Funding acquisition, M.G.; Investigation, M.G. and J.Z.; Methodology, Z.L.; Resources, S.J.; Software, Z.L.; Supervision, Z.L.; Writing—original draft, M.G.; Writing—review & editing, Z.L. All authors have read and agreed to the published version of the manuscript.

Funding: This research was funded by the Natural Science Foundation of Anhui Province (2108085QE208), the China Postdoctoral Science Foundation Funded Project (2021M700753), the Natural Science Research Project of Universities in Anhui Province (KJ2021A0463), and the National Natural Science Foundation of China (51779031, 41977219).

Institutional Review Board Statement: Not applicable.

Informed Consent Statement: Not applicable.

Data Availability Statement: The results reported in this study were obtained using our in-house code. The experimental data were obtained from measurements and experiments on RMT-150C. The data for producing the reported results will be made available upon request.

Conflicts of Interest: The authors declare no conflict of interest.

References

1. Donath, F.A. Strength variation and deformational behavior in anisotropic rock. In *State of Stress in the Earth Crust*; Judd, W.R., Ed.; Elsevier: Amsterdam, The Netherlands, 1964; pp. 281–300.
2. Amadei, B.; Rogers, J.D.; Goodman, R.E. Elastic constants and tensile strength of the anisotropic rocks. In Proceedings of the Fifth Congress of International Society of Rock Mechanics, Melbourne, Australia, 10–15 April 1983; Balkema: Rotterdam, The Netherlands, 1983; pp. 189–196.
3. Amadei, B. Importance of anisotropy when estimating and measuring in situ stresses in rock. *Int. J. Rock Mech. Min. Sci. Geomech. Abstr.* **1996**, *33*, 293–325. [CrossRef]
4. Gao, M.; Liang, Z.Z.; Jia, S.P.; Li, Y.C.; Yang, X.X. An equivalent anchoring method for anisotropic rock masses in underground tunnelling. *Tunn. Undergr. Space Technol.* **2019**, *85*, 294–306. [CrossRef]
5. Yagiz, S. Utilizing rock mass properties for predicting TBM performance in hard rock condition. *Tunn. Undergr. Space Technol.* **2008**, *23*, 326–339. [CrossRef]
6. Singh, T.N.; Pradhan, S.P.; Vishal, V. Stability of slopes in a fire-prone mine in Jharia Coalfield, India. *Arab. J. Geosci.* **2013**, *6*, 419–427. [CrossRef]
7. Mclamore, P.B.; Gray, K.E. The mechanical behaviour of anisotropic sedimentary rocks. *J. Eng. Ind.* **1967**, *89*, 62–76. [CrossRef]
8. Tien, Y.M.; Tsao, P.F. Preparation and mechanical properties of artificial transversely isotropic rock. *Int. J. Rock Mech. Min. Sci.* **2000**, *37*, 1001–1012. [CrossRef]
9. Jaeger, J.C.; Cook, N.G.W.; Zimmerman, R.W. *Fundamentals of Rock Mechanics*, 4th ed.; Blackwell Publishing: Malden, MA, USA, 2007.
10. Aadnoy, B.; Looyeh, R. *Petroleum Rock Mechanics: Drilling Operations and Well Design*; Gulf Professional Publishing: Oxford, UK, 2011.

11. Wu, N.; Liang, Z.Z.; Li, Y.C.; Li, H.; Li, W.R.; Zhang, M.L. Stress-dependent anisotropy index of strength and deformability of jointed rock mass: Insights from a numerical study. *Bull. Eng. Geol. Environ.* **2019**, *78*, 5905–5917. [CrossRef]
12. Guo, F.; Liang, Z.Z.; Gong, B.; Li, G. Tensile failure in stability analysis of rock and soil slopes. *Chin. J. Rock Mech. Eng.* **2017**, *36*, 3192–3205. (In Chinese)
13. Chen, Y.L.; Azzam, R.; Wang, M.; Xu, S.; Chang, L.Q. The uniaxial compressive and tensile tests of frozen saturated clay in Shanghai area. *Environ. Earth Sci.* **2011**, *64*, 29–36. [CrossRef]
14. ISRM. Suggested methods for determining tensile strength of rock materials. *Int. J. Rock Mech. Min. Sci. Geomech. Abstr.* **1978**, *15*, 99–103. [CrossRef]
15. Hobbs, D.W. Rock tensile strength and its relationship to a number of alternative measures of rock strength. *Int. J. Rock Mech. Min. Sci. Geomech. Abstr.* **1967**, *4*, 115–127. [CrossRef]
16. Li, D.Y.; Wong, L.N.Y. The Brazilian disc test for rock mechanics applications: Review and new insights. *Rock Mech. Rock Eng.* **2013**, *46*, 269–287. [CrossRef]
17. Wu, B.; Chen, R.; Xia, K. Dynamic tensile failure of rocks under static pretension. *Int. J. Rock Mech. Min. Sci.* **2015**, *80*, 12–18.
18. Bai, Q.S.; Tu, S.H.; Zhang, C. DEM investigation of the fracture mechanism of rock disc containing holes and its influence on tensile strength. *Theor. Appl. Fract. Mech.* **2016**, *86*, 197–216. [CrossRef]
19. Khosravi, A.; Simon, R.; Rivard, P. The shape effect on the morphplogy of the fracture surface induced by the Brazilian test. *Int. J. Rock Mech. Min. Sci.* **2016**, *93*, 201–209. [CrossRef]
20. Xia, K.; Yao, W.; Wu, B. Dynamic rock tensile strengths of Laurentian granite: Experimental observation and micromechanical model. *J. Rock Mech. Geotech. Eng.* **2017**, *9*, 116–124. [CrossRef]
21. Yuan, R.; Shen, B. Numerical modelling of the contact condition of a Brazilian disk test and its influence on the tensile strength of rock. *Int. J. Rock Mech. Min. Sci.* **2017**, *93*, 54–65. [CrossRef]
22. Zhang, S.W.; Shou, K.J.; Xian, X.F.; Zhou, J.P.; Liu, G.J. Fracture characteristics and acoustic emission of anisotropic shale in Brazilian tests. *Tunn. Undergr. Space Technol.* **2018**, *71*, 366–381. [CrossRef]
23. Wang, P.T.; Cai, M.F.; Ren, F.H. Anisotropic and directionality of tensile behaviours of a jointed rock mass subjected to numerical Brazilian tests. *Tunn. Undergr. Space Technol.* **2018**, *73*, 139–153. [CrossRef]
24. Barla, G.; Innaurato, N. Indirect tensile testing of anisotropic rocks. *Rock Mech.* **1973**, *5*, 215–230. [CrossRef]
25. Debecker, B.; Vervoort, A.; Napier, J.A.L. Fracturing in and around a natural discontinuity in rock: A comparison between boundary and discrete element models. In Proceedings of the 5th International Conference on Engineering Computational Technology, Las Palmas, Spain, 12–15 September 2006; p. 186.
26. Debecker, B.; Vervoort, A. A Two-dimensional discrete element simulations of the fracture behavior of slate. *Int. J. Rock Mech. Min. Sci.* **2013**, *61*, 161–170. [CrossRef]
27. Park, B.; Min, K.B. Discrete element modeling of transversely isotropic rock. In Proceedings of the 47th US Rock Mech/Geomech Symp ARMA, San Francisco, CA, USA, 23–26 June 2013; Volume 13, p. 490.
28. Dan, D.Q.; Konietzky, H.; Herbust, M. Brazilian tensile strength tests on some anisotropic rocks. *Int. J. Rock Mech. Min. Sci.* **2013**, *58*, 1–7. [CrossRef]
29. Dan, D.Q.; Konietzky, H. Numerical simulations and interpretations of Brazilian tensile tests on transversely isotropic rocks. *Int. J. Rock Mech. Min. Sci.* **2014**, *71*, 53–63. [CrossRef]
30. Liu, S.Y.; Fu, H.L.; Rao, J.Y.; Dong, H.; Cao, Q. Research on Barzilian disc splitting tests for anisotropy of slate under influence of different bedding orientations. *Chin. J. Rock Mech. Eng.* **2012**, *31*, 785–791. (In Chinese)
31. Vervoort, A.; Min, K.B.; Konietzky, H.; Cho, J.W.; Debecker, B.; Dinh, Q.D.; Fruhwirt, T.; Tavallali, A. Failure of transversely isotropic rock under Brazilian test conditions. *Int. J. Rock Mech. Min. Sci.* **2014**, *70*, 343–352. [CrossRef]
32. Chong, Z.H.; Li, X.H.; Hou, P.; Wu, Y.C.; Zhang, J.; Chen, T.; Liang, S. Numerical investigation of bedding plane parameters of transversely isotropic shale. *Rock Mech. Rock Eng.* **2017**, *50*, 1183–1204. [CrossRef]
33. Xu, G.W.; He, C.; Chen, Z.Q.; Su, A. Transversely isotropy of phyllite under Brazilian tests: Laboratory testing and numerical simulations. *Rock Mech. Rock Eng.* **2018**, *51*, 1111–1135. [CrossRef]
34. Kundu, J.; Mahanta, B.; Sarkar, K.; Singh, T.N. The effect of lineation on anisotropic in dry and saturated Himalayan schistose rock under Brazilian test conditions. *Rock Mech. Rock Eng.* **2018**, *51*, 5–21. [CrossRef]
35. Wijk, G. Some new theoretical aspects of indirect measurements of tensile strength of rocks. *Int. J. Rock Mech. Min. Sci. Geomech. Abstr.* **1978**, *15*, 149–160. [CrossRef]
36. *ASTM D3967-16*; Standard Test Method for Splitting Tensile Strength of Intact Core Specimens. ASTM International: West Conshohocken, PA, USA, 2001.
37. Aliha, M.R.M. Indirect tensile test assessment for rock materials using 3-D disc-type specimens. *Arab. J. Geosci.* **2014**, *7*, 4757–4766. [CrossRef]
38. Karaman, K.; Kesimal, A.; Ersoy, H. A comparative assessment of indirect methods for estimating the uniaxial compressive and tensile strength of rocks. *Arab. J. Geosci.* **2015**, *8*, 2393–2403. [CrossRef]
39. Youash, Y. Tension Tests on Layered Rocks. *Geol. Soc. Am. Bull.* **1969**, *80*, 303–306. [CrossRef]
40. Liao, J.J.; Tang, M.T.; Hsieh, H.Y. Direct tensile behavior of a transversely isotropic rock. *Int. J. Rock Mech. Min. Sci.* **1997**, *34*, 837–849. [CrossRef]

41. Barron, K. Brittle fracture initiation in and ultimate failure of rocks: Part II-Anisotropic rocks: Theory. *Int. J. Rock Mech. Min. Sci. Geomech. Abstr.* **1971**, *8*, 553–563. [CrossRef]
42. Nova, R.; Zaninetti, A. An investigation into the tensile behaviour of a schistose rock. *Int. J. Rock Mech. Min. Sci. Geomech. Abstr.* **1990**, *27*, 231–242. [CrossRef]
43. Lee, Y.K.; Pietruszczak, S. Tensile failure criterion for transversely isotropic rocks. *Int. J. Rock Mech. Min. Sci.* **2015**, *79*, 205–215. [CrossRef]
44. Ma, T.S.; Wu, B.S.; Fu, J.H.; Zhang, Q.G.; Chen, P. Fracture pressure prediction for layered formations with anisotropic rock strengths. *J. Nat. Gas Sci. Eng.* **2017**, *38*, 485–503. [CrossRef]
45. Ma, T.S.; Zhang, Q.B.; Chen, P.; Yang, C.H.; Zhao, J. Fracture pressure model for inclined wells in layered formations with anisotropic rock strengths. *J. Pet. Sci. Eng.* **2017**, *149*, 393–408. [CrossRef]
46. Shang, J.; Duan, K.; Cui, Y.; Handley, K.; Zhao, Z. Numerical investigation of the direct tensile behaviour of laminated and transversely isotropic rocks containing incipient bedding planes with different strengths. *Comput. Geotech.* **2018**, *104*, 373–388. [CrossRef]
47. Tang, C.A. Numerical simulation of progressive rock failure and associated seismicity. *Int. J. Rock Mech. Min. Sci.* **1997**, *34*, 249–261. [CrossRef]
48. Chen, C.S.; Pan, E.; Amadei, B. Determination of deformability and tensile strength of anisotropic rock using Brazilian tests. *Int. J. Rock Mech. Min. Sci.* **1998**, *35*, 43–61. [CrossRef]
49. Tavallali, A.; Vervoort, A. Failure of layered sandstone under Brazilian test conditions: Effect of micro-scale parameters on macro-scale behavior. *Rock Mech. Rock Eng.* **2010**, *43*, 641–653. [CrossRef]
50. Tavallali, A.; Vervoort, A. Effect of layered orientation on the failure of layered sandstone under Brazilian test conditions. *Int. J. Rock Mech. Min. Sci.* **2010**, *47*, 313–322. [CrossRef]
51. Tavallali, A.; Vervoort, A. Behaviour of layered sandstone under Brazilian test conditions: Layered orientation and shape effects. *J. Rock Mech. Geotech. Eng.* **2013**, *5*, 366–377. [CrossRef]
52. Khanlari, G.; Rafiei, B.; Abdilor, Y. An experimental investigation of the Brazilian tensile strength and failure patterns of Laminated sandstones. *Rock Mech. Rock Eng.* **2015**, *48*, 843–852. [CrossRef]
53. Tan, X.; Konietzky, H.; Fruhwirt, T.; Dan, D.Q. Brazilian tests on transversely isotropic rocks: Laboratory testing and numerical simulations. *Rock Mech. Rock Eng.* **2015**, *48*, 1341–1351. [CrossRef]
54. Cho, J.W.; Kim, H.; Jeon, C.K.; Min, K.B. Deformation and strength anisotropy of Asian gneiss, Boryeong shale, and Yeoncheon schist. *Int. J. Rock Mech. Min. Sci.* **2012**, *50*, 158–169. [CrossRef]
55. Simpson, N.D.J. An Analysis of Tensile Strength, Fracture Initiation and Propagation in Anisotropic Rock (Gas Shale) Using Brazilian Test Equipped with High Speed Video and Acoustic Emission. Master's Thesis, Norwegian University of Science and Technology, Trondheim, Norway, 2013.
56. Deng, H.F.; Zhang, X.J.; Zhang, H.B.; Wang, C.X.J.; Fang, J.C.; Xiao, Y. Analysis and discussion on Brazilian tests on layered rock tensile strength. *Rock Soil Mech.* **2016**, *37*, 309–316. (In Chinese)
57. Wang, P.T.; Yang, T.H.; Xu, T.; Yu, Q.L.; Liu, H.L. A model of anisotropic property of seepage and stress for jointed rock mass. *J. Appl. Math.* **2013**, *2013*, 420536. [CrossRef]

applied sciences

MDPI

Article

Tensile Characteristics and Fracture Mode of Frozen Fractured Rock Mass Based on Brazilian Splitting Test

Tingting Wang [1], Pingfeng Li [2,*], Chun'an Tang [3,4], Bingbing Zhang [2], Jiang Yu [1] and Tao Geng [5]

1 School of Resources and Civil Engineering, Northeastern University, Shenyang 110819, China
2 Hongda Blasting Engineering Group Co., Ltd., Guangzhou 510623, China
3 State Key Laboratory of Coastal and Offshore Engineering, Dalian University of Technology, Dalian 116024, China
4 State Key Laboratory of Frozen Soil Engineering, Northwest Institute of Eco-Environment and Resources, Chinese Academy of Sciences, Lanzhou 730000, China
5 School of Earth Sciences, China University of Geosciences, Wuhan 430074, China
* Correspondence: hdbplpf@163.com

Abstract: Frozen fractured rock mass is often encountered during the implementation of geotechnical engineering in cold regions. The tensile strength parameters of frozen rock play an important role in the construction of rock slopes involving tensile failure. In order to study the tensile characteristics of a frozen fractured rock mass in a cold region, original rock specimens were mined and processed in the Yulong Copper Mine, and artificial, frozen fractured marble specimens were made. The effects of different ice-filled crack angles, lengths, and widths on the force–displacement curve and the tensile strength of frozen rock were studied by laboratory Brazilian splitting experiments and RFPA^{3D}, and the evolution law of the tensile strength of frozen rock was revealed. At the same time, wing crack initiation and cracking mode after tensile failure were analyzed by high-speed camera; the whole process of the Brazilian splitting of frozen rock was reconstructed, and the development of microcrack initiation in frozen rock was analyzed. The following conclusions were drawn from the test results: the frozen rock specimens have typical brittle-failure characteristics. The tensile strength of frozen rock decreases gradually with the increase in the width and length of ice-filled cracks, and decreases first and then increases with the increase in the angle of the ice-filled crack. The ice-filled crack incurs damage first, and then the wing cracks start from the tip of the ice-filled crack and extend continuously. The tensile strength of frozen rock is significantly affected by the angle and length of ice-filled cracks.

Keywords: frozen rock mass; Brazilian splitting test; tensile strength; ice-filled crack ice; failure characteristics

Citation: Wang, T.; Li, P.; Tang, C.; Zhang, B.; Yu, J.; Geng, T. Tensile Characteristics and Fracture Mode of Frozen Fractured Rock Mass Based on Brazilian Splitting Test. *Appl. Sci.* **2022**, *12*, 11788. https://doi.org/10.3390/app122211788

Academic Editor: Arcady Dyskin

Received: 20 October 2022
Accepted: 18 November 2022
Published: 20 November 2022

1. Introduction

The Yulong Copper Mine in Tibet, China, is 4569–5118 m above sea level. The mining area is a semi-arid alpine region with a continental climate, and has rich and reliable copper resources. This region mainly adopts a strip mining method. In the process of resource development, the rock slope of the strip mine in such high-altitude and cold regions is exposed for a long time, and there are cracks, weak interlayers, stratification, and other structural surfaces (Figure 1). Under rainfall and groundwater conditions, the fissures will gather water to form a frozen rock mass. In the process of mineral resource development, a frozen rock mass is often encountered. In order to carry out safe and efficient engineering and mining, a deep understanding of the mechanical properties of frozen rock mass is of crucial guiding significance for improving the development efficiency of open-pit mineral resources and construction safety and prevention of accidents in high altitude cold regions [1–7].

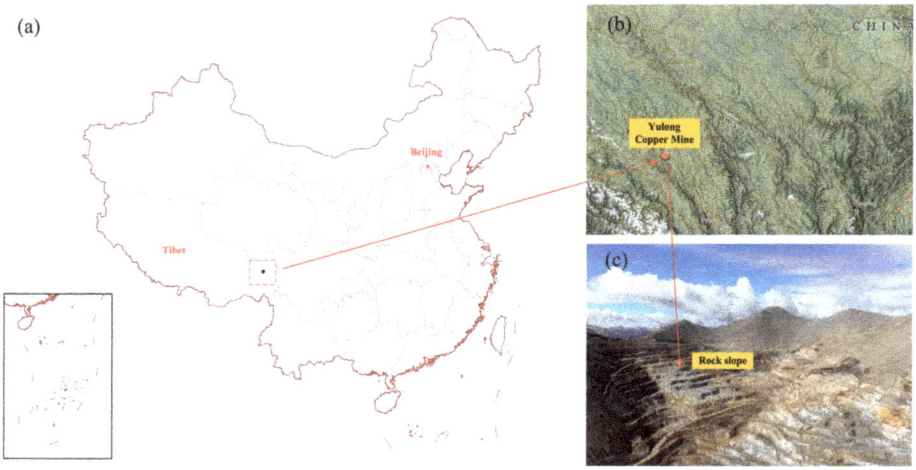

Figure 1. Yulong Copper Mine in Tibet, China. (**a**) Tibet Autonomous Region in Southwest China; (**b**) Satellite imagery of Yulong copper mine; (**c**) Rock slope in Yulong copper mine.

The tensile strength, compressive strength and shear strength of rock are important indicators for measuring the physical and mechanical properties of rock. The damage and failure of rock are usually controlled by the tensile strength. There are two main methods to test the tensile strength of rock, the first is direct stretching, and the second is indirect stretching. However, the direct tensile test requires specific laboratory equipment and complex operations, and the indirect tensile test has a more mature theoretical system. So, the indirect tensile test, namely the Brazilian splitting test, is generally used to test the tensile strength of rock [8,9].

Many scholars [10–13] have analyzed the changes of tensile strength, tensile sensitivity coefficient, radial strain, AE energy accumulation and release, crack initiation and fracture mode of rock specimens through the Brazilian splitting laboratory test and numerical simulation. Their research considered the prefabricated fissures, pores, different strain rates, degrees of water saturation, different particle sizes, different angles, high temperatures, filling of fissures, and other factors. At present, there are mature theoretical systems and methods for reference. The research on low-temperature rock mechanics mainly focuses on the mechanical properties of frozen intact rock [14–17] and the influence of the freeze–thaw cycle on the failure characteristics of the rock's mechanical properties [18–21]. The mechanical properties of frozen intact rock are mainly determined by freezing temperature, initial water content and loading rate. Aoki et al. [22] studied the mechanical properties of various rocks under dry or saturated conditions through uniaxial compression and Brazilian splitting tests. The results showed that the compressive and tensile strength and elastic modulus increased significantly at −160 °C, and the compressive and tensile strength and elastic modulus varied according to the type of rock and the degree of water saturation. Yamabe et al. [23] studied the mechanical properties of sandstone at low temperatures through uniaxial and triaxial experiments. They found that the tensile and compressive strength of rock increased with the decrease in temperature. Still, the elastic modulus was less affected by temperature in the section from −10 to −20 °C. Xu et al., Yang et al., Xi et al., and Shan et al. [24–27] studied the mechanical properties of a variety of rocks under negative temperature through uniaxial and triaxial experiments. The results showed that the compressive strength, internal friction angle, cohesion force and elastic modulus of the rocks all increased with the decrease in temperature. They attributed the increase in mechanical properties with decreasing temperature to the shrinkage of mineral particles at low temperatures, the increased strength of the ice itself, and the interaction of the frost swelling force with the rock. Li et al. [28] studied the strength characteristics of frozen limestone sandstone in different confining pressures under cyclic loading and unloading, and found that the peak strength increased under low confining pressures, while the peak strength 'weakened' under high confining pressures.

Some scholars have conducted compression experiments on frozen rock specimens with cracks at low temperatures. Bai et al. [29] studied the strength, deformation and crack evolution characteristics of frozen red sandstone with two ice-filled defects under different temperatures and sealing pressures. Yang et al. [30] studied the influence of crack inclination angle, length, width, confining pressure and temperature on the mechanical properties of single and double crack rock mass under low-temperature conditions. They found that Poisson's ratio and elastic modulus were less affected by confining pressure under low-temperature conditions, and rock mass strength increased with the decrease in temperature. The inclination angle had the most significant effect on the strength of rock mass, followed by length and temperature. The above studies used red sandstones or similar materials to discuss the fracture mode and mechanical properties of frozen rock mass after compression. They considered the relevant engineering problems in the construction process of the artificial freezing method. The mechanical properties of the frozen fracture rock mass with high density which is encountered in the blasting construction of the cold region strip mine are not clear.

Currently, relevant studies on frozen fracture rock mass mainly evaluate the compressive strength. In contrast, field sampling in cold regions and tensile strength tests of frozen fracture rock mass have not been involved. In particular, there has not been a comprehensive and systematic analysis of the influence of ice-filled crack space distribution on the tensile strength of frozen fracture rock mass and the failure process of wing crack propagation under load. At the same time, it is difficult to obtain the original rock in a high altitude cold region project. So, it would be of great value to find an effective numerical simulation method to analyze the mechanical properties of frozen rock mass.

This study focused on the tensile strength of frozen fractured rock mass in high altitude cold regions. The aim of the study was to reveal the tensile characteristics and fracture mode of frozen fractured rock mass during the Brazilian splitting test process. The original rock sampling and processing were carried out from the Yulong Copper Mine in Tibet, China, to make artificial ice-filled crack marble specimens. The Brazilian splitting test of frozen rock mass was carried out. Firstly, the influence of ice-filled cracks with different angles on the strength and fracture mode of the frozen rock mass was investigated experimentally, and the initiation and propagation of wing cracks were recorded by a high-speed camera. Furthermore, the effects of the width and length of the ice-filled cracks on the tensile properties of rock mass were discussed by RFPA^{3D}.

2. Test Preparation

2.1. Specimens Processing

All the marble rock specimens were taken from the Yulong Copper Mine in Tibet, China. The rock core was fabricated into a specimen, whose size was $\phi = 50 \times 25$ mm. The selected specimens were processed through the carver to form different angled cracks (0, 30, 45, 60, 90°). Two of each specification were made for the post-production of ice-filled rock specimens. The test results and similar fracture modes were selected as the final test results to reduce the test error.

In order to prevent moisture migration during freezing and ensure uniform moisture in the specimen, the prepared specimens were first placed in a constant low temperature environment of −20 °C for rapid freezing, and the freezing time was 48 h. Subsequently, a uniaxial compression testing machine was used as the power device to carry out the Brazilian splitting test on the frozen rock specimens (Figure 2). The constant deformation rate (0.001 mm/s) was loaded until the specimen was split. The tensile strength of the specimen was calculated using the formula proposed by Shloido [31]:

$$\sigma_t = \frac{2P}{\pi d t} \tag{1}$$

where σ_t is the failure load of the rock specimen, d is the diameter of the rock specimen, and t is the thickness of the rock specimen.

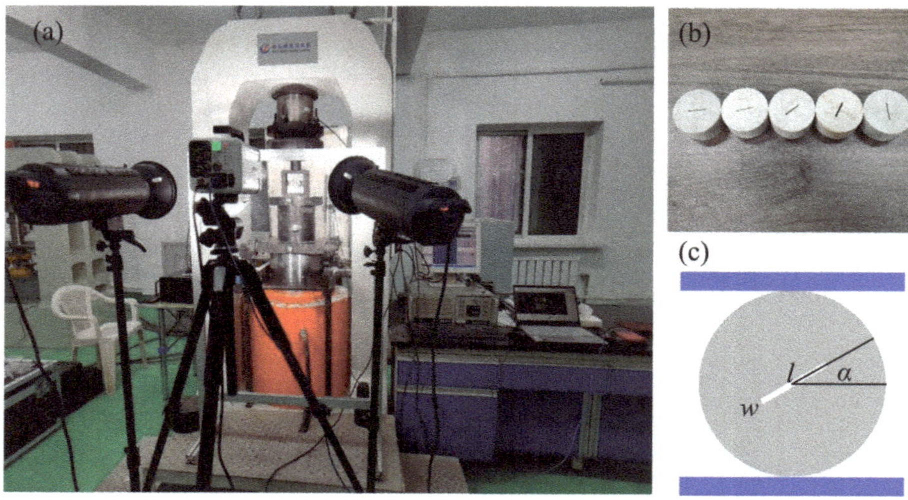

Figure 2. Design drawing of experimental device and specimen. (**a**) Photograph of experimental device; (**b**) Test specimens; (**c**) l is the ice-filled crack length, α is the ice-filled crack angle, and w is the ice-filled crack width.

The compressive and tensile strengths of intact frozen rock under natural water were measured. The uniaxial compressive force–displacement curve of the frozen rock specimen fell sharply when it increased to 131.02 kN (Figure 3). When the tensile strength of the intact specimen reached the peak value of 8.88 kN, the stress curve fell quickly (Figure 4).

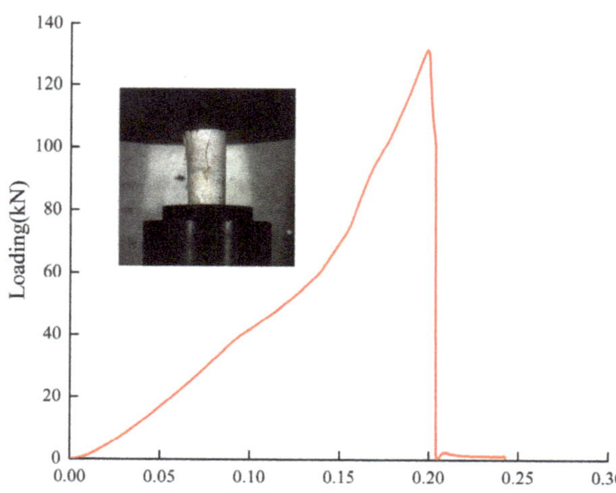

Figure 3. Uniaxial compression force–displacement curve of intact frozen rock specimen.

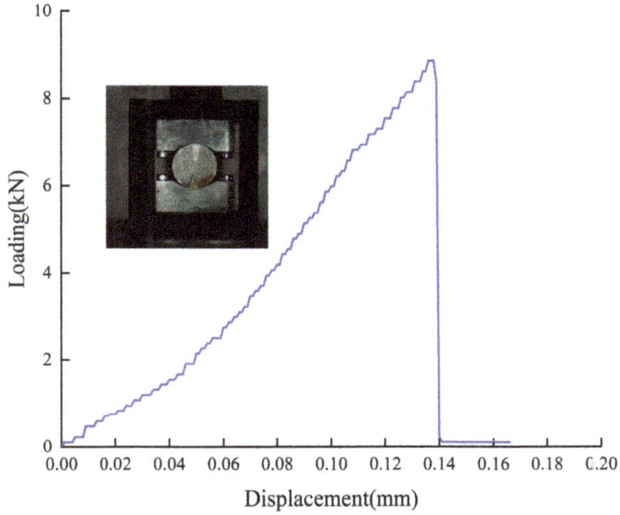

Figure 4. Brazilian splitting force–displacement curve of intact frozen rock specimen.

2.2. Test Results of Frozen Rock Specimens in Different Ice-Filled Crack Angles

2.2.1. Analysis of Tensile Strength

Figure 5 shows the Brazilian splitting force–displacement curve of frozen rock with different ice-filled crack angles. The study found that all the frozen rock specimens showed typical brittle failure characteristics. The tensile strength of the frozen rock specimens decreased gradually and then increased with the increase in the ice-filled crack angles. The tensile strength fell quickly after the peak, and the specimens fractured after producing a small displacement.

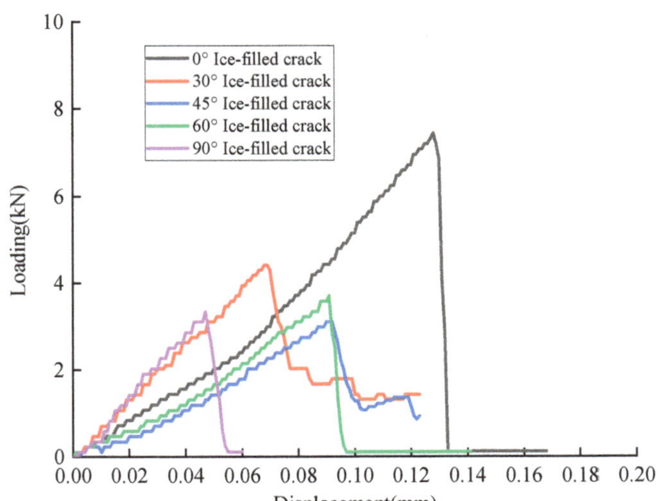

Figure 5. Brazilian splitting test force–displacement curve of frozen rocks with different ice-filled crack angles.

2.2.2. Analysis of Fracture Mode

Figures 6 and 7 show the splitting failure process of frozen rock specimens containing different ice-filled crack angles. The splitting failure is caused by the damage of ice during loading, the wing crack penetration of the ice-filled crack tip and the extension of the far-field crack. In the process of splitting, the ice-filled crack incurs damage first, and then the wing crack at the tip extends to the top and bottom of the specimen, and the macroscopic crack is gradually formed. When the ice-filled crack angle α is 0°, the ice-filled crack is perpendicular to the loading direction. The wing crack of the ice-filled crack tip cannot penetrate through the whole specimen along the ice-filled crack, and the tensile strength of the specimen is relatively high. When the ice-filled crack angle $\alpha > 0^\circ$, the ice crack's force bearing and transmission during loading gradually decrease with the increase in the ice-filled crack angle. Crack formation is hindered by the ice-filled crack. As the ice-filled crack angle α increases, the crack evolution is less hindered. As the angle between the ice-filled crack and the loading direction gradually decreases, the crack and the ice-filled crack are quickly connected, eventually leading to specimen fracture. When the ice-filled crack angle α is 90°, the ice-filled crack is parallel to the loading direction, and the fissure ice effect is minimized.

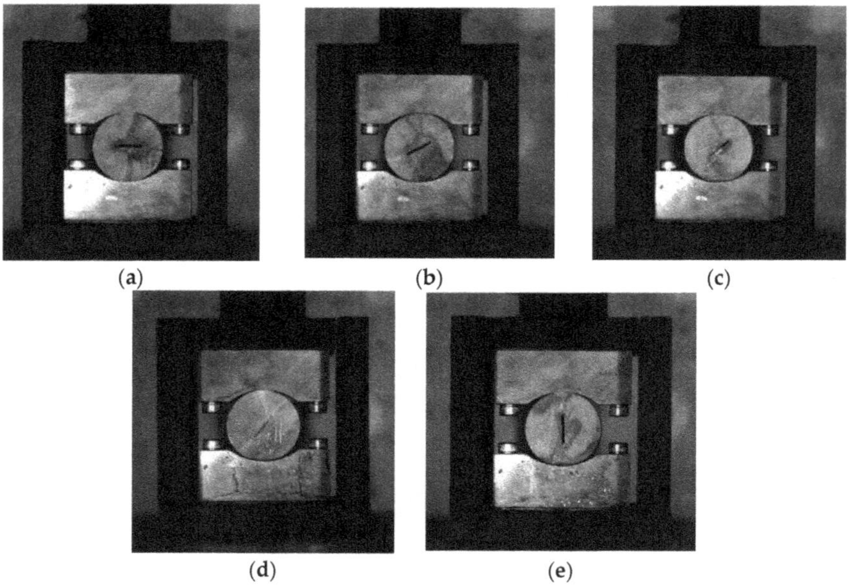

Figure 6. Fracture mode of ice-filled cracks at different angles in Brazilian splitting test: (**a**) 0°, (**b**) 30°, (**c**) 45°, (**d**) 60°, (**e**) 90°.

Figure 7. Fracture morphology process of a 60° ice-filled crack frozen rock specimen in Brazilian splitting test: (**a**) 0 s, (**b**) 90 s, (**c**) 100 s.

2.3. RFPA^{3D} Feasibility Verification

Rock is a heterogeneous material. The heterogeneity plays an important role in determining the crack initiation and fracture mode of rock mass. The influence of heterogeneity is pronounced in the progressive failure process. The Weibull distribution is used to consider the heterogeneity of rock in the RFPA method. RFPA is also suitable for the fracture analysis of brittle materials. Compared with ABAQUS and COMSOL, RFPA calculates crack initiation and propagation without considering the convergence problem. At present, a large number of papers have been published on numerical simulation analysis of rock materials using RFPA, and the principle can be found in the relevant papers [32–38].

RFPA^{3D} was used to perform the splitting experiments of the frozen rock specimens. The aim of the simulation was to obtain the internal microcrack initiation and extension, and the energy accumulation and release in the process of frozen rock failure. The size of the model was ϕ = 50 × 25 mm, and the mesh size was 0.25 × 0.5 × 0.25 mm. Cardboard was added at both ends to reduce the end effect. The thickness of the cardboard was 5 mm and the loading rate was 0.0005 mm/step. The material parameters of the model (Table 1) were obtained from the uniaxial compression of intact frozen rock and Brazilian splitting experiments, and converted by RFPA empirical formula [39,40].

Table 1. Material parameters of the model.

	Elasticity Modulus (MPa)	m	Compressive Strength (MPa)	m	Poisson Ratio	Friction Angle
Granite	32,000	5	147	5	0.25	30°
Ice	6000	10	8	10	0.35	26.5°
Cardboard	50,000	100	1 e8	100	0.3	30°

Figure 8 shows the numerical simulation results of the Brazilian splitting of intact frozen rock. Compared with the experimental results, the frozen rock fracture mode was similar, and macroscopic cracks were formed in the loading direction. The intact frozen rock specimen showed a typical brittle failure form. AE events increased sharply before macroscopic cracks occurred. The frozen rock specimen fractured instantly and released a large amount of energy (Figure 9). The load capacity of the intact frozen rock specimen was 8.15 kN in the numerical simulation and 8.88 kN in the experiment, which were similar to each other.

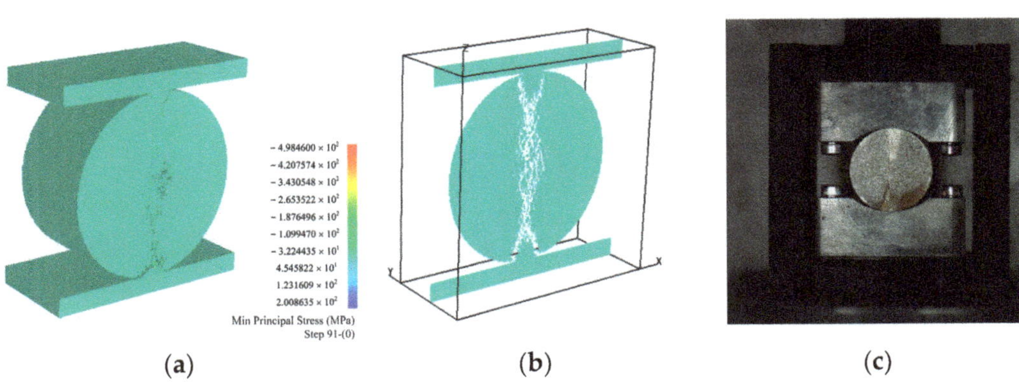

Figure 8. Numerical simulation results of Brazilian splitting of intact frozen rock: (**a**) minimum principal stress, (**b**) model slice, (**c**) fracture mode.

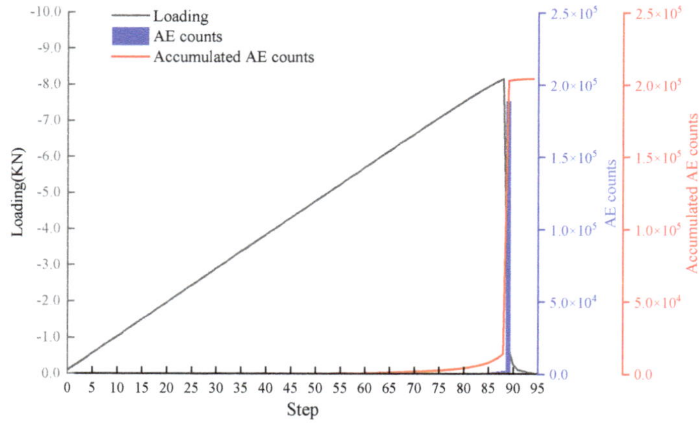

Figure 9. The relationship curve between load, AE counts and accumulated AE counts for intact frozen rock.

Figure 10 shows the numerical simulation results of the Brazilian splitting of frozen rock with different ice-filled crack angles. Red and blue circles in the AE distribution indicate the shear and tensile failure of the element. Compared with the experimental results (Figure 6), the splitting fracture modes of the frozen rock model were similar. The ice in the crack supported and transferred some force during the loading. However, the ice strength was lower than the rock strength, which induced the stress concentration at the ice-filled crack. When loaded to a certain extent, wing cracks occurred at the tip and penetrated through the whole model.

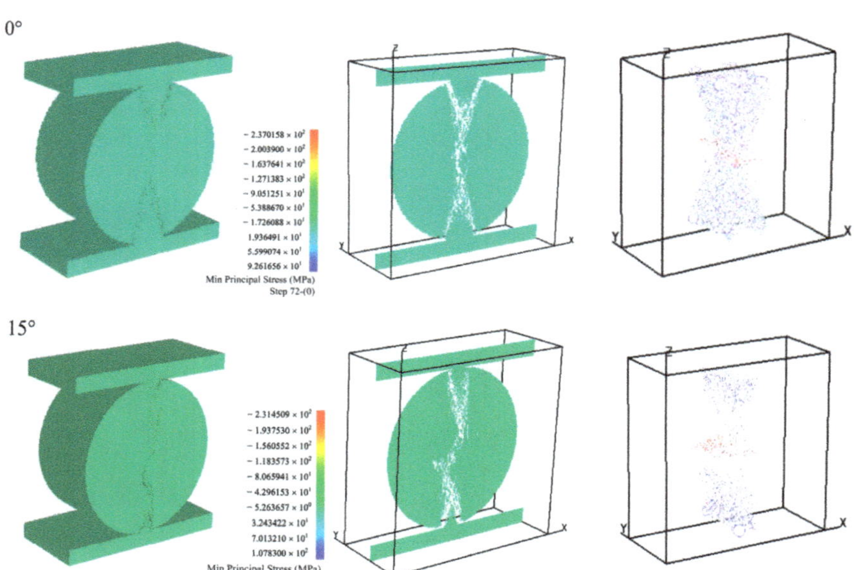

Figure 10. *Cont.*

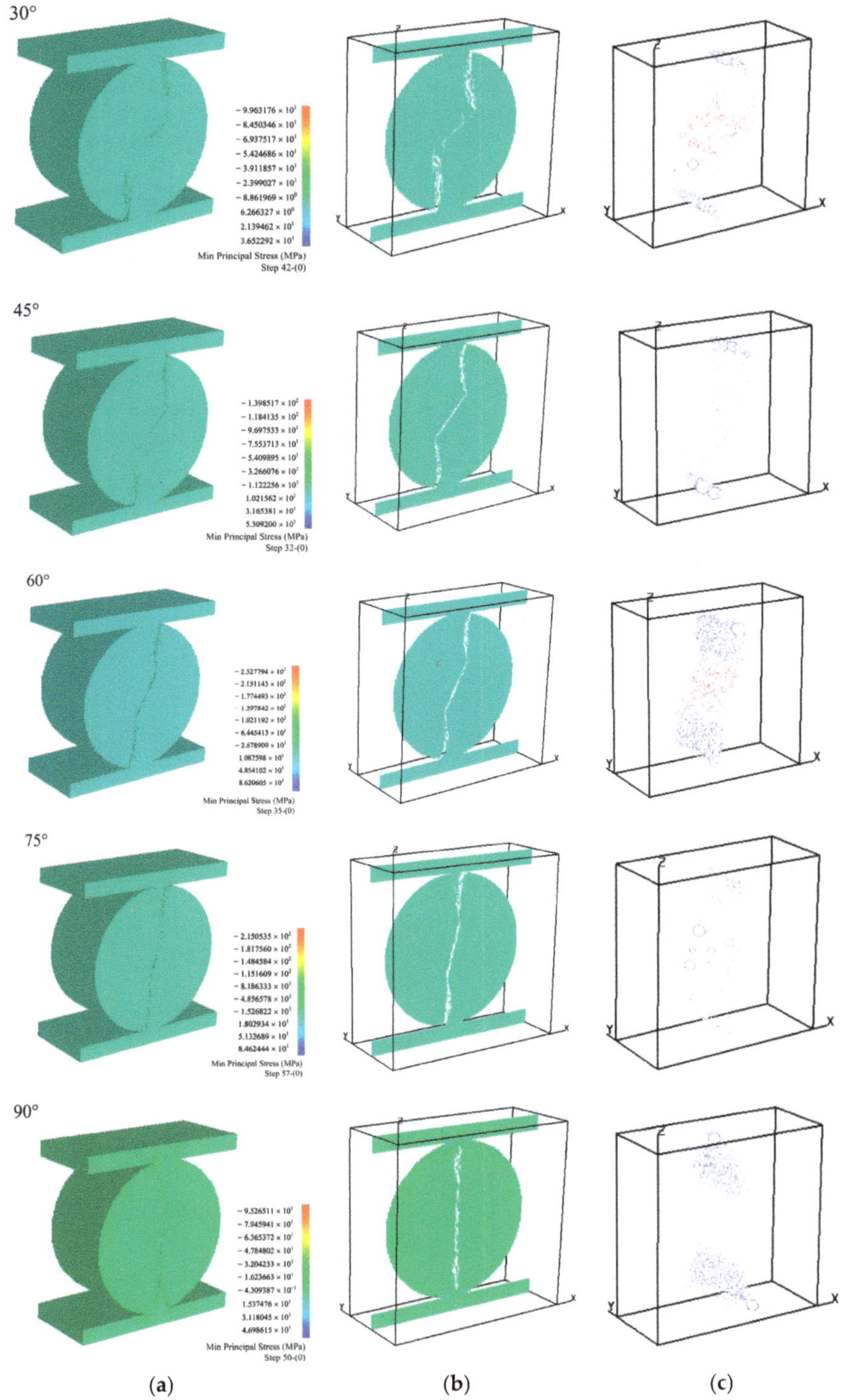

Figure 10. Numerical simulation results of Brazilian splitting of frozen rock in different ice-filled crack angles: (**a**) minimum principal stress, (**b**) model slice; (**c**) AE.

Figure 11 shows the relationship curve between the load, AE counts and accumulated AE counts for frozen rock with different ice-filled crack angles. The peak load of 0° ice-filled crack frozen rock was 7.62 kN, and splitting failure occurred at step 67. The peak load of 15° ice-filled crack frozen rock was 6.81 kN, and splitting failure occurred in step 87. The peak load of 30° ice-filled crack frozen rock was 4.11 kN, and splitting failure occurred at step 38. The peak load of the 45° fractured frozen rock specimen was 2.59 kN, and splitting failure occurred in step 29.

The frozen rock specimens with 0°, 15°, 30°, and 45° ice-filled crack showed typical brittle failure mode. At the initial stage of loading, the internal microcracks of the frozen rock specimen accumulated continuously. AE events were fewer with no macroscopic cracks on the surface. When an ice-filled crack incurred damage and small macroscopic cracks began to appear on the surface of the specimen, AE events increased and the damaged elements accumulated stably. When the tensile strength increased to the peak value, the damaged elements increased. At the same time, the AE events increased suddenly, and the macroscopic cracks on the surface of the model further extended. When the stress fell, there were still a small number of AE events existing, and the damaged elements increased continuously. The cracks on the surface of the model were aggravated and extended, and the specimen finally failed.

The peak loads of the frozen rock with 60, 75 and 90° angle ice-filled cracks were 2.73, 2.57 and 3.49 kN, respectively, and the splitting failures occurred at steps 30, 38, and 34. There were two peak regions of AE events in these specimens. The first peak was the energy released by the ice-filled cracks, and the second peak was the brittle failure of the specimen. The frozen rock specimens with 60, 75 and 90° ice-filled cracks also showed a typical brittle failure mode.

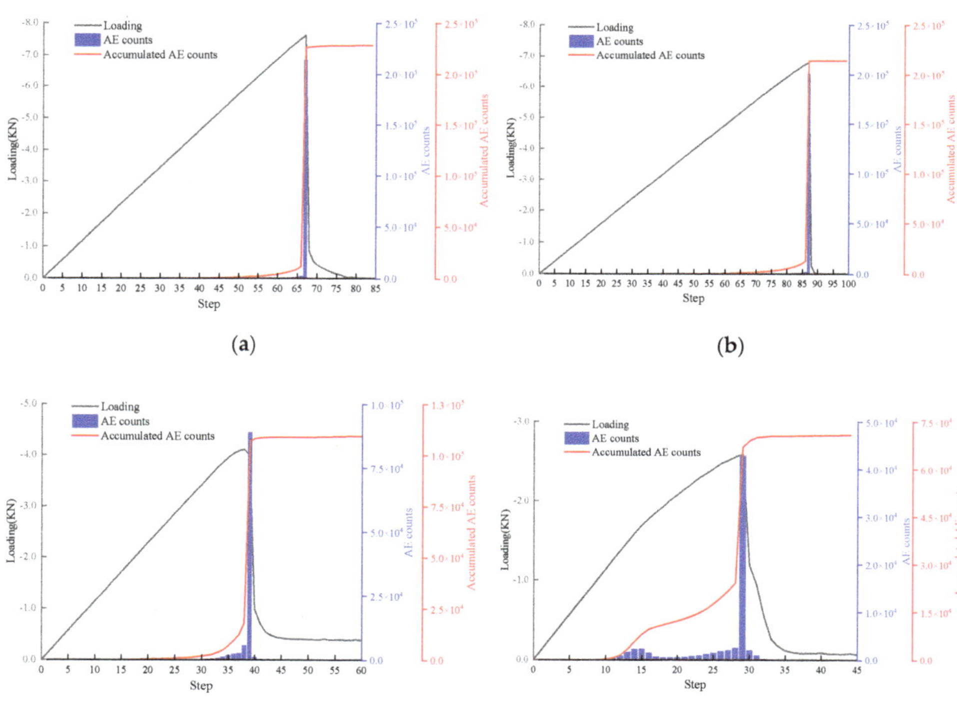

(a) (b) (c) (d)

Figure 11. *Cont.*

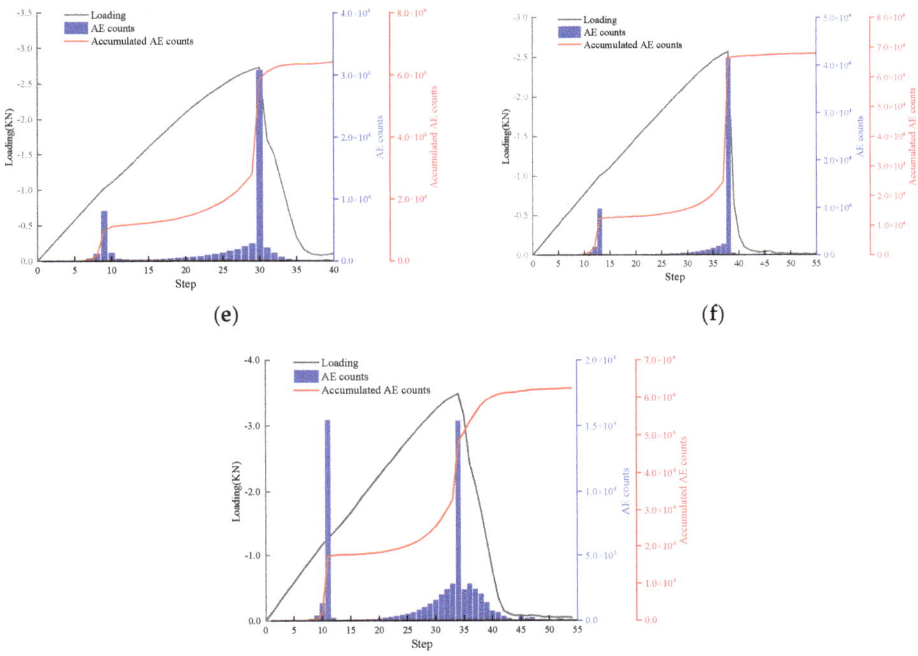

(e) **(f)**

(g)

Figure 11. The relationship curve between load, AE counts and accumulated AE counts for frozen rock with different ice-filled crack angles: (**a**) 0°, (**b**) 15°, (**c**) 30°, (**d**) 45°, (**e**) 60°, (**f**) 75°, (**g**) 90°.

Figure 12 shows that the numerical simulation results are close to the experimental results by comparing the tensile strength of frozen rock with different ice-filled angles. The tensile strength of the specimens decreased gradually and then increased with the increase in the angle of the ice-filled crack. By comparing the tensile strength and fracture mode of the frozen rock with the experimental results, it proves that RFPA^{3D} is feasible for simulating the splitting process of frozen rock mass.

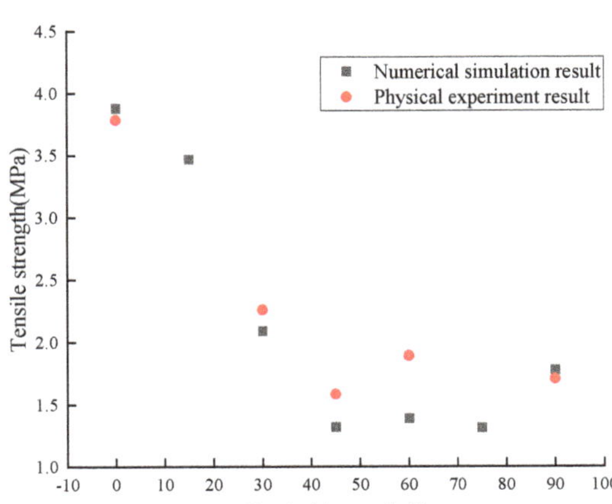

Figure 12. Comparison between experimental and simulated values of tensile strength of frozen rock.

3. Multi-Factor Discussion

In order to study the change of tensile strength, crack initiation and fracture mode of frozen rock under the influence of multiple factors, RFPA^{3D} was used to conduct numerical simulation analysis on specimens with different widths of ice-filled crack (l = 20 mm, $\alpha = 60°$, w = 2, 3, 4 mm) and lengths of ice-filled crack (w = 1.5 mm, $\alpha = 30°$, l = 10, 15, 25 mm). The material parameters are shown in Table 1.

3.1. Different Lengths for Ice-Filled Crack

Figure 13 shows that the fracture mode of the frozen rock specimens with different lengths of ice-filled crack started from the ice crack first, and then the wing crack was formed from the crack tip of the ice to the loading end.

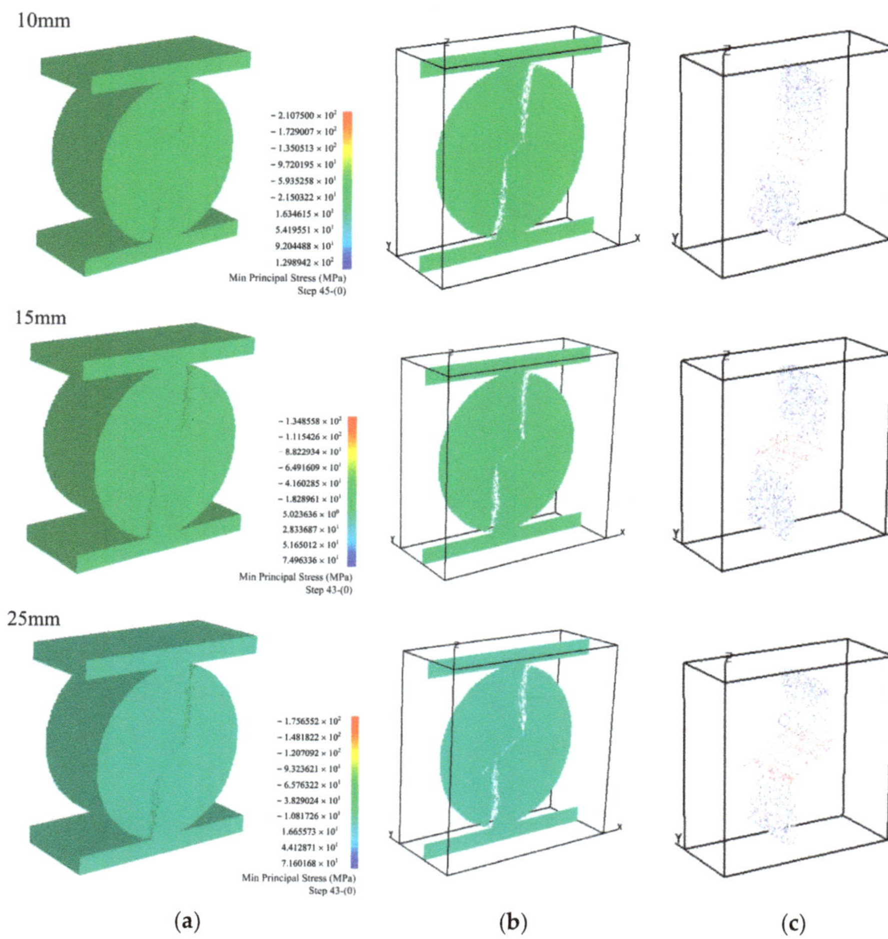

Figure 13. Brazilian splitting fracture mode of ice-filled crack frozen rock of different lengths. (**a**) Minimum principal stress, (**b**) model slice, (**c**) AE.

The peak loads of frozen rock with 10, 15, and 25 mm length ice-filled cracks were 4.63, 4.35, and 4.04 kN, respectively, and the splitting failure occurred at steps 42, 40, and 40 (Figure 14a–c). Combined with the calculation results for 20 mm length frozen fractured rock specimens in 2.3, it was found that the tensile strength of the frozen rock gradually decreased with the increase in the length of the ice-filled crack. The longer the ice-filled crack, the higher the ice content, and the lower the bearing capacity of the specimen.

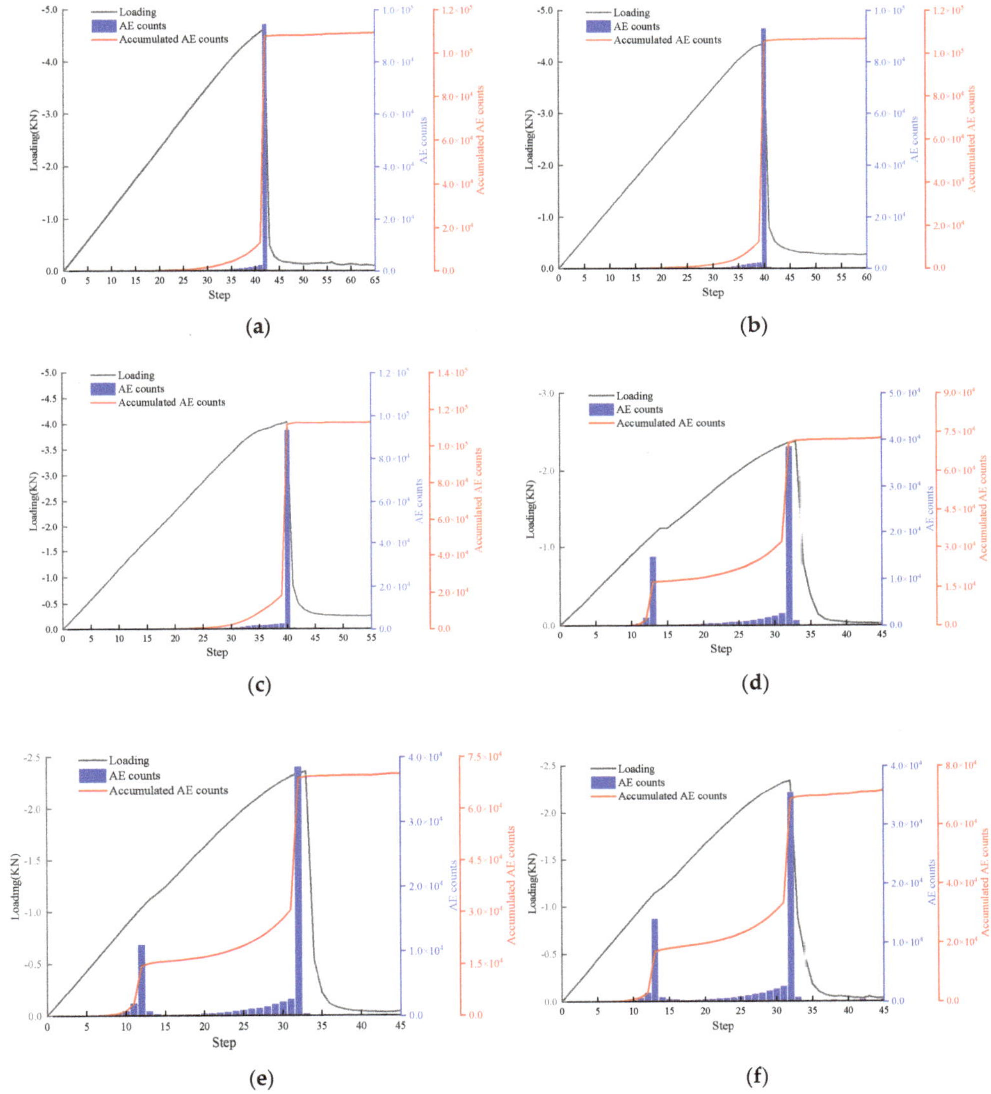

Figure 14. The relationship curve between load, AE counts and accumulated AE counts for ice-filled cracks of different lengths and widths of frozen rock. (**a**) l = 10 mm, (**b**) l = 15 mm, (**c**) l = 25 mm, (**d**) w = 2 mm, (**e**) w = 3 mm, (**f**) w = 4 mm.

3.2. Different Widths for Ice-Filled Cracks

Frozen rocks with different ice-filled crack widths also incur damage and failure elements from the ice. Then, the wing crack starts from the tip of the ice-filled crack, and gradually extends to the loading end (Figure 15). The peak loads of frozen rock with 2, 3, and 4 mm width ice-filled cracks were 2.38, 2.36, and 2.34 kN, respectively. Combined with the calculation results of 1.5 mm width frozen fractured rock specimens in 2.3, it was found that the tensile strength of frozen rock gradually decreased with the increase in the width of the ice-filled crack. There were also two peak regions of AE events (Figure 14d–e). The first peak was the energy released by the ice-filled crack, and the second peak was the brittle failure of the specimen.

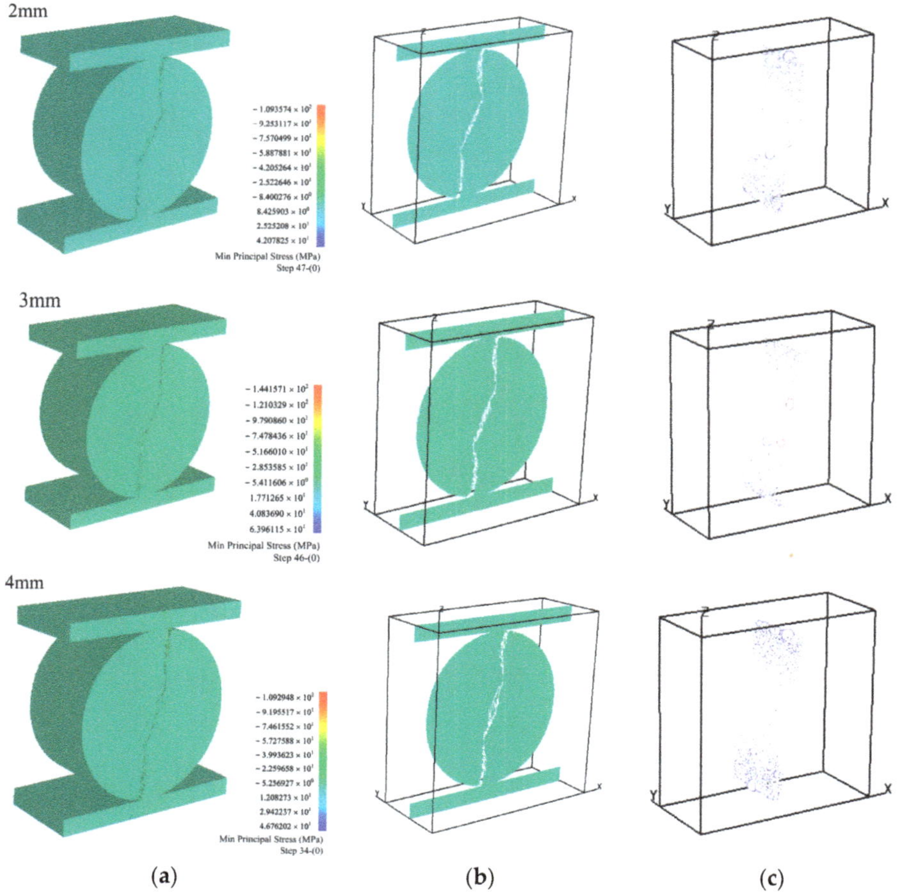

Figure 15. Brazilian splitting fracture mode of frozen rock with ice-filled cracks of different widths. (**a**) Minimum principal stress, (**b**) model slice, (**c**) AE.

4. Conclusions

(1). The tensile strength of frozen rock specimens decreased gradually and then increased with the increase in the ice-filled crack angle. When the ice-filled crack angle was 0°, the tensile strength of the ice-filled crack frozen rock specimen was similar to that of the intact frozen rock specimen. The tensile strength was the lowest when the ice-filled crack angle was 45°.

(2). The peak loads of frozen rock with 1.5, 2, 3, and 4 mm width ice-filled cracks were 2.73, 2.38, 2.36, and 2.34 kN, respectively. The peak loads of frozen rock with 10, 15, 20, and 25 mm length ice-filled cracks were 4.63, 4.35, 4.11 and 4.04 kN, respectively. The tensile strength of frozen rock gradually decreased with the increase in the length and width of the ice-filled cracks.

(3). Except for the 0° ice-filled cracks, the frozen fractured rock masses incurred damage elements at the ice-filled crack first, and then the wing crack started from the tip of the ice-filled crack and extended continuously, leading to the failure of the frozen rock specimens. When the angle of the ice-filled crack was 0°, the ice crack was perpendicular to the loading direction, and the effects on the frozen rock specimen were small.

(4). The frozen rock specimens showed typical brittle failure characteristics. With the loading process, the model experienced four stages: internal microcrack initiation, surface macroscopic crack generation, macroscopic crack propagation, and complete failure.

Author Contributions: Conceptualization, T.W. and P.L.; methodology, T.W., C.T. and P.L.; software, T.W., J.Y. and C.T.; validation, T.W., T.G., J.Y. and B.Z.; formal analysis, T.W. and T.G.; investigation, T.W. and B.Z.; resources, T.W., C.T. and P.L.; data curation, T.W. and P.L.; writing—original draft preparation, T.W.; writing—review and editing, C.T. and P.L.; visualization, T.W.; supervision, C.T. and P.L.; project administration, C.T.; funding acquisition, C.T. and P.L. All authors have read and agreed to the published version of the manuscript.

Funding: This research was funded by the National Natural Science Foundation of China (Grant No. 42050201) and the Open Fund of State Key Laboratory of Frozen Soil Engineering (Grant No. SKLFSE202013).

Institutional Review Board Statement: Not applicable.

Informed Consent Statement: Informed consent was obtained from all subjects involved in the study.

Data Availability Statement: The data presented in this study are available on request from the corresponding author. The data are not publicly available due to privacy concerns.

Acknowledgments: Hongda Blasting Engineering Group Co., Ltd. are appreciated for their assistance.

Conflicts of Interest: The authors declare they have no conflict of interest to this work.

References

1. Huang, S.B.; Liu, Y.Z.; Guo, Y.L.; Zhang, Z.; Cai, Y. Strength and failure characteristics of rock-like material containing single crack under freeze-thaw and uniaxial compression. *Cold Reg. Sci. Technol.* **2019**, *162*, 1–10. [CrossRef]
2. Liu, N.F.; Li, N.; Li, G.F.; Song, Z.P.; Wang, S.J. Method for Evaluating the Equivalent Thermal Conductivity of a Freezing Rock Mass Containing Systematic Fractures. *Rock Mech. Rock Eng.* **2022**, *55*, 7333–7355. [CrossRef]
3. Wang, T.T.; Tang, C.A.; Li, P.F.; Tang, S.; Liu, M.H.; Zhang, B.B. Frost-Heaving Cracking Sensitivity of Single-Flaw Rock Mass Based on a Numerical Experimental Method. *Geofluids* **2021**, *2021*, 3436119. [CrossRef]
4. Ma, D.D.; Xiang, H.S.; Ma, Q.Y.; Kaunda, E.E.; Huang, K.; Su, Q.Q.; Yao, Z.M. Dynamic damage constitutive model of frozen silty soil with prefabricated crack under uniaxial load. *J. Eng. Mech.* **2021**, *147*, 04021033. [CrossRef]
5. Xu, S.H.; Li, N.; Wang, X.D.; Xu, Z.G.; Yuan, K.K.; Tian, Y.Z.; Wang, L.L. Damage test and degradation model of saturated sandstone due to cyclic freezing and thawing of rock slopes of open-pit coal mine. *Chin. J. Rock Mech. Eng.* **2016**, *35*, 2561–2571.
6. Wu, N.; Liang, Z.Z.; Zhang, Z.H.; Li, S.H.; Lang, Y.X. Development and Verification of Three-Dimensional Equivalent Discrete Fracture Network Modelling Based on the Finite Element Method. *Eng. Geol.* **2022**, *306*, 106759. [CrossRef]
7. Wan, W.; Li, C.C. Microscopic and Acoustic Interpretations of the Physics of Rock Burst and the Difference in Fracturing Patterns in Class I and Class II Rocks. *Rock Mech. Rock Eng.* **2022**, *55*, 6841–6862. [CrossRef]
8. Yang, S.Q.; Yin, P.F.; Huang, Y.H. Experiment and Discrete Element Modelling on Strength, Deformation and Failure Behaviour of Shale under Brazilian Compression. *Rock Mech. Rock Eng.* **2019**, *52*, 4339–4359. [CrossRef]
9. Wang, M.; Ping, C. Experimental study on the validity and rationality of four Brazilian disc tests. *Geothch. Geol. Eng.* **2018**, *36*, 63–76. [CrossRef]
10. Haeri, H.; Shahriar, K.; Marji, M.F.; Moarefvand, P. Experimental and numerical study of crack propagation and coalescence in pre-cracked rock-like disks. *Int. J. Rock Mech. Min. Sci.* **2014**, *67*, 20–28. [CrossRef]
11. Gao, M.; Liang, Z.Z.; Jia, S.P.; Zou, J.Q. Tensile Properties and Tensile Failure Criteria of Layered Rocks. *Appl. Sci.* **2022**, *12*, 6063. [CrossRef]
12. Zhou, J.; Zeng, Y.J.; Guo, Y.T.; Chang, X.; Liu, L.W.; Wang, L.; Hou, Z.K.; Yang, C.H. Effect of natural filling fracture on the cracking process of shale Brazilian disc containing a central straight notched flaw. *J. Pet. Sci. Eng.* **2021**, *196*, 107993. [CrossRef]
13. Zhao, Z.H.; Liu, Z.N.; Pu, H.; Li, X. Effect of thermal treatment on Brazilian tensile strength of granites with different grain size distributions. *Rock Mech. Rock Eng.* **2018**, *51*, 1293–1303. [CrossRef]
14. Liu, B.; Sun, Y.D.; Wang, B.; Han, Y.H.; Zhang, R.H.; Wang, J.X. Effect of water content on mechanical and electrical characteristics of the water-rich sandstone during freezing. *Environ. Earth Sci.* **2020**, *79*, 236–249. [CrossRef]
15. Jia, H.L.; Zi, F.; Yang, G.S.; Li, G.Y.; Shen, Y.J.; Sun, Q.; Yang, P.Y. Influence of pore water (ice) content on the strength and deformability of frozen argillaceous siltstone. *Rock Mech. Rock Eng.* **2020**, *53*, 967–974. [CrossRef]
16. Kodama, J.; Goto, T.; Fujii, Y.; Hagan, P. The effects of water content, temperature and loading rate on strength and failure process of frozen rocks. *Int. J. Rock Mech. Min. Sci.* **2013**, *62*, 1–13. [CrossRef]
17. Wang, Y.; Yi, Y.F.; Li, C.H.; Han, J.Q. Anisotropic fracture and energy characteristics of a Tibet marble exposed to multi-level constant-amplitude (MLCA) cyclic loads: A labscale testing. *Eng. Fract. Mech.* **2021**, *244*, 107550. [CrossRef]

18. Jia, H.L.; Ding, S.; Zi, F.; Dong, Y.H.; Shen, Y.J. Evolution in sandstone pore structures with freeze-thaw cycling and interpretation of damage mechanisms in saturated porous rocks. *Catena* **2020**, *195*, 104915. [CrossRef]
19. Zhou, X.P.; Fu, Y.H.; Wang, Y.; Zhou, J.N. Experimental study on the fracture and fatigue behaviors of flawed sandstone under coupled freeze-thaw and cyclic loads. *Theor. Appl. Fract. Mech.* **2022**, *119*, 103299. [CrossRef]
20. Zhang, H.M.; Xia, H.J.; Yang, G.S.; Zhang, M.J.; Peng, C.; Ye, W.J.; Shen, Y.J. Experimental research of influences of freeze-thaw cycles and confining pressure on physical-mechanical characteristics of rocks. *J. China Coal Soc.* **2018**, *43*, 441–448.
21. Bayram, F. Predicting mechanical strength loss of natural stones after freeze-thaw in cold regions. *Cold Reg. Sci. Technol.* **2012**, *83–84*, 98–102. [CrossRef]
22. Aoki, K.; Hibiya, K.; Yoshida, T. Storage of refrigerated liquefied gases in rock caverns: Characteristics of rock under very low temperatures. *Tunn. Undergr. Space Technol.* **1990**, *5*, 319–325. [CrossRef]
23. Yamabe, T.; Neaupane, K.M. Determination of some thermo-mechanical properties of Sirahama sandstone under subzero temperature conditions. *Int. J. Rock Mech. Min. Sci.* **2001**, *38*, 102–1034. [CrossRef]
24. Xu, G.M.; Liu, Q.S.; Peng, W.W.; Chang, X.X. Experimental study on basic mechanical behaviors of rocks under low temperatures. *Chin. J. Rock Mech. Eng.* **2006**, *25*, 2502–2508. (In Chinese)
25. Yang, G.S.; Xi, J.M.; Li, H.J.; Cheng, L. Experimental study of rock mechanical properties under triaxial compressive and frozen conditions. *Chin. J. Rock Mech. Eng.* **2010**, *29*, 459–464. (In Chinese)
26. Xi, J.M.; Yang, G.S.; Pang, L.; Lv, X.T.; Liu, F.L. Experimental study on basic mechanical behaviors of sandy mudstone under low freezing temperature. *J. China Coal Soc.* **2014**, *39*, 1262–1268.
27. Shan, R.L.; Yang, H.; Guo, Z.M.; Liu, X.; Song, L. Experimental study of strength characters of saturated red sandstone on negative temperature under triaxial compression. *Chin. J. Rock Mech. Eng.* **2014**, *33*, 3657–3664.
28. Li, T.; Ma, Y.J.; Liu, B.; Sheng, H.L.; He, P. Strength characteristics and elastic modulus evolution of frozen gray sandstone under cyclic loading. *J. China Coal Soc.* **2018**, *43*, 2438–2443.
29. Bai, Y.; Shan, R.L.; Ju, Y.; Wu, Y.X.; Tong, X.; Han, T.Y.; Dou, H.Y. Experimental study on the strength, deformation and crack evolution behaviour of red sandstone samples containing two ice-filled fissures under triaxial compression. *Cold Reg. Sci. Technol.* **2020**, *174*, 103061. [CrossRef]
30. Yang, H.; Shan, R.L.; Zhang, J.X.; Wu, F.M.; Guo, Z.M. Mechanical properties of frozen rock mass with two diagonal intersected fractures. *Int. J. Min. Sci. Technol.* **2018**, *28*, 631–638. [CrossRef]
31. Shloido, G.A. Determining the tensile strength of frozen ground. *Hydrotech. Constr.* **1968**, *2*, 238–240. [CrossRef]
32. Wang, S.Y.; Sloan, S.W.; Tang, C.A. Three-dimensional numerical investigations of the failure mechanism of a rock disc with a central or eccentric hole. *Rock Mech. Rock Eng.* **2014**, *47*, 2117–2137. [CrossRef]
33. Fan, N.; Wang, J.R.; Deng, C.B.; Fan, Y.P.; Wang, T.T.; Guo, X.Y. Quantitative Characterization of Coal Microstructure and Visualization Seepage of Macropores Using CT-Based 3D Reconstruction. *J. Nat. Gas Sci. Eng.* **2020**, *81*, 103384. [CrossRef]
34. Liang, Z.; Wu, N.; Li, Y.; Li, H.; Li, W. Numerical Study on Anisotropy of the Representative Elementary Volume of Strength and Deformability of Jointed Rock Masses. *Rock Mech. Rock Eng.* **2019**, *52*, 4387–4402. [CrossRef]
35. Khoojine, A.S.; Shadabfar, M.; Tabriz, Y.E. A Mutual Information-Based Network Autoregressive Model for Crude Oil Price Forecasting Using Open-High-Low-Close Prices. *Mathematics* **2022**, *10*, 10173172.
36. Zhu, W.C.; Tang, C.A. Numerical simulation of Brazilian disk rock failure under static and dynamic loading. *Int. J. Rock Mech. Min. Sci.* **2006**, *43*, 236–252. [CrossRef]
37. Wu, N.; Liang, Z.Z.; Li, Y.; Qian, X.K.; Gong, B. Effect of confining stress on representative elementary volume of jointed rock masses. *Geomech. Eng.* **2019**, *18*, 627–638.
38. Shadabfar, M.; Cheng, L.S. Probabilistic Approach for Optimal Portfolio Selection Using a Hybrid Monte Carlo Simulation and Markowitz Model. *Alex. Eng. J.* **2020**, *59*, 3381–3393. [CrossRef]
39. Tang, C.A. Numerical simulation of progressive rock failure and associated seismicity. *Int. J. Rock Mech. Min. Sci.* **1997**, *34*, 249–261. [CrossRef]
40. Liao, Z.Y.; Zhu, J.B.; Tang, C.A. Numerical investigation of rock tensile strength determined by direct tension, Brazilian and three-point bending tests. *Int. J. Rock Mech. Min. Sci.* **2019**, *115*, 21–32. [CrossRef]

applied sciences

MDPI

Article

Study on the Dynamic Splitting Mechanical Properties of Annular Sandstone Specimens with Temperature–Water Coupling in a Coal Mine

Qi Ping [1,2,3,*], Qi Gao [2,3], Yulin Wu [2,3], Chen Wang [2,3], Kaifan Shen [2,3], Shuo Wang [2,3], Shiwei Wu [2,3] and Yijie Xu [2,3]

1 State Key Laboratory of Mining Response and Disaster Prevention and Control in Deep Coal Mine, Anhui University of Science and Technology, Huainan 232001, China
2 Engineering Research Center of Mine Underground Projects, Ministry of Education, Anhui University of Science and Technology, Huainan 232001, China; 2020200277@aust.edu.cn (Q.G.); 2020200404@aust.edu.cn (Y.W.); 2020200327@aust.edu.cn (C.W.); 2020200341@aust.edu.cn (K.S.); 2020200372@aust.edu.cn (S.W.); 2021200374@aust.edu.cn (S.W.); 2021200476@aust.edu.cn (Y.X.)
3 School of Civil Engineering and Architecture, Anhui University of Science and Technology, Huainan 232001, China
* Correspondence: ahpingqi@163.com or qping@aust.edu.cn; Tel.: +86-139-5645-9398

Featured Application: The basic physical parameters and dynamic properties of annular and intact sandstone specimens treated by temperature–water coupling are compared and analyzed.

Abstract: With the gradual deepening of mine excavation depth, the strong disturbance of deep strata becomes more and more obvious. Rock's failure under blasting mainly depends on its dynamic tensile strength. The changes in rock's dynamic properties are obviously affected by temperature and water. In order to study the dynamic tensile properties of annular sandstone specimens under the influence of temperature and water, deep sandstone was drilled, followed by water bath tests at eight temperatures (25~95 °C). It can be seen from the analysis of test results that the mass and volume growth rates of the annular and the intact sandstone specimens first increased and then decreased, while the density growth rate first decreased and then increased. The mass and volume growth rates of the annular sandstone specimens were smaller, but the density growth rate was larger. Because of the increase in water temperature, the dynamic compressive strength first increased and then decreased. The dynamic tensile strength of the annular sandstone specimen was lower. The average strain rate and peak strain also showed a quadratic function relationship of first decreasing and then increasing with the increase in water temperature. The average strain rate of the annular sandstone specimen was smaller, but the peak value changed greatly. The Brazilian disc validity condition is applicable to two failure conditions of sandstone specimens. Through XRD and SEM analysis, we found that the changes in the dynamic properties of sandstone specimens were not due to their own material composition, but to the damage to their structure caused by the temperature–water coupling effect.

Keywords: rock impact dynamics; sandstone annular specimen; SHPB; Brazilian disc split test; temperature–water coupling

Citation: Ping, Q.; Gao, Q.; Wu, Y.; Wang, C.; Shen, K.; Wang, S.; Wu, S.; Xu, Y. Study on the Dynamic Splitting Mechanical Properties of Annular Sandstone Specimens with Temperature–Water Coupling in a Coal Mine. *Appl. Sci.* **2022**, *12*, 4608. https://doi.org/10.3390/app12094608

Academic Editors: Zhengzhao Liang, Nuwen Xu and Bei Jiang

Received: 4 April 2022
Accepted: 30 April 2022
Published: 3 May 2022

1. Introduction

With the continuous increase in the human demand for mineral resources, shallow resources are not enough to meet human demand, so the depth of mine excavation is gradually increasing. However, with the increase in depth, the excavation becomes more difficult, and the underground strata environment becomes more complicated. Therefore, it is necessary to study the mechanical properties of ore at the bottom of mines, and how they are affected by various strong disturbances.

Temperature and groundwater are common influencing factors in the process of mine excavation. Ping [1] applied temperature–water coupling treatment to sandstone, and carried out dynamic compression tests to analyze the dynamic properties of sandstone, such as peak strength, peak strain, average strain rate, and elastic modulus. Ping [2] conducted dynamic splitting tests on the temperature–water coupling of sandstone with different impact pressures, and found that the dynamic properties were affected by strain rate. Wang et al. [3] found that the dynamic properties of sandstone are affected by water bonding force and the Stefan effect. Roy [4] treated different types of sandstone for different saturation times, and concluded that their mechanical properties and fracture toughness decreased with the increase in saturation. Wang et al. [5] investigated the mechanical properties and failure law of two kinds of coal rock with different moisture content. Teng et al. [6] researched the degradation mechanism of shale under different water-bearing states. Deng et al. [7] considered the influence of five different water contents on the splitting tensile strength of layered sandstone. Wang et al. [8] conducted splitting tensile strength tests of sandstone Brazilian discs under different temperatures and water contents, providing basic reference data for the standardization of splitting strength and engineering applications. Ping et al. [9,10] conducted SHPB impact tests on sandstone and limestone at different temperatures to analyze their dynamic properties. It was concluded that different effects of high temperature made the peak strength of the rock first increase, and then decreases. Zhang [11] used an SHPB device to study the dynamic properties and damage characteristics of sandstone at different temperatures. Qin et al. [12] analyzed the influence of mesoscopic changes on the uniaxial compressive strength of high-temperature sandstone via acoustic damage, X-ray diffraction, and scanning electron microscopy. Xu et al. [13] combined and predicted the rate-dependent effect of dynamic tensile strength of two igneous rocks by using incubation time as a criterion. Ke et al. [14,15] studied the dynamic and static mechanical properties of rocks under freeze–thaw cycles. Zheng et al. [16] enriched the detection methods of water-induced mine disasters by studying the influence of water on the induction signals of rock failure charge. Chang et al. [17] studied the tensile strength characteristics of sandstone, mudstone, coal, and medium sandstone under different water content states. Geng [18] tested dynamic damage to sandstones using acoustic emission technology, and concluded that water absorption reduces brittleness but increases ductility. Taiki [19] carried out uniaxial compression tests on clay with cracks of different prefabricated lengths and different water contents, and concluded that the deformation modulus depends on the water content, but has nothing to do with crack length. Li [20] studied the energy storage and acoustic emission characteristics of rock under different water content states. Wang [21] studied the influence of moisture content on the characteristic strength of soft coal, and found that the shear strength reached the maximum when the moisture content was 4.23%. Zhao [22] studied the surface characteristics, porosity, and permeability of limestone treated with different temperatures by dry-ice cooling. Liu [23] conducted SHPB tests on limestone treated at different temperatures under different confining pressures, and the test results showed that 400 °C was an inflection point of dynamic properties.

The strength and other mechanical properties of deep sandstone are affected not only by temperature and groundwater, but also by its own inhomogeneity and size [24]. Scholars at home and abroad have studied the mechanical properties and failure modes of annular specimens with holes by simulating their own cracks and internal channels. Wang et al. [25] carried out an experimental study on radial compression of annular granite under various temperature and humidity conditions, and found that the maximum tensile strain could be used as a failure judgment parameter for annular granite. Wu et al. [26] used annular sandstone specimens with different inner diameter to conduct Brazilian disc splitting tests. Yang et al. [27] constructed a discrete element model for Brazilian disc splitting testing of a single-hole disc, and studied the influence of aperture and eccentricity on specimen deformation and strength characteristics. You et al. [28] conducted triaxial compression tests on marble samples with two different channels to analyze their bearing capacity and deformation characteristics. It was concluded that the different characteristics of marble affect the deformation and damage to marble specimens. Ying [29] conducted impact loading tests on a tunnel's surrounding rock in different directions to analyze its fracture toughness and other characteristics, which was of great value to tunnel engineering research.

It can be seen that the rock dynamic properties of temperature–water coupling are worthy of further study. Since the rock's inhomogeneity is also the main factor affecting its mechanical properties, this test drilled the sandstone specimen and then carried out temperature–water coupling treatment at different temperatures (25 °C~95 °C), so as to study its physical properties, dynamic properties, and failure mode. Annular sandstone specimens can be used not only to simulate the influence of inhomogeneity on mechanical properties, but also to simulate the failure law of shaft walls affected by transverse load in the process of shaft excavation. Because rock's failure mainly depends on its tensile strength, but it is difficult to directly measure the tensile strength, the Brazilian disc splitting test was adopted in this experiment [30].

2. Materials and Methods

2.1. Sandstone Specimen Processing and Preparation

According to the requirements of the test specification [31,32] and the dynamic test specification [33], the sandstone required for the test was cored, cut, and polished. The height and diameter of the processed sandstone specimen were 25 mm and 50 mm, respectively. In the preliminary experiment, the failure characteristics and data processing of the 10 mm aperture specimen were more representative, which was convenient for the final results' analysis. Therefore, a 10 mm drill bit was used for drilling in this experiment. The center of the circle was found on the specimen and marked with a marker. Then, a drilling machine was used for drilling. In order to prevent the influence of friction heating on the strength of the drill bit, water was used to cool and lubricate the drill bit during drilling. The effect diagram of drilling the specimen is shown in Figure 1.

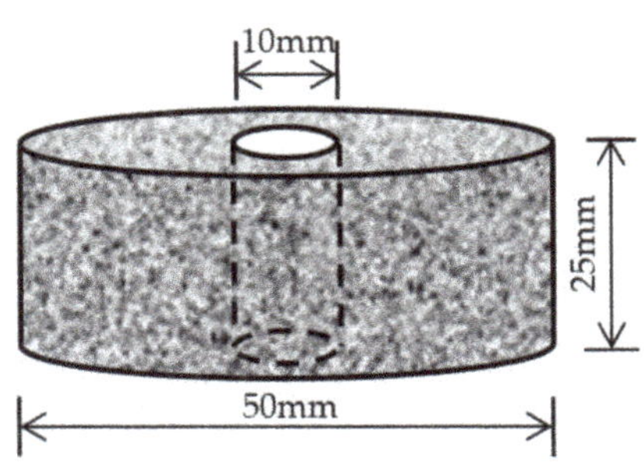

Figure 1. Drilling effect drawing.

2.2. Experimental Plan

The sandstone in this paper was taken from the roadway sandstone of a coal mine, and the depth of the mine's bottom is about 600 m. Combined with the local geological and climatic conditions, we improved the test scheme. Because the temperature of the bottom of the mine is kept above 25 °C all year round, and the water temperature reaches 100 °C, boiling occurs. The annular and intact sandstone specimens were subjected to temperature–water coupling treatment at eight temperatures (25 °C~95 °C), followed by SHPB impact testing. It was found that when the impact pressure was 0.3 MPa, the waveform obtained from the test was better and more conducive to data processing. Therefore, the impact pressure was set to 0.3 MPa in this test. At the end of the test, the test data were processed, and the dynamic properties and failure modes of the annular sandstone specimens were analyzed and compared with those of the intact sandstone specimens, so as to analyze the influence rule of the annular sandstone's dynamic properties.

After processing, the two specimens were treated with thermostatic water baths at different temperatures. In order to ensure that the sandstone specimens met the water saturation condition, each temperature gradient was treated with a 48h constant-temperature water bath. After the temperature–water coupling effect, the surface of the sandstone specimen was dried, its mass and its internal and external diameter were measured, and then the specimen was placed on the SHPB pressure bar for dynamic splitting tensile testing after different water bath temperatures.

2.3. SHPB Test Device

The pressure rod device (SHPB) used in this study is shown in Figure 2.

Figure 2. SHPB test device.

Brazilian disc splitting tests were carried out on the annular and intact sandstone specimens treated by temperature–water coupling with eight temperature gradients. Immediately after the impact, the test data were saved and the post-impact fragments were collected. The clamping mode of the specimen is shown in Figure 3.

Figure 3. Clamping method of the specimen.

Through the data collected with the signal collection device, the data were calculated and processed, and the dynamic parameters required by the test—such as impact load *P*(*t*), dynamic strain ε_s, and average strain rate $\dot{\varepsilon_s}$—were obtained. The calculation principle [34,35] is shown in Formula (1).

$$\left.\begin{array}{c} P(t) = E_0 A_0[\varepsilon_I(t) - \varepsilon_R(t)] = E_0 A_0 \varepsilon_T(t) \\ \varepsilon(t) = \frac{2C_0}{D}\int_0^{\tau}[\varepsilon_I(t) - \varepsilon_T(t)]dt = \frac{2C_0}{D}\int_0^{\tau}\varepsilon_R(t)dt \\ \dot{\varepsilon}(t) = -\frac{2C_0}{D}[\varepsilon_I(t) - \varepsilon_T(t)] = -\frac{2C_0}{D}\varepsilon_R(t) \end{array}\right\} \quad (1)$$

where E_0 is the elastic modulus, and A_0 is the cross-sectional area of the press bar material.
ε_I is the incident strain, ε_R is the reflected strain, and ε_T is the projected strain.
C_0 is the longitudinal wave velocity of the pressure bar device;
ρ_0 is the material density of the pressure rod device;
D is the diameter of the sandstone specimen;
τ is the duration of the stress waves.

3. Analysis of Test Data

3.1. Physical Properties Analysis

Photos of sandstone specimens treated with different water temperatures are shown in Figure 4

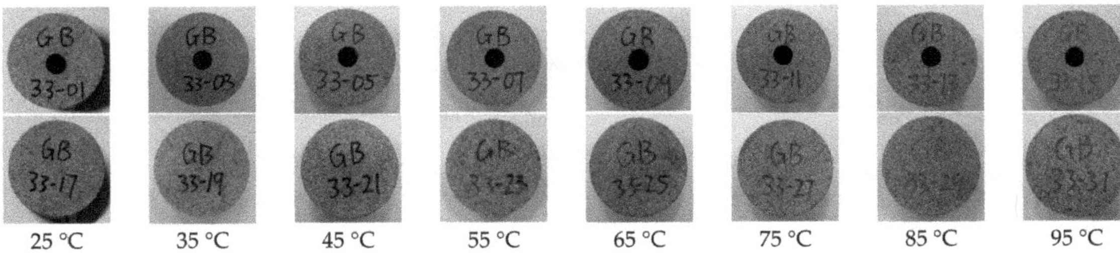

Figure 4. The apparent morphology of sandstone specimens after the effect of temperature–water coupling.

As can be seen from the picture, the increase in the water temperature lightens the surface color of the specimen. The specimen numbers also gradually became blurred under the temperature–water coupling effect. When touching the specimens' surface gently with the hand, the higher the water temperature, the stronger the surface frosting feeling, and the more granular fine sand floating on the specimen surface.

Before and after the temperature–water coupling effect, the physical basic data of the sandstone specimens were calculated, and the results are shown in Table 1.

Table 1. Comparison of physical data before and after temperature–water coupling.

Temperature (°C)	Specimen Number	Before Temperature–Water Coupling			After Temperature–Water Coupling		
		Mass (g)	Volume (mm³)	Density (g/mm³)	Mass (g)	Volume (mm³)	Density (g/mm³)
25	GB33-01	122.98	47.05	2.61	123.20	47.10	2.62
	GB33-02	123.25	47.24	2.61	123.63	47.28	2.62
35	GB33-03	123.81	47.51	2.61	124.12	47.54	2.61
	GB33-04	123.58	47.09	2.62	123.93	47.17	2.63
45	GB33-05	122.31	47.23	2.59	122.61	47.30	2.59
	GB33-06	123.51	47.14	2.62	123.89	47.18	2.63
55	GB33-07	124.03	47.30	2.62	124.37	47.30	2.63
	GB33-08	121.11	46.91	2.58	121.44	47.01	2.58
65	GB33-09	121.45	47.42	2.56	121.83	47.48	2.57
	GB33-10	123.96	47.31	2.62	124.24	47.33	2.62
75	GB33-11	123.76	47.41	2.61	124.10	47.44	2.62
	GB33-12	124.21	47.51	2.61	124.52	47.56	2.62
85	GB33-13	124.13	47.61	2.61	124.41	47.62	2.61
	GB33-14	124.12	47.60	2.61	124.45	47.62	2.61
95	GB33-15	122.34	47.29	2.59	122.61	47.29	2.59
	GB33-16	121.98	47.11	2.59	122.22	47.12	2.59
25	GB33-17	133.23	48.85	2.73	133.73	49.07	2.73
	GB33-18	126.55	48.80	2.59	127.32	49.16	2.59
35	GB33-19	127.25	48.89	2.60	127.91	49.21	2.60
	GB33-20	125.96	49.00	2.57	126.60	49.39	2.56
45	GB33-21	129.60	49.45	2.62	130.13	49.42	2.63
	GB33-22	127.84	49.54	2.58	128.76	49.31	2.61
55	GB33-23	126.77	48.94	2.59	127.31	49.00	2.60
	GB33-24	129.52	49.65	2.61	130.40	50.37	2.59
65	GB33-25	129.27	48.56	2.66	129.74	48.95	2.65
	GB33-26	134.65	49.26	2.73	135.55	49.55	2.74
75	GB33-27	125.92	48.89	2.58	126.49	49.14	2.57
	GB33-28	127.46	48.92	2.61	128.12	49.16	2.61
85	GB33-29	129.17	49.29	2.62	129.83	49.49	2.62
	GB33-30	129.15	48.93	2.64	129.60	49.05	2.64
95	GB33-31	127.22	48.85	2.60	127.72	48.87	2.61
	GB33-32	128.66	49.04	2.62	129.09	49.05	2.63

The basic physical properties of sandstone before and after temperature–water coupling were analyzed according to the data in the table.

The curve of the mass growth rate is shown in Figure 5.

$$\left.\begin{aligned} m_1' &= 0.166 + 0.004t - 3.857 \times 10^{-5}t^2\left(R^2 = 0.9731\right) \\ m_2' &= 0.308 + 0.010t - 9.538 \times 10^{-5}t^2\left(R^2 = 0.9780\right) \end{aligned}\right\} \quad (2)$$

where m_1' is the annular sandstone specimen's mass growth; m_2' is the intact sandstone specimen's mass growth.

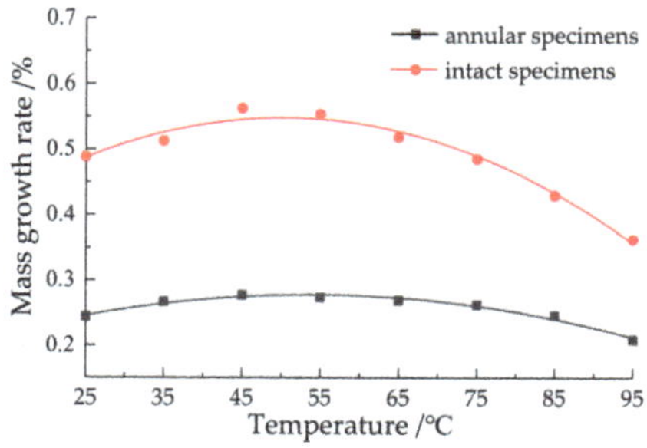

Figure 5. Mass growth rate varies with water temperature.

The mass growth rate curve satisfies the quadratic function relation, and an inflection point appeared at 45 °C. The mass growth rate of the intact sandstone specimens was larger. The correlation coefficients of the two specimens' fitting curves were 0.9731 and 0.9780, respectively, showing a strong correlation. Analysis of the reasons shows that the overall volume of the intact sandstone specimen was larger, and more water could be absorbed, which entered the internal cracks of the sandstone and increased its mass. At the same time, the increase in water temperature enlarged the cracks in the sandstone specimen, enabling it to absorb more water, so its mass increased first. However, with the increase in water temperature, some granular sand appeared and flaked off on the surface of the specimen, resulting in a decrease in quality. When the temperature exceeded 45 °C, the amount of degradation damage exceeded the amount of water absorption, so the mass growth rate decreased; thus, the mass growth rate first increased and then decreased.

The curve of the volume growth rate is shown in Figure 6.

$$\left.\begin{aligned} V_1' &= 0.040 + 0.003t - 4.032 \times 10^{-5}t^2\left(R^2 = 0.9894\right) \\ V_2' &= 0.061 + 0.036t - 3.696 \times 10^{-5}t^2\left(R^2 = 0.9860\right) \end{aligned}\right\} \quad (3)$$

where V_1' is the annular sandstone specimen's volume growth; V_2' is the intact sandstone specimen's volume growth.

The volume growth rate first increased and then decreased with the increase in water temperature, and the correlation coefficient reached 0.9894 and 0.9860, respectively, showing a strong correlation. The volume growth rate of the intact sandstone specimen was greater, because the volume of the intact sandstone specimen was larger and could absorb more water, resulting in the increase in internal cracking, so the volume growth rate was higher. Similarly, the increase in water temperature not only increased the volume, but also caused the deterioration and damage of the specimen, with part of the surface falling off, leading to a decrease in the overall volume, so the trend of first increasing and then decreasing appeared.

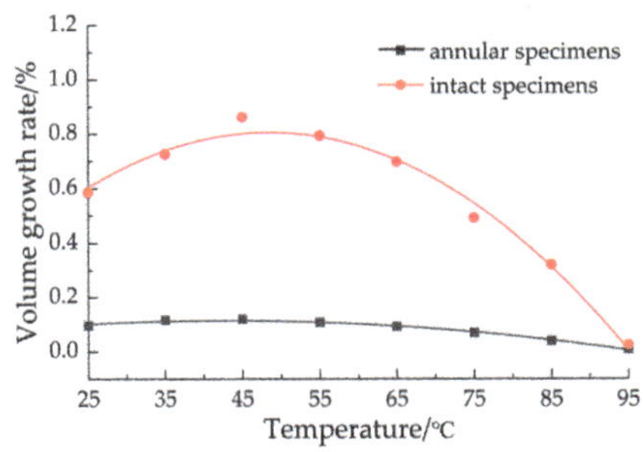

Figure 6. The volume growth rate varies with water temperature.

The density growth rate presents a law under mass and volume changes, as shown in Figure 7.

$$\left.\begin{array}{l}\rho_1' = 0.243 - 0.003t + 3.242 \times 10^{-5}t^2\left(R^2 = 0.9515\right) \\ \rho_2' = 0.362 - 0.025t + 2.649 \times 10^{-5}t^2\left(R^2 = 0.9914\right)\end{array}\right\} \quad (4)$$

where $\rho_1{}'$ is the annular sandstone specimen's density growth; $\rho_2{}'$ is the intact sandstone specimen's density growth.

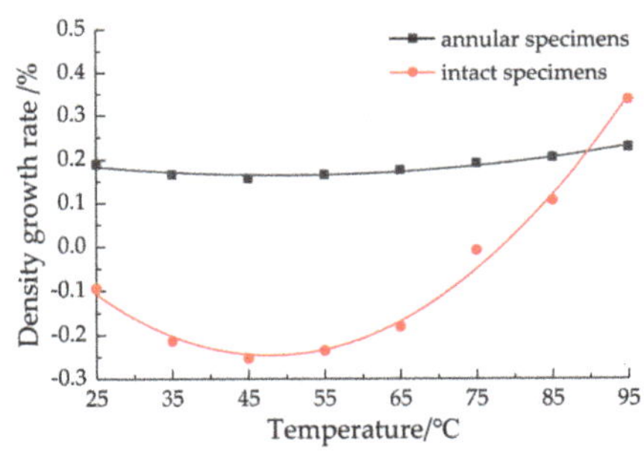

Figure 7. Density growth rate varies with water temperature.

Because volume growth is greater than mass growth, density first shows a decreasing trend. When the temperature exceeds 45 °C, the mass growth rate is greater than the volume growth rate, so the density growth rate tends to increase. Due to the better integrity and greater volume and mass of intact sandstone specimens, their density growth rate changes more obviously under the temperature–water coupling.

3.2. Dynamic Performance Analysis

When studying the dynamic properties of the two sandstone specimens, applying Vaseline to the surface of the specimen and pressure bar can reduce friction and end-face resistance, so as to improve the reliability of the test data. The impact test data have a large dispersion, so we prepared five sandstone specimens of each type. In order to make the test results more convincing, we selected two data within the error range of 15% and took their average value.

After processing the test data, the results are shown in Table 2.

Table 2. SHPB test data.

Temperature (°C)	Specimen Number	Annular Sandstone Specimens: Peak Stress (MPa)	Average Strain Rate (s^{-1})	Peak Strain ($\times 10^{-3}$)	Specimen Number	Intact Sandstone Specimens: Peak Stress (MPa)	Average Strain Rate (s^{-1})	Peak Strain ($\times 10^{-3}$)
25	GB33-01	17.06	84.50	2.86	GB33-17	25.26	81.30	3.62
	GB33-02	15.29	84.40	2.81	GB33-18	26.19	81.70	3.54
35	GB33-03	17.21	83.50	2.64	GB33-19	26.91	80.80	3.43
	GB33-04	16.69	83.30	2.73	GB33-20	26.25	80.70	3.49
45	GB33-05	16.48	83.60	2.50	GB33-21	27.31	80.70	3.03
	GB33-06	17.54	83.10	2.40	GB33-22	26.60	79.90	3.38
55	GB33-07	15.10	83.40	2.60	GB33-23	26.62	80.30	3.58
	GB33-08	17.52	83.80	2.88	GB33-24	26.32	80.90	3.56
65	GB33-09	16.73	84.60	3.24	GB33-25	26.41	79.70	4.02
	GB33-10	14.51	84.00	2.81	GB33-26	26.10	83.20	4.09
75	GB33-11	15.48	84.60	3.89	GB33-27	25.73	82.00	4.52
	GB33-12	15.08	85.30	3.97	GB33-28	25.96	81.30	4.54
85	GB33-13	15.00	85.90	4.93	GB33-29	25.01	83.20	5.21
	GB33-14	13.26	86.20	4.68	GB33-30	25.33	83.10	5.25
95	GB33-15	13.80	85.60	5.46	GB33-31	21.97	85.20	5.61
	GB33-16	12.95	88.00	5.02	GB33-32	25.36	83.30	5.64

3.2.1. Stress–Strain Relationship Analysis

The stress–strain relationship of the two sandstone specimens is shown in Figure 8.

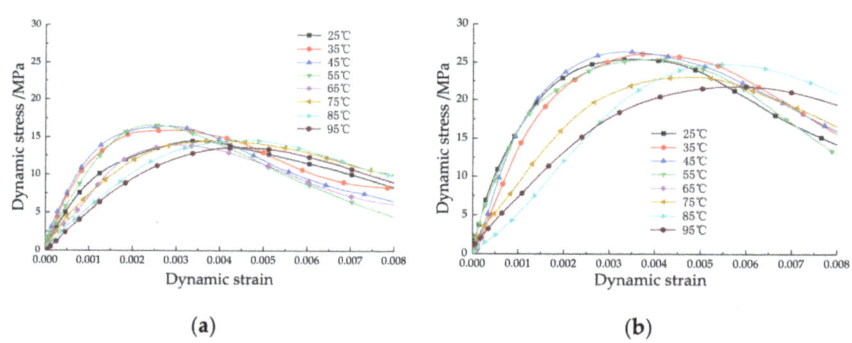

Figure 8. Stress–strain curves: (**a**) stress–strain curves of annular sandstone specimens; (**b**) stress–strain curves of intact sandstone specimens.

By comparing and analyzing the stress–strain curves of the two sandstone specimens with different structures, their dynamic properties can be preliminarily obtained. The dynamic tensile strength of annular sandstone is obviously lower than that of intact sandstone, due to the influence of pores. The slope of the curves of the two sandstone specimens also changes with the increase in temperature under the temperature–water coupling effect. The slope of the curves of the two sandstone specimens also changes with the increase in temperature under the temperature–water coupling effect.

3.2.2. Dynamic Tensile Strength Analysis

The tensile strength of the specimen is consistent with the peak stress, and the instability failure occurs when the sandstone specimen reaches the tensile strength. After testing and data processing, the peak of the dynamic stress–strain curve is the dynamic tensile strength.

Dynamic tensile strength is shown in Figure 9.

$$\left.\begin{array}{l} P_1(t) = 15.045 - 0.085t + 0.001 \times 10^{-5}t^2 \left(R^2 = 0.9598\right) \\ P_2(t) = 23.133 - 0.145t + 0.001 \times 10^{-5}t^2 \left(R^2 = 0.9682\right) \end{array}\right\} \quad (5)$$

where $P_1(t)$ is the annular sandstone specimen's dynamic tensile strength; $P_2(t)$ is the intact sandstone specimen's dynamic tensile strength.

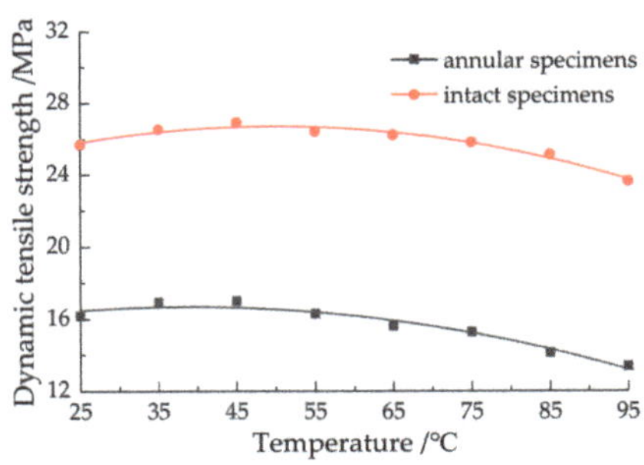

Figure 9. Relationship between dynamic tensile strength and water temperature.

According to the analysis, the peak tensile strength of the two sandstone specimens changed at 45 °C. Through quadratic function fitting, the correlation coefficients reached 0.9598 and 0.9682, respectively, showing strong correlation. In the process of testing, water entered the internal void. Considering the influence of water bonding force and the Stefan effect [36], the strength of the specimens was enhanced to a certain extent. After 45 °C, the effects of temperature and water on the damage to the specimen are greater than its enhancement. Partial spalling occurred on the surface of the specimens, internal cracks increased, and the specimens were more prone to instability failure. Therefore, the peak strength of both specimens gradually decreased. Because the annular sandstone specimen had lower integrity, its peak tensile strength was lower than that of the annular sandstone specimen.

3.2.3. Average Strain Rate Analysis

The failure difficulty of specimens can be reflected by the average strain rate.

The variation in the average strain rate with water temperature is shown in Figure 10.

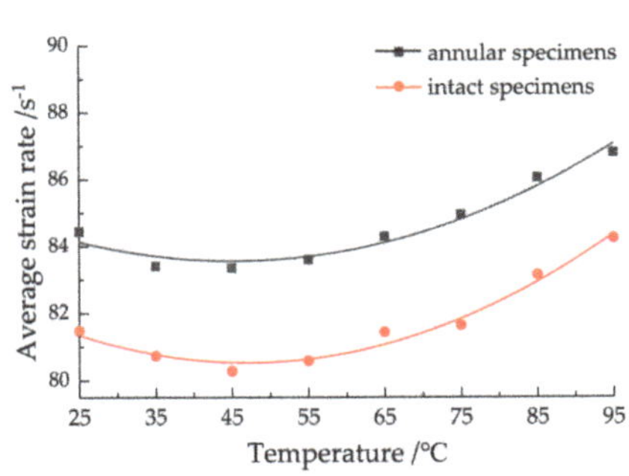

Figure 10. Relationship between average strain rate and water temperature.

Average strain rate first decreased but then increased with the change in water temperature, and an inflection point appeared at 45 °C. The correlation coefficients reached 0.9614 and 0.9750, respectively, by quadratic function fitting, indicating obvious correlation. This was related to the dynamic tensile strength. Under the influence of water bonding force and the Stefan effect, the dynamic tensile strength of the specimens increased, so the failure difficulty of the specimens increased and the average strain rate decreased accordingly. When the water temperature exceeded the inflection point, the damaging effects of temperature and water were more obvious, so the specimens were easier to destroy and the average strain rate increased accordingly. Due to the influence of its structure, the dynamic tensile strength of the annular sandstone specimen was low, and this specimen was more prone to failure. Therefore, the average strain rate of the annular sandstone specimen was lower than that of the intact sandstone specimen.

$$\left.\begin{array}{l}\dot{\varepsilon}_1 = 86.454 - 0.128t + 0.001 \times 10^{-5}t^2 \left(R^2 = 0.9614\right) \\ \dot{\varepsilon}_2 = 84.258 - 0.157t + 0.002 \times 10^{-5}t^2 \left(R^2 = 0.9750\right)\end{array}\right\} \quad (6)$$

where $\dot{\varepsilon}_1$ is the annular sandstone specimen's average strain rate; $\dot{\varepsilon}_2$ is the intact sandstone specimen's average strain rate.

3.2.4. Peak Strain Analysis

The peak strain can also reflect the degree of failure and breakage to a certain extent.

The peak strain first decreased and then increased with the increase in temperature, and reached the inflection point at 45 °C, while the correlation coefficient reached 0.9713 and 0.9720, respectively, indicating an obvious correlation. Consistent with the dynamic tensile strength and average strain rate, the water bonding force and Stefan effect increased the specimens' strength, so the peak strain decreased. As the water temperature increased, damage was the main effect, making the specimens easier to damage and increasing the peak strain. According to the figure, the integrity, peak strength, and peak strain of intact sandstone specimens are higher than those of annular sandstone specimens. In the impact test, the annular hole played a buffer role, and the inner ring was squeezed first when it was impacted by the pressure bar, so the peak strain of the annular sandstone specimen was slightly lower than that of the intact sandstone specimen.

The variation of peak strain with water temperature is shown in Figure 11.

$$\left.\begin{array}{l}\varepsilon_{T1} = 4.101 - 0.076t + 9.554 \times 10^{-5}t^2 \left(R^2 = 0.9713\right) \\ \varepsilon_{T2} = 4.343 - 0.051t + 7.021 \times 10^{-5}t^2 \left(R^2 = 0.9720\right)\end{array}\right\} \quad (7)$$

where ε_{T1} is the annular sandstone specimen's peak strain; ε_{T2} is the intact sandstone specimen's peak strain.

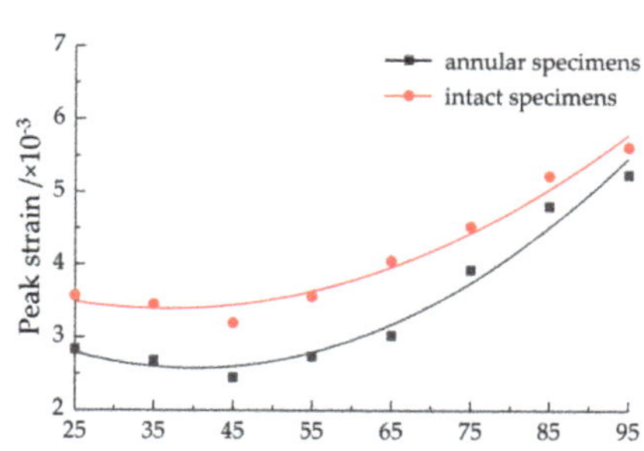

Figure 11. Relationship between peak strain and water temperature.

3.3. Failure Mode Analysis

Failure modes can also reflect the dynamic characteristics of sandstone specimens to a certain extent.

The failure modes of each water temperature gradient are shown in Figure 12.

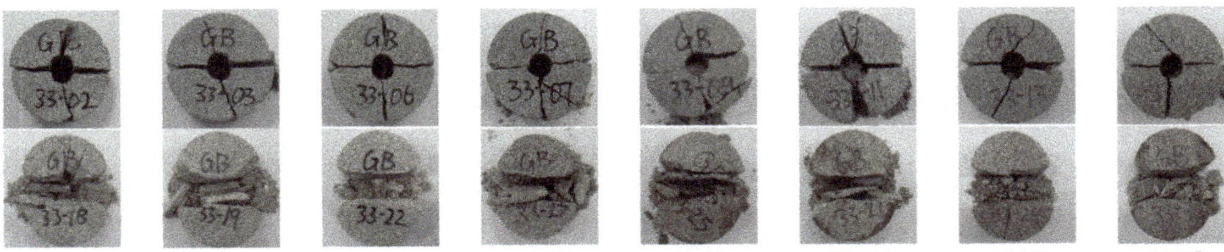

25 °C 35 °C 45 °C 55 °C 65 °C 75 °C 85 °C 95 °C

Figure 12. Fracture morphology of sandstone specimens.

Through observation and analysis, we found that the annular sandstone specimen under impact load was mainly divided into four parts, accompanied by some small fragments spalling. The specimen was first cracked along the axis of the incident bar, and then the upper and lower parts were extruded, cracked from the edge of the specimen, and destroyed along the direction of the internal fissure. The failure mode was type I failure. When the water temperature was 45 °C, the fragments were mainly divided into four parts of similar size, which were broken apart along the central axis and the vertical central axis. When the temperature was greater than or less than 45 °C, the strength of the specimen was low, and the size of fragments was not uniform. It can be concluded that the failure of the upper and lower parts of the specimen was mainly along the internal cracks, thus showing irregularity. Therefore, at 45 °C, the integrity of the sandstone specimens was better. The fracture morphology of the complete sandstone specimen was split along the central axis, leaving two fan-shaped fragments, small fragments, and powder. At the same time, we found that the fan-shaped area of the sandstone specimen was larger at 45 °C, indicating that its strength was higher. The failure modes of the two sandstone specimens meet the validity conditions of the Brazilian disc test.

The fracture morphology can be verified laterally, based on the dynamic tensile strength, average strain rate, and peak strain variation scale of the sandstone specimens. It was proven that 45 °C is an inflection point for the dynamic parameters of sandstone specimens with temperature–water coupling.

3.4. XRD Pattern and SEM Photo Analysis

After the impact test, an X-ray diffractometer and an electron microscope were used to conduct XRD and SEM tests on sandstone specimens, respectively, so as to measure their XRD patterns and SEM images. The measured data are shown in Figures 13 and 14, respectively.

After the analysis of the XRD patterns, no new substances were found in the sandstone after the temperature–water coupling effect. The composition of sandstone is mainly SiO_2, with a small amount of $K_2(PtCl)_4$ and $CaCuV_2O_7$. SEM images show the morphological characteristics of the sandstone fracture surface. As the temperature of the temperature–water coupling increased gradually, the fracture surface became coarser, and angular grains appeared at the fracture surface (the coarsest at 45 °C). The XRD patterns and SEM image analysis show that the dynamic properties of sandstone are mainly determined by the internal structural changes after the temperature–water coupling effect.

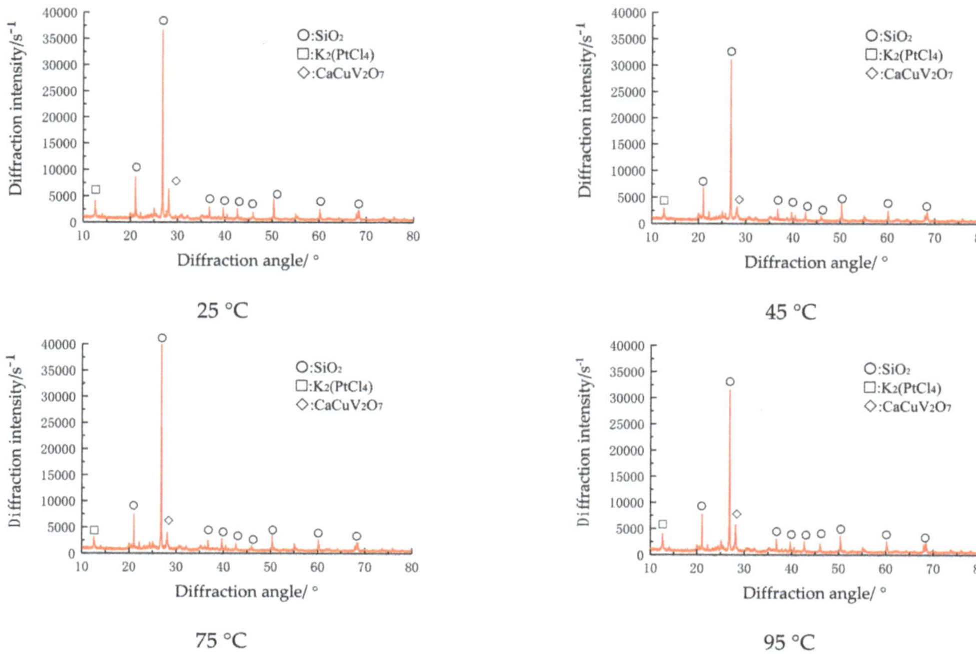

Figure 13. XRD patterns of sandstone specimens treated with different water bath temperatures.

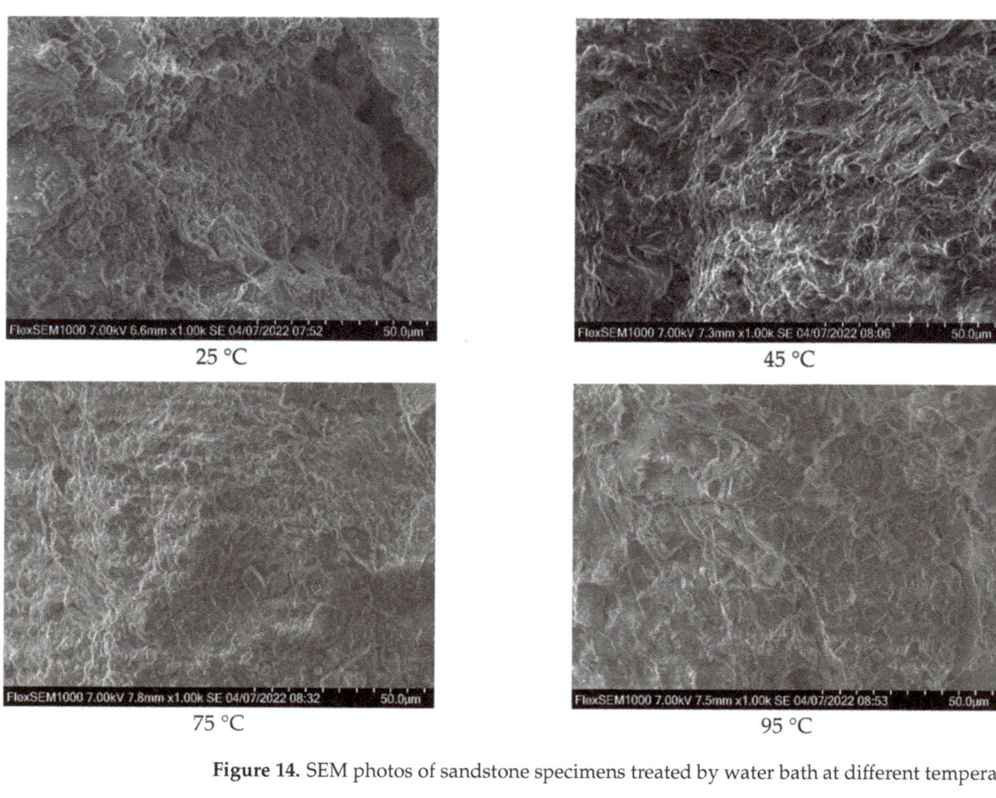

Figure 14. SEM photos of sandstone specimens treated by water bath at different temperatures.

4. Conclusions

In this study, eight temperature gradients (25 °C~95 °C) were used to conduct temperature–water coupling treatment for annular sandstone specimens and intact sandstone specimens, followed by testing of their dynamic splitting mechanical properties (SHPB). The following conclusions were obtained:

(1) After the water bath, the growth rate of the mass and volume of the sandstone specimens first increased and then decreased with the increase in temperature, and the growth rate of density first decreased and then increased under the joint influence of mass and volume changes, both of which met the quadratic function relationship.
(2) The peak strength, average strain rate, and peak stress of the two sandstone specimens all met the quadratic function relation, the correlation was very strong, and the law of increase and decrease at 45 °C changed.
(3) Due to the influence of its structure, the strength of annular sandstone is obviously lower than that of intact sandstone. However, the prefabricated holes in the annular sandstone specimens have a certain buffer effect on the impact load, so the strain is lower.
(4) The fracture morphology of the two sandstone specimens was consistent with the Brazilian disc splitting model. In addition, the fragmentation morphology changed with temperature to different degrees, and tended to be complete at 45 °C.
(5) The changes in the basic physical parameters and dynamic properties of sandstone samples after temperature–water coupling mainly depend on their internal structural damage, and their material composition does not change with the increase in the water bath temperature.

Author Contributions: Conceptualization, Q.P.; Data curation, Q.G.; Formal analysis, C.W.; Funding acquisition, Q.P.; Investigation, Q.G., Y.W. and Y.X.; Methodology, Q.P.; Project administration, S.W. (Shiwei Wu); Resources, Q.P. and K.S.; Software, K.S.; Supervision, Y.W.; Validation, S.W. (Shuo Wang); Writing—original draft, Q.G.; Writing—review & editing, Q.P. All authors have read and agreed to the published version of the manuscript.

Funding: This research was funded by the National Natural Science Foundation of China (no. 52074005, no. 52074006), Anhui University of Science and Technology Graduate Innovation Fund Project (no. 2021CX2032), and the National College Student Innovation and Entrepreneurship Training Program (no. 202110361022, no. 2021103661027, no.202110361032).

Institutional Review Board Statement: Not applicable.

Informed Consent Statement: Informed consent was obtained from all subjects involved in the study.

Data Availability Statement: The data used to support the findings of this study are available from the corresponding author upon request.

Acknowledgments: We would like to thank Anhui University of Science and Technology for providing the experimental conditions.

Conflicts of Interest: The authors declare no conflict of interest. The funders had no role in the design of the study; in the collection, analyses, or interpretation of data; in the writing of the manuscript, or in the decision to publish the results.

References

1. Ping, Q.; Qi, D.Z.; Diao, Q.; Zhang, C.L.; Gao, Q.; Wu, Y.L.; Wu, B.B. Experimental Study on Physical and Dynamic Mechanical Properties of Temperature-Water Coupled Sandstone. *Shock Vib.* **2021**, *2021*, 6568436. [CrossRef]
2. Ping, Q.; Gao, Q.; Wang, C.; Wu, Y.L.; Zhang, C.L.; Qi, D.Z.; Diao, Q.; Zang, L.; Su, Z.Q. Study on physical and dynamic splitting mechanical properties of temperature-water coupled sandstone. *IOP Conf. Ser. Earth Environ. Sci.* **2021**, *861*, 042077. [CrossRef]
3. Wang, B.; Li, X.B.; Yin, T.B.; Ma, C.D.; Yin, Z.Q.; Li, Z.G. SHPB test on dynamic strength of saturated sandstone. *J. Rock Mech. Eng.* **2010**, *29*, 1003–1009. [CrossRef]
4. Roy, D.G.; Singh, T.N.; Kodikara, J.; Das, R. Effect of Water Saturation on the Fracture and Mechanical Properties of Sedimentary Rocks. *Rock Mech. Rock Eng.* **2017**, *50*, 2585–2600. [CrossRef]

5. Wang, K.; Jing, Y.F.; Xu, C. Mechanical properties and damage statistical Model of coal with different water content under uniaxial compression. *J. Rock Mech. Eng.* **2018**, *37*, 1070–1079. [CrossRef]
6. Teng, J.Y.; Tang, J.X.; Zhang, Y.N.; Duan, J.C.; Wang, J.B. Damage process and characteristics of lower layer water-bearing shale under uniaxial compression. *Rock Soil Mech.* **2017**, *38*, 1629–1638. [CrossRef]
7. Deng, H.F.; Zhang, Y.C.; Li, J.L.; Wang, W.; Zhi, Y.Y. Effect of water content on splitting tensile strength of layered sandstone. *J. Rock Mech. Eng.* **2017**, *36*, 2778–2787. [CrossRef]
8. Wang, C.; Lai, Y.M.; You, Z.M.; Yang, Y.; Yu, D.H.; Liu, G.H. Experimental study on the influence of temperature and water content on the splitting strength of rock. *Glacial Permafr.* **2016**, *38*, 1317–1324.
9. Ping, Q.; Su, H.P.; Ma, D.D.; Zhang, H.; Zhang, C.L. Experimental study on physical and dynamic characteristics of limestone under different high-temperature conditions. *Rock Soil Mech.* **2021**, *42*, 932–942. [CrossRef]
10. Ping, Q.; Wu, M.J.; Zang, H.; Yuan, P. Experimental study on dynamic mechanical properties of sandstone at high temperature. *J. Undergr. Space Eng.* **2019**, *15*, 691–698.
11. Zhang, R.R. Test and analysis of dynamic mechanics and damage characteristics of deep sandstone treated at different temperatures. *J. Rock Mech. Eng.* **2018**, *37*, 3879–3890. [CrossRef]
12. Qin, N.; Ge, Q.; Liang, Z.H.; Sun, J.B.; Wang, Y.Y. Prediction of macro-mesoscopic damage of sandstone under high temperature and uniaxial strength by BP neural network. *Exp. Mech.* **2021**, *36*, 105–113. [CrossRef]
13. Xu, X.; Chi, L.Y.; Yang, J.; Yu, Q. Experimental Study on the Temporal and Morphological Characteristics of Dynamic Tensile Fractures in Igneous Rocks. *Appl. Sci.* **2021**, *11*, 11230. [CrossRef]
14. Bo, K.; Jian, Z.; Deng, H.G.; Yang, X.G. Dynamic Characteristics of Sandstone under Coupled Static-Dynamic Loads after Freeze-Thaw Cycles. *Appl. Sci.* **2020**, *10*, 3351. [CrossRef]
15. Bo, K.; Zhang, C.Y.; Liu, C.J.; Ding, L.M.; Zheng, Y.; Li, N.; Wang, Y.X.; Lin, H. An experimental study on characteristics of impact compression of freeze-thawed granite samples under four different states considering moisture content and temperature difference. *Environ. Earth Sci.* **2021**, *80*, 661. [CrossRef]
16. Zheng, W.H.; Shi, T.W.; Pan, Y.H.; Luo, H.; Lv, X.F. Effects of moisture content on rock charge induction signal research. *Rock Soil Mech.* **2022**, *43*, 659–668. [CrossRef]
17. Chang, H.M.; Zhao, D.; Cai, T.Y.; Su, Y.F.; Yuan, J. Law and mechanism of influence of water content on tensile strength of deep rock. *Min. Res. Dev.* **2022**, *42*, 94–100. [CrossRef]
18. Geng, J.H.; Cao, L.W. Failure analysis of water-bearing sandstone using acoustic emission and energy dissipation. *Eng. Fract. Mech.* **2020**, *231*, 107021. [CrossRef]
19. Taiki, S.; Chisato, S.; Ukyo, U.; Ryota, I.; Yutaka, F. Effect of water contents and initial crack lengths on mechanical properties and failure modes of pre-cracked compacted clay under uniaxial compression. *Eng. Geol.* **2022**, *301*, 106593. [CrossRef]
20. Li, C.M.; Liu, N.; Liu, W.R.; Feng, R.M. Study on Characteristics of Energy Storage and Acoustic Emission of Rock under Different Moisture Content. *Sustainability* **2021**, *13*, 1043. [CrossRef]
21. Wang, C.; Gao, Y.W.; Guo, C.Y.; Wei, K. Experimental study on the influence of moisture content on soft coal strength characteristics. *Int. J. Min. Miner. Eng.* **2021**, *12*, 163. [CrossRef]
22. Zhao, Y.Y.; Sun, Q.; Wang, S.F.; Zhang, L.W.; Jiang, Z.Q. Pore characteristics and permeability changes of high-temperature limestone after rapid cooling by dry ice. *Heat Mass Transf.* **2022**, 1–14. [CrossRef]
23. Liu, L.; Li, R.; Qin, H.; Sun, W. Experimental SHPB Study of Limestone Damage under Confining Pressures after Exposure to Elevated Temperatures. *Metals* **2021**, *11*, 1663. [CrossRef]
24. Liang, Z.Z.; Zhang, Y.B.; Tang, S.B.; Li, L.C.; Tang, C.N. Effect of rock mass size on rock mass characteristics. *J. Rock Mech. Eng.* **2013**, *32*, 1157–1166.
25. Wang, C.; Wang, H.B.; Xiong, Z.Q.; Wang, C.; Cheng, L.P.; Zhan, S.F. Experimental study on radial compression mechanical characteristics of circular granite under cyclic temperature and humidity. *J. Rock Mech. Eng.* **2020**, *39*, 3260–3270. [CrossRef]
26. Wu, Q.H.; Zhao, F.J.; Li, X.B.; Wang, S.M.; Wang, B.; Zhou, Z.H. Mechanical properties of annular sandstone samples under radial compression. *Rock Soil Mech.* **2018**, *39*, 3969–3975. [CrossRef]
27. Yang, S.Q.; Li, Y.; Huang, Y.H.; Tian, W.L. Particle Flow analysis of macroscopic and meso-mechanical properties of single-hole disc Splitting Test. *J. China Univ. Min. Technol.* **2019**, *48*, 984–992. [CrossRef]
28. You, M.Q.; Su, C.D.; Gou, Y. Experimental Study on strength and deformation characteristics of marble channel specimens. *J. Rock Mech. Eng.* **2007**, *2007*, 2420–2429. [CrossRef]
29. Ying, P.; Li, W.J.; Zhu, Z.M.; Li, X.H.; Gao, W.T.; Shu, Y. Influence of impact loading orientations on the mechanical behaviour of rocks around a tunnel. *Int. J. Rock Mech. Min. Sci.* **2022**, *152*, 105071. [CrossRef]
30. Bieniawski, Z.T.H. Suggested methods for determining tensile strength of rock materials. *Int. J. Rock Mech. Min. Sci. Geomech. Abstr.* **1978**, *15*, 99–103. [CrossRef]
31. *GB50021-2019*; Code for Investigation of Geotechnical Engineering. China Architecture and Building Press: Beijing, China, 2019.
32. Ulusay, R. *The ISRM Suggested Methods for Rock Characterization, Testing and Monitoring: 2007–2014*; Springer International Publishing: Cham, Switzerland, 2015; pp. 51–68.
33. *T/CSRME001-2019*; Technical Specification for Testing Method of Rock Dynamic Properties. Chinese Society of Rock Mechanics and Engineering: Beijing, China, 2019.

34. Ping, Q. Dynamic and static tensile mechanical properties of sandstone and comparison analysis. *J. Undergr. Space Eng.* **2013**, *9*, 246–252.
35. Li, W.; Xie, H.P.; Wang, Q.Z. SHPB experimental study on dynamic splitting tension of marble. *Explos. Impact* **2006**, *2006*, 12–20. [CrossRef]
36. Zhao, J.P.; Wang, M.H.; Zhao, Y.H. Effect of water content on dynamic tensile strength of sandstone. *Gold Sci. Technol.* **2019**, *27*, 216–222. [CrossRef]

applied sciences

MDPI

Article

Seepage Characteristics and Failure Prediction during the Complete Stress–Strain Process of Limestone under High Water Pressure

Chunyan Bao [1,*], Yong Yin [1], Shibin Tang [2], Annan Jiang [3] and Hong Li [2]

1 College of Civil Engineering, Shaoxing University, Shaoxing 312000, China; yinyong19970202@163.com
2 Center for Rock Instability and Seismicity Research, Dalian University of Technology, Dalian 116024, China; tang_shibin@126.com (S.T.); hong.li@dlut.edu.cn (H.L.)
3 Highway and Bridge Engineering Institute, Dalian Maritime University, Dalian 116026, China; jiangannan@163.com
* Correspondence: fengniaobcy@163.com

Citation: Bao, C.; Yin, Y.; Tang, S.; Jiang, A.; Li, H. Seepage Characteristics and Failure Prediction during the Complete Stress–Strain Process of Limestone under High Water Pressure. *Appl. Sci.* **2022**, *12*, 6041. https://doi.org/10.3390/app12126041

Academic Editor: Muhammad Junaid Munir

Received: 6 May 2022
Accepted: 10 June 2022
Published: 14 June 2022

Abstract: The seepage characteristics during the complete stress–strain process of limestone under high water pressure were simulated via an experimental study of limestone post-peak penetrating behavior, and an approach to predict the formation of the seepage channel, namely, acoustic emission positioning technology, is proposed. The results showed that (1) whether in the experiment or in the numerical simulation, the sudden drop in the stress–strain curve after peaking indicated the full formation of shear fractures and seepage paths. (2) By using acoustic emission positioning technology in the simulation, the entire stress–strain process of limestone, from microfracture initiation and compaction to transfixion, could be monitored to observe the dynamic and real-time development of the microfractures. (3) The combination of acoustic emission technology with seepage monitoring revealed the real-time location and growth direction of micro ruptures and predicted the depth of penetration. The developed approach can improve forecast accuracy for landslides involving a low-permeability rock mass with cracks. In this study, limestone post-peak seepage characteristics were analyzed, and a method to forecast the formation of rock seepage paths before transfixion is provided. This work could provide a reference and guiding elements for ensuring the safety of slopes with high internal water pressure.

Keywords: fractured limestone; hypertonic pressure; complete stress–strain process; permeability characteristics; acoustic emission; damage prediction

1. Introduction

Energy shortage is one of the bottlenecks that restrict national economic development. China ranks first in the world in terms of water resources, most of which are distributed in Southwestern Sichuan, Yunnan, and Tibet Provinces where the terrain is challenging. The construction of large and extra-large water conservancy and hydropower projects in these areas faces the stability problem of high and steep slopes, resulting in considerable personnel safety concerns and possible property losses [1]. Although the permeability of rock engineering slopes is low [2], such slopes have many joints and cracks that provide channels for water infiltration. In China, especially in slopes in the southern region, the existence of water reduces the effective stress and changes the saturation of the slopes, leading to instability of slopes subjected to engineering work.

Studying the seepage law of water in rocks in essential to analyze the influence of water on the stability of rock slopes. Since Darcy's law was put forward 150 years ago, researchers have achieved fruitful results in the study of rock seepage. Theoretical and experimental research results for single-fracture seepage under the condition of saturated seepage are abundant and include the influence of fracture surface geometry and extension characteristics on the seepage law of fractures, the applicable conditions and correction of the seepage cubic theorem, and the change in the seepage characteristics and seepage law of fractures under deformation or load conditions [3–6]. Studies on the seepage characteristics and seepage movement of fractured rock mass have mainly adopted equivalent continuum, discrete network, and mixed models [7–12]. Many scholars [12–15] studied the evolution of the permeability characteristics of rock blocks in the process of deformation and progressive failure, and their results revealed a strong permeability scale effect. However, due to the complexity of rock structures and the differences in physical experimental conditions and environments, the test results are not generalizable and thus are limited. Effective geophysical research methods must be developed to study the influence of slope rock mass damage and fracture on rock mass permeability in the process of rock mass excavation and loading [16].

Rock permeability has two components [17]. One is the permeability of the primary pores and cracks in rocks, and the other is the permeability change caused by rock failure. Experimental and numerical research on the permeability of primary pores and cracks in rocks is extensive, but studies on the entire process of permeability change caused by rock failure are relatively scarce. Li Shiping [13] conducted many sandstone total stress–strain–permeability experiments and obtained the rock permeability–strain relationship equation. This article mainly studied the influence of the stress level on rock permeability and the permeability–strain equation of rock. It studied the permeability characteristics of fractured limestone under high osmotic pressure (>1 MPa). A method combining acoustic emission location with seepage measurement was proposed to determine the formation of seepage channel. Yang Tianhong et al. [18] studied the relationship between rock permeability and damage evolution (fracture) from the perspective of seepage–stress–damage coupling by using a numerical method. Although these studies deepened our understanding of the effect of rock fracture on rock permeability, most of them focused on rocks with good permeability (e.g., sandstone), whereas minimal attention has been devoted to rocks with low permeability [19] (e.g., limestone and granite). Rocks with low permeability are difficult to observe at the initial stage of loading. Only when a rock is cracked under a load and broken through will permeability increase sharply, which is extremely unfavorable to the stability of the slope. Given that the seepage and failure processes of rocks are time-dependent, studying the seepage characteristics of rocks in relation to time and fracture is also essential. Such investigations can improve stability and rock slope predictions by identifying the seepage characteristics of low-permeability rocks during the fracture process.

In this study, a non-Darcy steady-state seepage test was conducted on limestone samples after the peak by using the triaxial seepage test system produced by TOP INDUSTRIE Company in France. The relationship between the seepage velocity and pressure gradient and the curve of axial strain and water head pressure under different hydrostatic pressures after the peak were obtained, and regular results were achieved. Unlike in the numerical test method, in the physical test method, the specimen is wrapped in a closed cylinder. The entire failure process of the specimen cannot be observed, and even the occurrence of failure cannot be judged by the naked eye. Therefore, this study used a special numerical simulation method to establish a plane model of non-uniform rock sample seepage, and the whole failure process was investigated from the meso perspective. The specific research contents are as follows. (1) Limestone is dense in texture, and its permeability is mainly fracture permeability. Therefore, prefabricated weak surface and prefabricated weak zone models were adopted in the current study, and the seepage channel was formed by axial loading. The stress state change of the rock samples and the evolution law of permeability in the process of damage and failure under high osmotic pressure were studied. The analysis focused on the variation of the stress–strain–permeability coefficient during the whole process and of the permeability law during crack initiation, propagation, and coalescence. (2) The prefabricated weak surface was similar to the crack formed by artificial splitting, so the biting effect of the crack surface was worse than that of the natural crack, and the roughness and fluctuation of the natural crack surface could not be effectively simulated. Therefore, the shear crack formed by the force of the complete model was also regarded as the seepage channel to simulate the permeability characteristics of the natural crack. (3) Rock as a brittle material exhibits the acoustic emission phenomenon during its failure process. The failure of rock samples can be predicted by monitoring and collecting acoustic emission information during the failure process. By collecting acoustic emission information during the failure process of the three models mentioned above, the failure and seepage catastrophe of the samples in this study were predicted. (4) Through numerical simulation, the whole stress process, micro-fracture distribution, crack development direction, and permeability of the rock samples were revealed to vividly and concretely reflect the crack situation. The results can provide a reference for the real-time control and prevention of accidents.

2. Calculation Method of Seepage Velocity in the Rock Total Stress–Strain Process

The key task in rock seepage–stress coupling is to study the variation in the permeability coefficient during deformation and failure. Experimental studies play an irreplaceable role in examining the seepage characteristics of rock fractures. Limestone is dense in texture and it is permeable mainly in cracks [20]. Permeability changes in limestone specimens during deformation are directly caused by the initiation, opening, and connection of cracks [21].

Many studies have shown that for a fractured rock mass, the coupling of stress and seepage fields is based on the fluid–solid coupling problem of a single fracture. The cubic law is the basis for studying the seepage of a single fracture [22]. When a fractured surface exists in a completely impervious rock mass, it is completely parallel and smooth. Poiseuille derived the theoretical formula of fluid motion in a smooth parallel plate gap under the condition of homogeneous constant motion for a viscous incompressible fluid, as follows [23]:

$$u = \frac{gb^2}{12v} J or q = \frac{gb^3}{12v} J \quad (1)$$

where u is the average velocity (m/s) of the fluid in a fracture with opening b, g is the gravitational acceleration (m/s^2), v is the viscous coefficient of fluid motion, J is the hydraulic gradient, and q is the flow rate (m^2/s).

He Yulong et al. [24]. applied the cubic criterion to rock fracture seepage and found that it is effective, as shown in Figure 1. In the implementation process of this test scheme, when the water pressure difference between the two ends of the rock sample is large, the fissure water pressure causes additional deformation of the rock surface on both sides of the fissure runoff, which increases the width of the fissure. Hence, the cubic criterion is still valid for open rough joints. The following formula is used to calculate seepage velocity:

$$u = \frac{Q}{A} \quad (2)$$

where u is the seepage velocity, Q is the flow, and A is the cross-sectional area of the specimen.

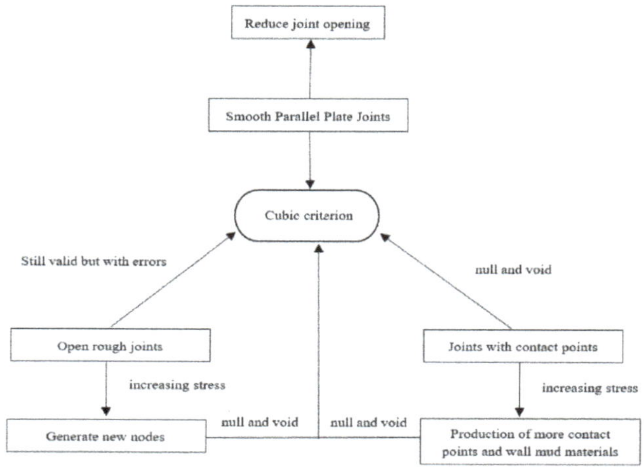

Figure 1. Validity of the cubic law for different fractured joints.

3. Brief Introduction to the Physical Test for Limestone Specimens

3.1. Test System and Procedures (The Numerical Simulation Method Uses the FEM)

The triaxial rheological test system developed by Lille University of Science and Technology and produced by TOP INDUSTRIE Company in France (Figure 2) was used in this experiment. The test system has three independent loading devices for confining pressure, axial pressure, and pore water pressure. It has the following functions and characteristics. (1) The specially designed water pressure application system has a high flow rate and stable flow and pressure and can directly output up to 70 MPa of water pressure, thus meeting the requirements of the steady-state seepage test under high permeability pressure. (2) The specially designed fluorinated rubber sleeve ensures the complete isolation of pore water pressure and confining pressure. (3) The high confining pressure (up to 70 MPa) and high axial pressure (up to 100 MPa) can realize strain and stress loading control, and the control precision is ±0.01 MPa. (4) Data collection is fully automatic.

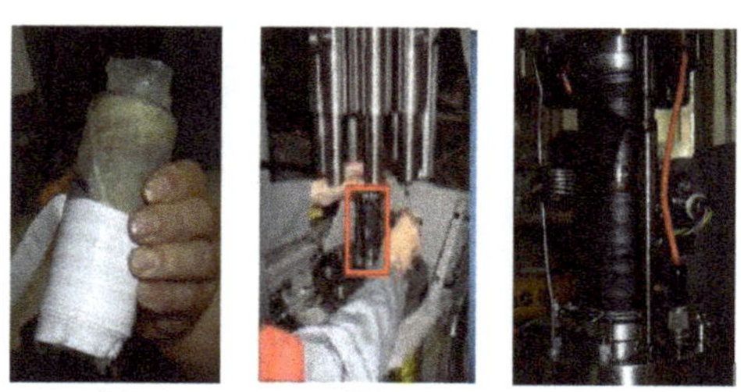

Figure 2. Experimental system and specimen installation. (**a**) Air exhaust. (**b**) Entering the loading room. (**c**) Installation specimen.

The specific test steps implemented in this study were as follows. First, a 2 MPa net confining pressure was applied to the rock sample, followed by axial displacement (0.05 mm/min). Second, a net confining pressure was applied, and stress control was implemented and maintained at 3 MPa. The first-order water pressure was 0.3 MPa (the classification standard is 1/10 points of the confining pressure at all levels) at the inlet, and the outlet was maintained at atmospheric pressure. Third, the first-order water pressure was maintained for a certain period. After the deformation of the specimen stabilized, the deformation of the specimen was recorded, and the seepage velocity and axial and circumferential strains were measured. Fourth, the next level of water pressure was applied.

After the application of all levels of water pressure under the net confining pressure of each level was completed, the second level of confining pressure and all water pressure levels under it were applied.

The deformation of rock mass under the action of stress and humidity includes the deformation ε_e caused by stress and the deformation ε_{sh} caused by humidity; the total deformation is

$$\varepsilon_{sh} = \alpha_{sh}\Delta h = \alpha_{sh}h_0$$

In the formula α_{sh} is the contraction coefficient; h_0 is the initial relative humidity (100%).

When the relative humidity is higher than 50%, the relationship between the strain of rock mass and the variation of relative humidity can be approximately considered as linear:

$$\sigma = \lambda I t_r\varepsilon + 2\mu\varepsilon - (3\lambda + 2\mu) + \varepsilon_{sh}I$$

In the formula. λ and μ are Latin constants, i.e., $\mu = \frac{E}{2(1+v)}$, $\lambda = \frac{2v\mu}{1-2v}$.

3.2. Test Results and Analysis

The test results showed that when the hydrostatic pressure was simulated by a 2 MPa net confining pressure, and 1 MPa seepage pressure was applied to the specimen under uniaxial loading, the failure of the specimen was manifested in shear cracks, which were closer to the natural cracks than those formed by artificial splitting (Figure 3). Owing to the limitation of the test methods, Darcy's law is generally used in tests of rock permeability. It requires the setting of a stable water pressure difference and seepage velocity, and the seepage velocity is converted by the amount of seepage water in a certain period. Given the limitation in the number of tests and the length of the article, the following shows only the test results for Specimen 2#. The stress–strain curve for Specimen 2# during uniaxial compression under hydrostatic pressure is shown in Figure 4a. The stress–strain curve began to decline after the peak value, and the amplitude was large. Given that the rock sample was located inside the instrument and could not be observed with the naked eye, the method for judging shear cracks was as follows. First, a sudden decrease in stress meant that the rock sample could be damaged. Second, because the limestone used in this study had a dense texture and poor permeability, the permeability of the intact rock block relative to the permeability of the fracture could be ignored. Therefore, in the axial loading process, the pore water pressure remained unchanged, and the outlet valve was opened. At this time, the amount of seepage water could be observed. When the outlet pore water pressure decreased then slowly increased, the shear fracture had penetrated; otherwise, the test failed.

Figure 3. Shear damaged limestone samples. (**a**) 1# test specimen, (**b**) 4# test specimen, (**c**) 8# test specimen, (**d**) 12# test specimen.

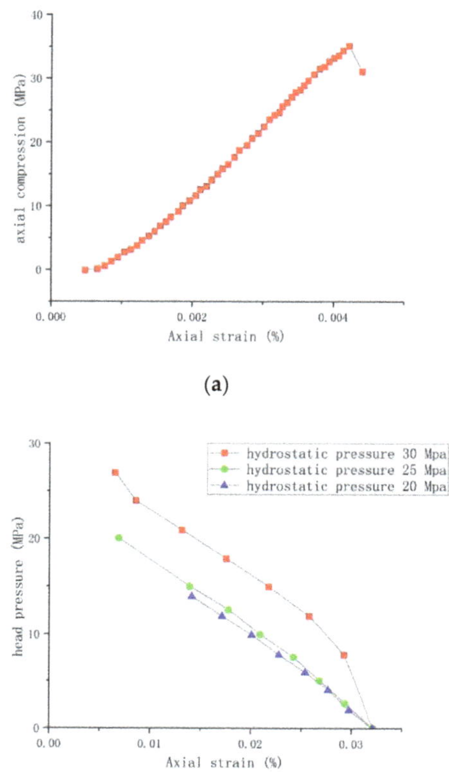

(a)

(b)

Figure 4. Experiment results for Sample 2#. (**a**) Stress–strain curve. (**b**) Post-peak strain–head pressure curve.

Figure 4b shows the experimental curve of axial strain and head pressure under different hydrostatic pressures after the peak. When the stress–strain reached its peak value, shear cracks were completely formed. With the further increase in the axial strain of limestone, the formed fracture pressure closed, the fracture width decreased, and the permeability decreased. This phenomenon can be seen in Figure 4b. Under the three hydrostatic pressures, the head pressure decreased with the increase in strain. Similarly, with the increase in hydrostatic pressure, the width of the cracks and the amount of seepage decreased. Moreover, under the three hydrostatic pressures, the strain and head pressure test curves of the specimens were basically consistent.

4. Numerical Simulation of Permeability in the Whole Process of Rock Fracture under High Permeability Pressure

4.1. Numerical Calculation Method and Numerical Model

The seepage–failure characteristics of limestone can be obtained from test results. However, due to the limitations of the physical test itself, observing the whole process of crack initiation and propagation in the specimen and revealing the flow process of water in the shear cracks are difficult. Therefore, this study used a numerical method to simulate the whole process of seepage–damage–failure of limestone. The sample size used in the numerical model was consistent with that adopted in the test, namely, H × W = 100 mm × 50 mm. The model was divided into 400 × 200 units, with pads at both ends. In this calculation, the permeability coefficient of the cushion plate was assumed to be larger than that of the sample so that the water could penetrate into the rock sample through the cushion plate as soon as possible. The model size and loading are shown in Figure 5, Section 1-1. The mechanical parameters of the sample are shown in Table 1.

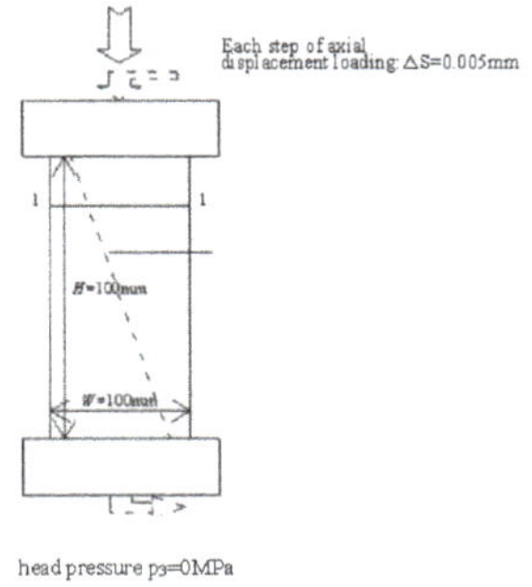

(a) Model 1: prefabricated weak surface

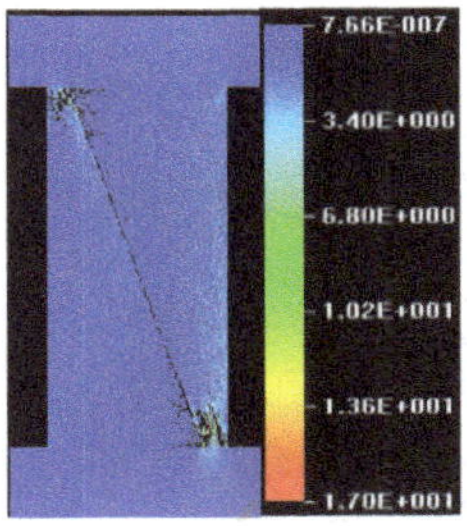

(a) Model 1

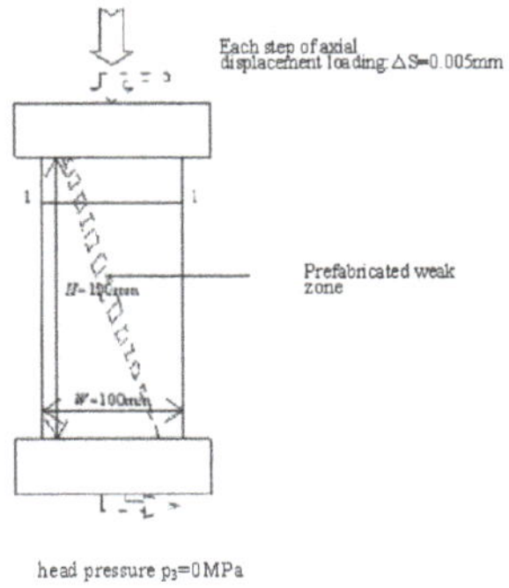

(b) Model 2: prefabricated weak zone

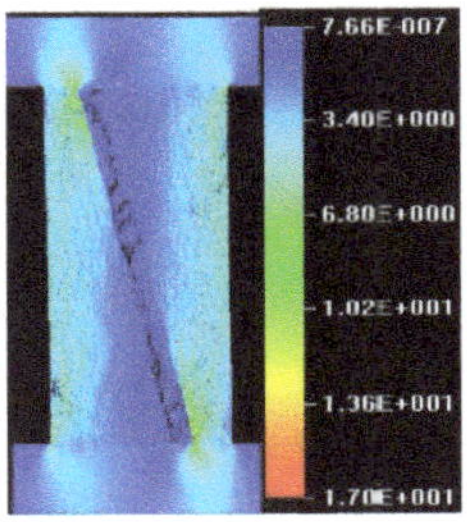

(b) Model 2

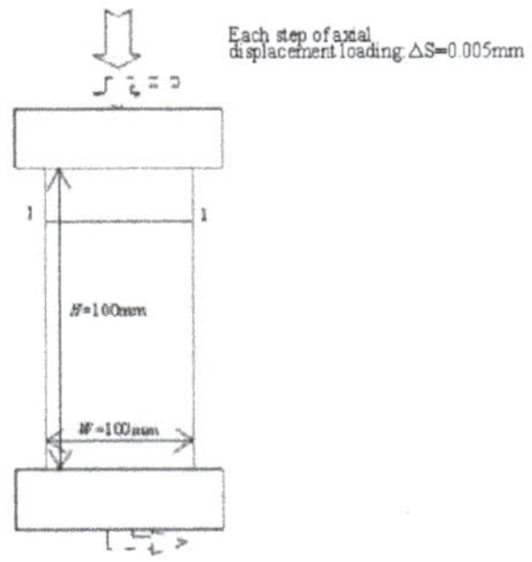

(c) Model 3: complete model

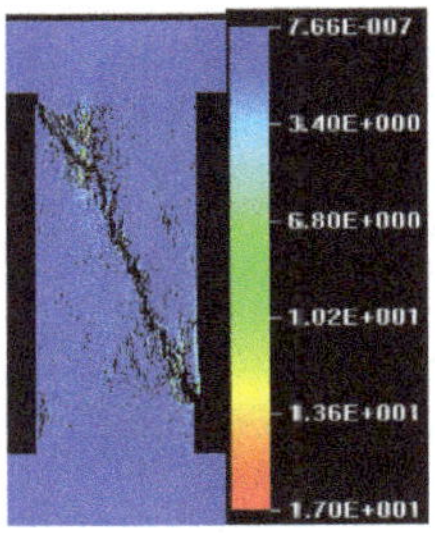

(c) Model 3

Figure 5. Loading diagram of the numerical simulation test and image of the model failure stress (MPa).

Table 1. Sample mechanical properties (average value).

	Parameter Name	Unit	Numerical Value
Parameters of the specimen (m = 2)	Mean elastic modulus E Intensity Poisson ratio μ Permeability coefficient	MPa MPa m/d	50,000 100 0.25 1.00×10^{-8}
	Parameter Name	**Unit**	**Numerical Value**
Pad parameters (homogeneous)	Mean elastic modulus E Intensity Poisson ratio μ Permeability coefficient	MPa MPa m/d	150,000 300 0.2 1
	Parameter Name	**Unit**	**Numerical Value**
Prefabricated weak surface (weak zone) parameters (m = 3)	Mean elastic modulus E Intensity Poisson ratio μ Permeability coefficient	MPa MPa m/d	20,000 20 0.25 1.00×10^{-8}

4.2. Evolution Law of the Permeability Coefficient in the Whole Process of Model Stress-Strain

The final failure modes of the prefabricated weak surface model (Model 1), prefabricated weak zone model (Model 2), and complete model (Model 3) are shown in Figure 6, and the stress–strain–permeability curves are shown in Figures 7a, 8a and 9a, respectively. Comparison of the post-peak curves in Figure 6a–c with those in Figures 7a, 8a and 9a indicated that when macroscopic shear cracks appeared in Models 1 and 3, the overall stress level decreased, and only the local area underwent high stress, which was due to the existence of some contact points in the shear failure zone. Both models had a certain shear slip after failure. However, the overall stress level of Model 2 did not decrease significantly because, although a shear failure zone formed in the weak area, the rocks on both sides of the failure zone were still intact and could continue to be loaded, and the model did not exhibit interlayer dislocation. Usually, sandstone has large pores high permeability. With axial displacement loading, the pores of sandstone closed. Therefore, the permeability coefficient began decreased slightly from the initial value at first. However, compared with sandstone, limestone has a dense texture, a small permeability coefficient, and almost no water permeability. Therefore, in accordance with the permeability characteristics of limestone, the permeability process of limestone can be divided into the following stages.

1. First stage: With axial displacement loading, some elements were destroyed because the load reached its failure strength. At this time, a micro-fracture occurred and formed a small seepage channel, and the permeability coefficient of the model increased.
2. Second stage: The axial displacement loading continued to increase. When the microcracks reached a certain number, the previous microcracks closed, and the permeability coefficient decreased slightly.
3. Third stage: With the increase in external load stress, failure occurred rapidly, and the cracks interconnected to form a shear failure zone. The penetrating cracks provided a favorable channel for seepage, and the permeability coefficient was greatly improved at this stage.
4. Fourth stage: The shear failure zone was gradually compacted under axial load, and the permeability coefficient decreased. This change in permeability coefficient could be clearly seen in both the experimental results (Figure 4b) and the numerical results. The prefabricated weak zone model (Figure 5b) was not compacted because the intact rock on both sides of the failure zone could continue to withstand the external load, and the decrease in the permeability coefficient was not obvious.

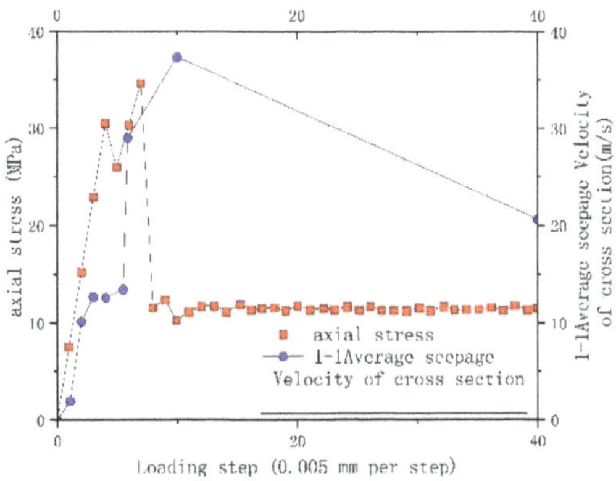

(a)

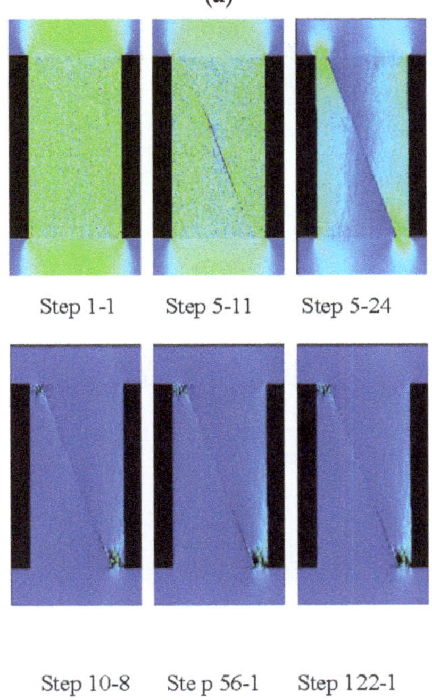

(b)

Figure 6. *Cont.*

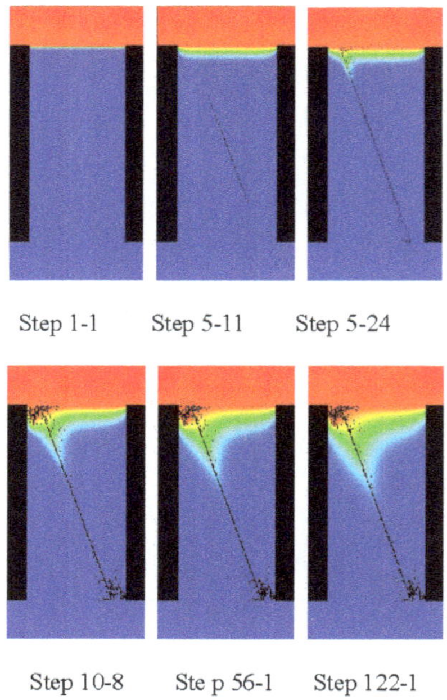

(**c**)

Figure 6. Numerical results of the cracked mode. (**a**) Whole process curve of load step-stress-permeability coefficient. (**b**) Stress results of the model failure process (MPa). (**c**) Results of water pressure distribution in the model failure process (m).

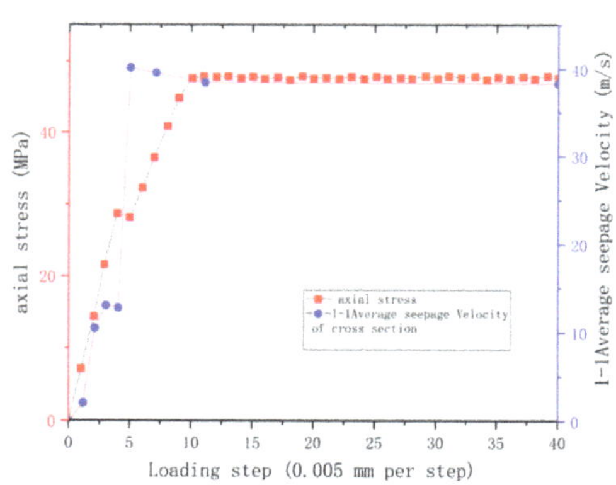

(**a**)

Figure 7. *Cont.*

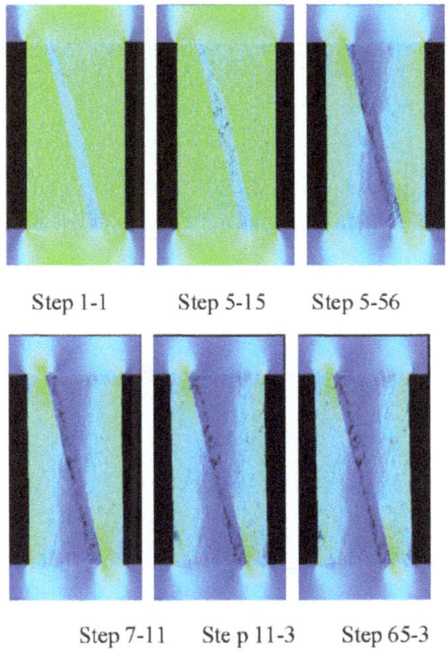

(b)

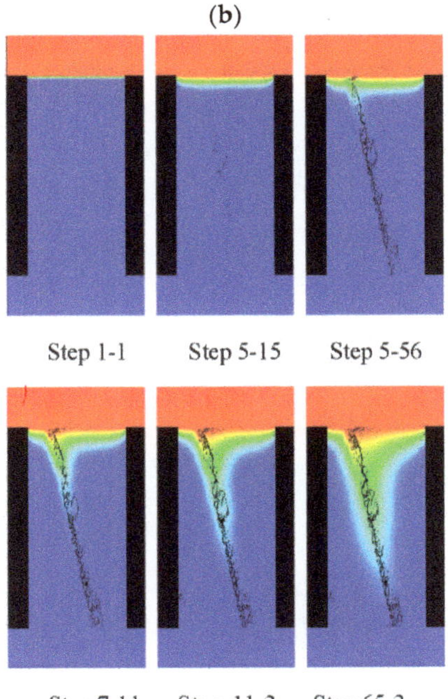

(c)

Figure 7. Numerical results of the model with a weak zone. (**a**) Whole process curve of load step-stress-permeability coefficient. (**b**) Presupposed weak zone model failure process stress results (MPa). (**c**) Results of water pressure distribution during crack formation (m).

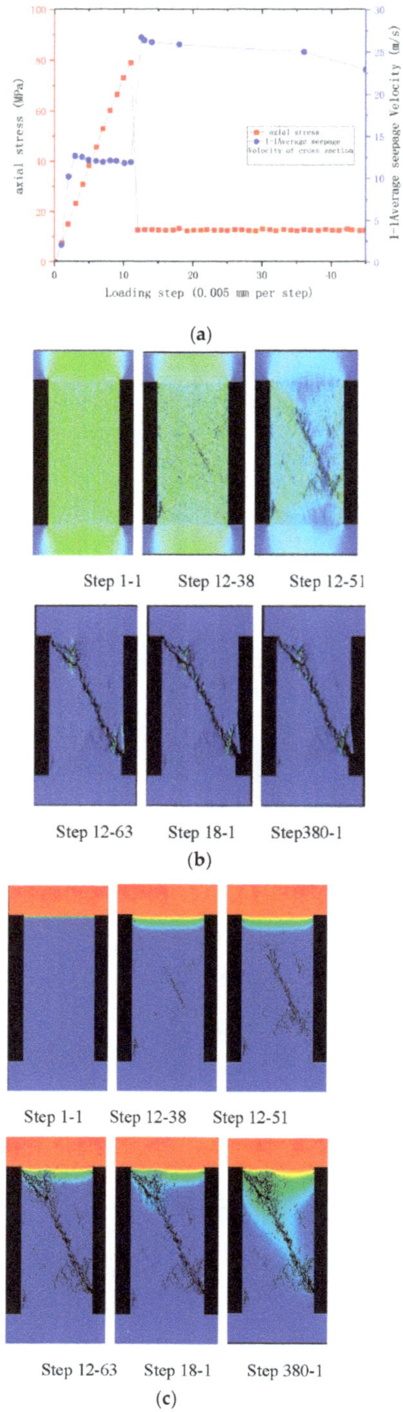

Figure 8. Numerical results for Sample 3. (**a**) Whole process curve of load step-stress-permeability coefficient. (**b**) Stress results for the intact specimen during the failure process (MPa). (**c**) Results of water pressure distribution during crack formation (m).

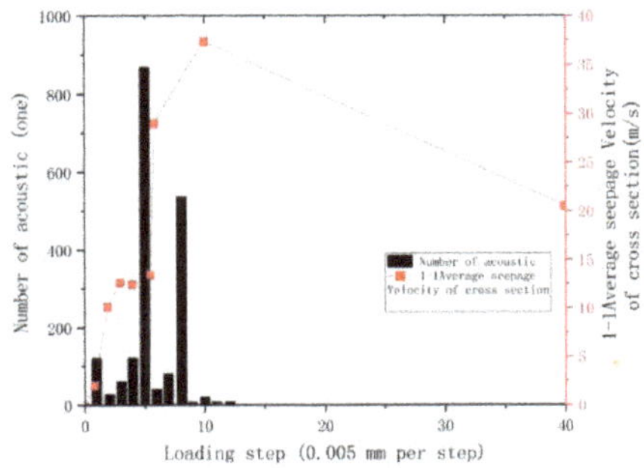

(a)

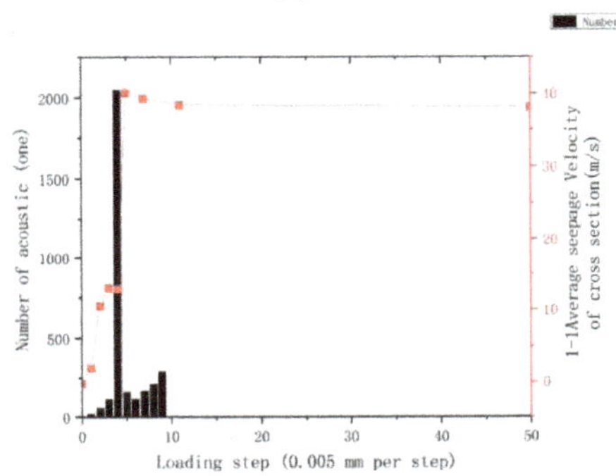

(b)

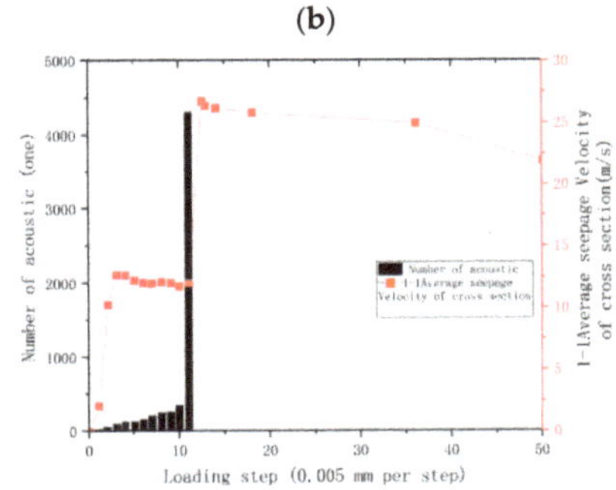

(c)

Figure 9. Numerical load step-stress-acoustic emission curve. (**a**) Prefabricated weak surface model. (**b**) Prefabricated weak zone model. (**c**) Complete model.

The four stages are shown in Figures 6a, 7a and 8a. With the increase in load step, the stress increased at the same time, a microfracture slowly initiated, and the permeability coefficient increased slowly from 0, which was the value at the first stage. As the load step continued to increase, the microfracture closed, and the step stage appeared, which was the second stage. After the step stage, the crack and permeability coefficient suddenly increased rapidly, which was the third stage. Then, due to the closed fracture permeability coefficient exhibiting a small decline, the fourth stage occurred.

Figures 6a and 8a indicate that the specimen slipped along the prefabricated joint surface, so the stress decreased sharply. The permeability coefficient increased sharply because the crack had penetrated. Figure 7a shows that no obvious stress drop process occurred, so the development of cracks could not be judged by the stress–strain curve. Figures 6b, 7b and 8b fully display the crack formation process, allowing our timely understanding of where the crack extended, whether it went through, and when it went through, which were difficult to determine in the experiment. In addition, the water pressure distribution results in the process of model failure are illustrated in different colors in Figures 6c, 7c and 8c to reflect the occurrence of seepage. Red represents high water pressure, and blue represents low water pressure, that is, water permeated from top to bottom and was sealed around. These figures fully show the process of water flowing from top (high water pressure) to bottom (low water pressure) along the crack. The head pressure at the crack was the largest, and a certain water pressure distribution formed near the crack area. The area that the seepage could reach was limited due to the low permeability of limestone itself, and almost no water pressure was observed outside the crack.

The microfractures in Model 1 were concentrated near the prefabricated joints, so the width of the crack was small, and the water penetration rate was low. Loading to Step 122–1 (load step 122, step 1; when element failure occurred in each loading step, step-in-step cycle calculation was applied, but when no failure occurred in this step, the next loading step was initiated) could not fully penetrate the entire model (Figure 6c). Given that the microfractures in Model 2 (Figure 5b) were mainly distributed in the weak region, only a few microfractures occurred in the other regions, resulting in a wide seepage channel in the weak region and a high penetration rate. By Step 65–3 (Figure 7c), the water flow penetrated into almost the entire model height. The seepage velocity of the intact rock block model (Figure 5c) was relatively low because the crack formation time was long, and the microfracture had the characteristics of dispersion and disorder and was randomly distributed in the whole model. Therefore, loading at Step 18–1 (Figure 8c) did not penetrate into half of the height of the model.

Figures 6a, 7a and 8a show that the seepage velocity of Model 2 was the largest, followed by those of Models 1 and 3 (the smallest). The reason was that the joints and weak zones of Model 1 and 2 connected the high water head area at the top, and water could easily pass through this area. The weak zone of Model 2 was wider than that of Model 1, and the permeability was stronger. This is the reason why the seepage velocity of Model 2 was larger than that of Model 1. The new cracks in Model 3 did not penetrate the upper and lower interfaces, and no effective seepage channel connecting the high water head area was formed. Water could only penetrate the cracks from the high water pressure through the microfracture and then through the cracks.

For the complete model and the prefabricated weak surface model, the stress sudden drop method can be used to determine whether a model seepage channel is formed. However, this method does not allow judging whether a seepage channel is formed in the prefabricated weak zone model. The prefabricated weak zone model can only judge whether the seepage channel is formed based on the results of the water pressure distribution. Given that the results of water pressure distribution in physical experiments cannot be vividly displayed as in numerical calculation, the seepage channel must be judged through other means.

4.3. Prediction of Rock Permeability during Crack Propagation

The analysis of the stress–strain–permeability curve revealed that the failure and seepage of the model could be judged by the stress drop. However, with hysteresis, passivity, and failure, predicting the catastrophic failure of the model was impossible. In fact, if the microfracture of the model is located and monitored, and crack coalescence is identified before microfracture coalescence, then crack coalescence and catastrophic seepage can be prevented effectively. When rock micro-fracture occurs under load, some strain energy is released in the form of elastic waves. These elastic waves can be monitored by the acoustic emission technology. Acoustic emission positioning technology is used to determine the location, time of occurrence, and strength of microfractures. In this way, we can analyze the mechanical behavior of the rock fracture process in terms of time, space, and strength and understand the defects of the current medium, both of which are crucial to obtain information about the precursors of the main fracture. Collection and analysis of the data obtained through the numerical model of acoustic emission (AE) monitoring showed that AE can effectively reflect a microfracture and that dense limestone materials mainly show permeability in the cracks. Therefore, if the development of cracks can be effectively predicted, then the occurrence of seepage can be effectively controlled. The underlying specific principle has been explained in [25].

Figure 9 shows the load step–stress–acoustic emission curve of the three models. With loading, the AE number increased slowly. Before the seepage velocity increased, AE suddenly occurred, and the permeability coefficient increased significantly. The relationship curve in Figure 10 indicates that AE showed certain advantages when judging the seepage characteristics. The AE location information in Figure 10 can well reflect the occurrence and penetration of microfractures and predict the permeability characteristics. In Figure 10, a circle represents an AE event. The size of the circle refers to the relative energy of AE. The white circle represents compressive failure, the red one represents shear failure, and the black one represents a historical AE event. Under axial load, oblique cracks were produced along the direction of the prefabricated weak plane in the prefabricated weak plane model, which was predictable. This phenomenon can also be seen clearly from the AE location information in Figure 10a. At Step 2–4, the model mainly exhibited minimal pressure damage, randomly distributed throughout the model. By Step 3–5, the compressive failure was still randomly distributed in the whole model, but the red tensile failure had begun to concentrate along a certain line of the model (the direction of the prefabricated weak surface). By Step 5–3, the small tensile failures formed a shear failure zone. Nevertheless, the failure point did not penetrate the upper and lower parts of the specimen, so the water at the upper end could not penetrate into the specimen in a wide range. Seepage along the crack occurred only when the crack passed through the whole sample at Step 8–25. Model 2 was mainly destroyed along the prefabricated weak zone. Figure 10b shows the AE location information of the prefabricated weak zone model. The AE distribution of the prefabricated weak zone model was basically consistent with that of the prefabricated weak surface model. At Step 5–37, the crack penetrated, and the microfractures were mainly distributed along the prefabricated weak zone.

Comparison of Figure 10a,b showed that in the early stage of loading (Figure 10c), the microfractures of the model were randomly distributed throughout the model, and no obvious shear microfracture occurred until Step 12–30, during which some concentrated red damage points emerged in the middle. Once the macrocracks were produced, surrounding unit damage was quickly induced, especially rapid damage at both ends of the crack, and a penetrating crack formed, resulting in a shear failure zone. The failure diagram shows that the failure mode from the numerical calculation was consistent with the experimental result (Figure 4), and both showed a high-angle shear failure.

Figure 11 shows the distribution of the permeability coefficient of each element in the 1-1 section and the change with loading step during the loading process of the complete model. At Step 1, the permeability coefficient did not fluctuate much and followed an almost horizontal line. At Step 11, the permeability coefficient began to fluctuate around the horizontal line. At Step 12, a large amplitude peak emerged near the section element number of 30, which can be judged as a crack through the site. When loaded to Step 153, the fracture surface was contacted and compacted by the load, resulting in a decrease in permeability and seepage velocity at the peak position. This process is clearly reflected in the variation of seepage velocity in the 1-1 section in Figure 11.

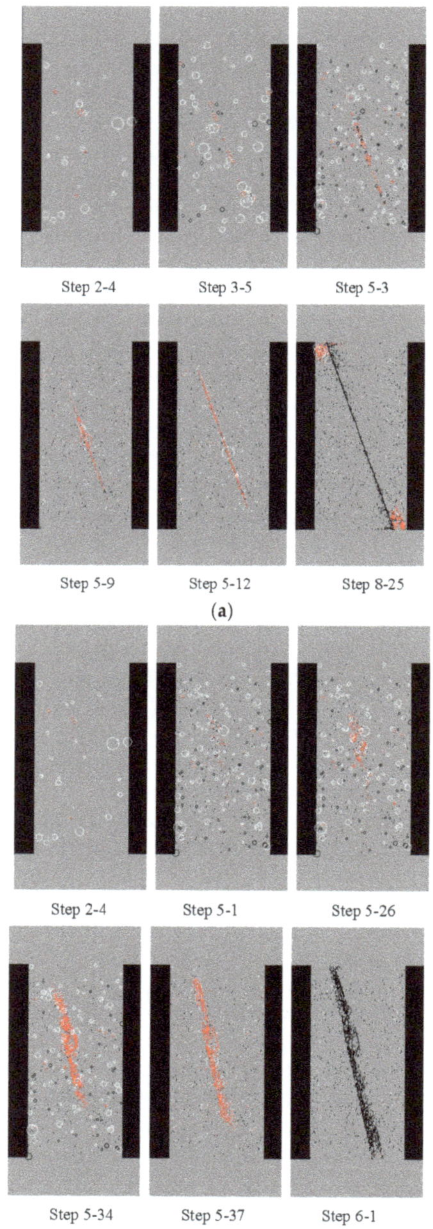

Figure 10. *Cont.*

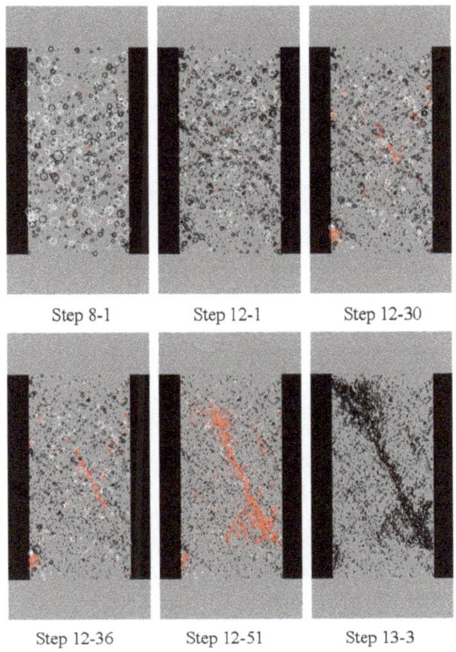

(c)

Figure 10. Numerical time distribution sequence of AE positioning in the sample loading process. (**a**) Acoustic emission results of Model 1's loading process. (**b**) Acoustic emission results of Model 2's loading process. (**c**) Acoustic emission results of Model 3's loading process.

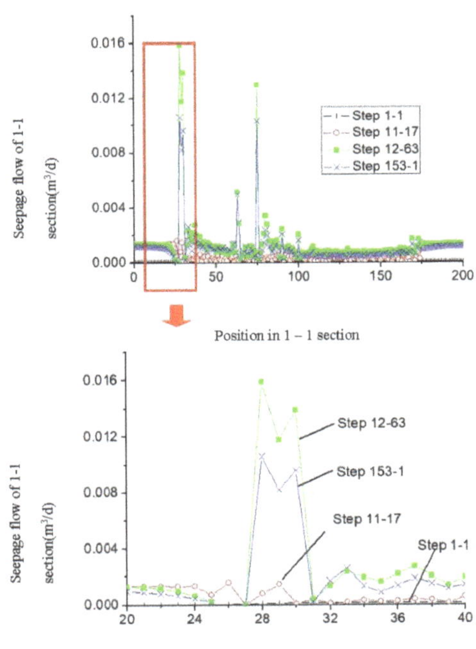

Figure 11. Numerical time distribution sequence of AE positioning in the specimen loading process.

The AE location information of the specimen loading process indicated that before the high osmotic pressure emerged, that is, before the crack went through, crack initiation underwent a process. Predicting the development direction of cracks by using the AE positioning method to monitor cracks is thus effective.

5. Discussion

In this paper, the seepage characteristics and the formation of seepage channels of limestone under high seepage pressure were studied through physical experiments and numerical simulation. Compared with previous studies, the rock permeability selected in this paper was low. Such rock seepage phenomenon is difficult to observe. Under the action of load, cracks were generated. When the cracks penetrated, the permeability increased sharply, which is extremely unfavorable to the slope safety. Finally, this paper puts forward a new idea to predict landslides by using acoustic emission to locate the position of micro cracks.

6. Conclusions

In this study, the post-peak limestone seepage test and numerical simulation were combined to study the seepage characteristics of limestone and the prediction method of seepage channel formation under high seepage pressure. The results showed that it is effective to judge the formation of the seepage channel based on the sudden stress drop for the complete model and prefabricated weak area model, but this method is invalid for the prefabricated weak area model.

In the three models in this study, the development of cracks began in the middle of the model, then developed toward both ends. At the beginning of crack initiation, no upward nor downward crack development was observed in the channel. However, as the loading continued, a crack developed toward both ends until, finally, it penetrated. Water was instantly squeezed into the seepage channel, and high osmotic pressure was generated instantaneously. This kind of sudden inflow of seepage exerts a destructive effect on slope stability and increases the difficulty of predicting catastrophic landslides with high seepage pressure.

The prediction accuracy for low-permeability jointed rock landslides can be improved to a certain extent by combining AE characteristics and seepage head changes. The AE location of microfractures in the model can reflect the location and development direction of microfractures in real time, and the seepage depth can be predicted by measuring the seepage head. If the AE location is combined with measured water head distribution results in practical engineering, then crack opening can be determined to effectively reflect and prevent crack propagation in real time. As a result, the main seepage channel will not run through, sudden seepage will not occur, and seepage pressure will not be transferred.

Author Contributions: Conceptualization, C.B. and S.T.; methodology, C.B.; software, RFPA; validation, C.B., S.T. and Y.Y.; formal analysis, A.J.; investigation, H.L.; resources, H.L.; data curation, C.B.; writing—original draft preparation, C.B.; writing—review and editing, Y.Y.; visualization, S.T.; supervision, H.L.; project administration, C.B.; funding acquisition, C.B. All authors have read and agreed to the published version of the manuscript.

Funding: NSFC-funded project (No. 51474046); National Natural Science Foundation of China—China Petroleum and Natural Gas Corporation Petrochemical Joint Fund Project (U1562103); Liaoning University Excellent Talents Support Project (LJQ2015076).

Informed Consent Statement: Informed consent was obtained from all subjects involved in the study.

Conflicts of Interest: The authors declare no conflict of interest.

References

1. Zhou, C.B. Review of life cycle safety control of high and steep slope in hydropower engineering. *J. Rock Mech. Eng.* **2013**, *32*, 1081–1093.
2. Wang, B.; Zhao, R.; Li, Y.S.; Xu, M.; Liu, X.H.; Hu, A.K. Experimental study on variation characteristics of rock permeability under different confining pressures in western sichun plateau-Taking the Balangshan tunnel as an example. *Secur. Environ. Eng.* **2021**, *28*, 179–186. [CrossRef]
3. Louis, C. *A Study of Groundwater Flow in Jointed Rock and Its Influence on the Stability of Rock Masses*; Rock Mechanics Research Report, No. 10; Imperial College of Science and Technology: London, UK, 1971.
4. Su, B.Y.; Zhan, M.L.; Zhao, J. The model test of the flow in Smooth fracture and the study of its mechanism. *ShuliXuebao* **1994**, *5*, 19–24. (In Chinese)
5. Tang, S.B.; Yu, Q.L.; Li, H.; Yu, C.Y.; Bao, C.Y.; Tang, C.A. Mesomechanical model of moisture diffusion and shrinkage cracking in building material–Model development. *Constr. Build. Mater.* **2013**, *47*, 511–529. [CrossRef]
6. Esaki, T.; Du, S.; Mitani, Y.; Ikusada, K.; Jing, L. Development of a shear-flow test apparatus and determination of coupled properties for a single rock joint. *Int. J. Rock Mech. Min. Sci.* **1999**, *36*, 641–650. [CrossRef]
7. National Research Council. *Rock Fractures and Fluid Flow: Contemporary Understanding and Applications*; National Academy Press: Washington, DC, USA, 1996.
8. Snow, D.T. Anisotropie Permeability of Fractured Media. *Water Resour. Res.* **1969**, *5*, 1273–1989. [CrossRef]
9. Oda, M. Permeability tensor for discontinuous rock masses. *Geotechnique* **1985**, *35*, 483–495. [CrossRef]
10. Liang, Z.; Wu, N.; Li, Y.; Li, H.; Li, W. Numerical Study on Anisotropy of the Representative Elementary Volume of Strength and Deformability of Jointed Rock Masses. *Rock Mech. Rock Eng.* **2019**, *52*, 4387–4402. [CrossRef]
11. Zhou, C.B.; Ye, Z.T.; Han, B. A preliminary study on unsaturated permeability of rock joints. *Chin. J. Geotech. Eng.* **1998**, *20*, 1–4. (In Chinese)
12. Hsieh, P.A.; Neuman, S.P.; Stiles, G.K.; Simpson, E.S. Field determination of the three-dimensional hydraulic conductivity tensor of anisotropic media: 1. theory. *Water Resour. Res.* **1985**, *21*, 1655–1665. [CrossRef]
13. Li, S.P.; Li, Y.S.; Wu, Z.Y. The permeability—Strain equations relating to complete stress—Strain path of the rock. *Chin. J. Geotech. Eng.* **1995**, *17*, 13–19. (In Chinese)
14. Peng, S.; Qu, H.; Luo, L.; Wang, L.; Duan, Y. An experimental study on the penetrability of sedimentary rock during the complete stress-strain path. *J. China Coal Soc.* **2000**, *25*, 113–116. (In Chinese)
15. Han, B.P. Study on permeability test of dolomite reservoir of wumishan formation in renqiu oilfield. *Chin. J. Geol.* **2000**, *35*, 396–403. (In Chinese)
16. Zhu, W.S.; Zheng, W.H.; Wang, W.T. Numerical simulation of a new damage rheology model for jointed rock mass. *J. Geotech. Eng.* **2010**, 32, 1011–1016.
17. Zhang, R.; Yan, L.; Qian, Z.W.; Sun, X.W.; Liu, W. Study on the evolution relationship of the deformation mechanism and permeability of underground rock. *J. Undergr. Space Eng.* **2022**, *18*, 129–135.
18. Yang, T.H.; Tang, C.A.; Xu, T.; Yang, T.H.; Tang, C.A.; Xu, T.; Rui, Y.Q. *Seepage Characteristic in Rock Failure-Theory, Model and Applications*; Science Press: Beijing, China, 2004; pp. 98–116. (In Chinese)
19. Zhao, Y.X.; Cao, B.; Zhang, T. Experimental study on influence of permeability of axial pressures and pentrative pressures on broken rocks. *J. Min. Sci.* **2018**, 3, 434–441. [CrossRef]
20. Tang, C.A.; Wang, J.M. Rock and seismic monitoring and prediction—The feasibility and the preliminary practice. *Rock Mech. Eng. News* **2010**, *89*, 43–55. (In Chinese)
21. Cui, W.; Wang, L.X.; Jiang, Z.; Wang, C.; Wang, L.H.; Zhang, S.R. Numerical Simulation of grouting process in rock mass with rough fracture network based on corrected cubic law. *Geotech. Mech.* **2021**, *42*, 2250–2258. [CrossRef]
22. Yang, X.R.; Jiang, A.N.; Zhang, F.R.; Zhang, Q. Seepage characteristics of fractured limestone under osmitic pressure-stress coupling. *J. Coal* **2019**, *44*, 101–109. [CrossRef]
23. Zhu, Z.D.; Guo, H.Q. *Hydraulic Foundation of Fractured Rock Mass*; Science Press: Beijing, China, 2007.
24. He, Y.L.; Tiao, Y.J.; Yang, L.Z. Experimental research on hydraulic behaviors in a single joint with various values of JRC. *Rock Mech. Eng.* **2010**, *29*, 3235–3240.
25. He, Y.L.; Yang, L.Z. Testing study on variational characteristics of rockmass permeability under loading-unloading of confining pressure. *Chin. J. Rock Mech. Eng.* **2004**, 23, 415–419. (In Chinese)

applied sciences

MDPI

Article

Experimental Study on the Mechanical Properties and Acoustic Emission Characteristics of Deep Soft Rocks under Low-Frequency Dynamic Disturbance

Ling Dong [1], Dong Wang [1,2,*], Xiaoming Sun [2], Yujing Jiang [1], Hengjie Luan [1,3], Guangchao Zhang [1], Guanglei Zhou [1] and Bin Liang [1]

1 State Key Laboratory of Mining Disaster Prevention and Control Co-Founded by Shandong Province and the Ministry of Science and Technology, Shandong University of Science and Technology, Qingdao 266590, China
2 State Key Laboratory for Geomechanics and Deep Underground Engineering, China University of Mining and Technology, Beijing 100083, China
3 Inner Mongolia Shanghaimiao Mining Co., Ltd., Ordos 016299, China
* Correspondence: wdwinter@163.com; Tel.: +86-156-2102-9309

Abstract: The strong dynamic disturbance in deep mines seriously affects the safe and efficient mining of deep resources. In this study, we used the creep disturbance impact loading system and acoustic emission system to conduct experiments on soft siltstone specimens under a combination of dynamic and static loads. Based on the failure characteristics and waveform signals, the mechanical properties and acoustic emission characteristics of soft rocks under different dynamic disturbances were quantitatively revealed. The experimental results show that: (1) Under the dynamic disturbance, the deformation of the siltstone specimens increases as the initial average stress increases. When the axial stress exceeds the upper stress threshold, cracks continue to propagate, resulting in the destabilization of the specimen. (2) The magnitude of the initial average stress is closely related to the degree of damage and failure mode of the siltstone. With the increase in the initial average stress, the failure mode of the siltstone specimens gradually changes. As the initial average stress increases, the maximum load first decreases, then increases, and finally decreases, and the fitted curve is polynomial. We used the RFPA2D cyclic loading module to analyze the effect of the elastic modulus of each loading step on the damage evolution of the specimen under dynamic disturbance. The waveform characteristics during the evolution of the damage of the specimens were analyzed by extracting signals at the key points.

Keywords: dynamic disturbance; soft rock; cyclic loading; mechanical properties; acoustic emission characteristics

Citation: Dong, L.; Wang, D.; Sun, X.; Jiang, Y.; Luan, H.; Zhang, G.; Zhou, G.; Liang, B. Experimental Study on the Mechanical Properties and Acoustic Emission Characteristics of Deep Soft Rocks under Low-Frequency Dynamic Disturbance. *Appl. Sci.* **2023**, *13*, 6399. https://doi.org/10.3390/app13116399

Academic Editor: Ricardo Castedo

Received: 23 March 2023
Revised: 19 May 2023
Accepted: 21 May 2023
Published: 24 May 2023

1. Introduction

Of the current energy and mineral reserves, coal is a strategic energy source supporting China's economic and social development [1,2]. At present, deep mining has become the new normal for coal resource development in China. Safety issues during deep mining and major disaster prevention and control have been written into China's "14th Five-Year Plan" [2]. Under complex geological conditions, deep mines are strongly disturbed by the mining dynamics; therefore, large deformation hazards and ground control in deep soft rock mines have become the focus of research in the field of deep mining [1,3,4]. The study of mechanical properties of soft rocks at depth under low-frequency dynamic disturbance is of important practical significance for understanding the deformation destabilization mechanism and stability control of soft rock mines [5,6].

Regarding the dynamic disturbance to coal and rocks at depth, scholars have conducted extensive studies on experimental mechanical properties and acoustic emission characteristics of rocks under dynamic and static loading conditions.

Since 2012, dozens of rock bursts or suspected rock bursts have occurred in the Asher copper mine in Xinjiang, with the most serious one occurring during the anchor network spray concrete support, where the surrounding rock was flaky and ejected over a large area for a short period of time [7]; the Chener gold mine in Shaanxi explored at a depth of about 1000 m, and rock bursts occurred from time to time after entering deep mining, resulting in flakes falling off the top of the roadway and the two gangs, and fragments rapidly flying away [8]. By studying mechanical properties under combined dynamic and static loads, experiments were performed using the split Hopkinson pressure bar (SHPB) to examine the damage characteristics and the energy dissipation behavior of strain-type rock bursts at depth [9–11]. Mechanical experiments under different dynamic perturbations were carried out to investigate the energy-based failure mechanism of rocks, the evolution of fracture characteristics, and the effect of roof dynamic disturbance on the stress of the floor [12–15]. By studying rock bursts near a deeply buried tunnel fault under static and dynamic loads, static and dynamic loads were used to simulate rock damage. A combined finite element method was used to simulate the damage process of an underground cavern, revealing the damage mechanism of deep hard rock influenced by the amplitude of dynamic stress wave, disturbance direction and dip angle of the structural surface [16,17]. Dynamic disturbance, and cyclic loading is related to the fatigue crack expansion of the material, and when the capacity of the material is exceeded, the material undergoes damage [18–20].

In studying acoustic emission (AE) characteristics of coal and rocks at depth, the rock failure process analysis (RFPA) code was used to analyze the influencing factors on AE characteristics of rock-like specimens at different positions, and the evolution of permeability of fractured rock and its anisotropy under hydrodynamic coupling conditions [21,22]. Zhang et al. [23] analyzed the evolution of acoustic emission and deformation field during crack propagation by establishing a mechanistic model for directional crack propagation. Researchers have carried out shear experiments on granite and triaxial compression experiments on coal specimens of different sizes to explore the variations of transverse and longitudinal AE wave signals and the evolution of energy [24–28]. Acoustic emission signals of rocks during damage and the influence of different fracture dip angles on the macroscopic mechanical properties of rocks were investigated by using uniaxial compression experiments with AE monitoring [29,30]. The torsional properties and damage development of long and hollow axisymmetric composite cylinders with different opening sizes were investigated experimentally. The acoustic emission (AE) of prismatic marble specimens during uniaxial compression loading was analyzed using the non-extensive statistical mechanics (NESM) method. Acoustic emission (AE) was used to record the microcracks during the impact of granite specimens [31–33]. Different materials can be studied using acoustic emission and digital image correlation techniques to investigate the specific characteristics of the material damage process under different loading effects in order to follow the evolution of the material under different damage modes and to determine the strain [34–37].

In summary, most of these experimental studies on the mechanical properties of coal and rock under combined dynamic and static loads were focused on single, high-frequency impact loading using SHPB and drop hammers. By applying the average value of different cyclic loading stress values, the siltstone is damaged under the action of dynamic perturbation, and the deformation damage characteristics of the dynamic perturbation are necessarily different from those of uniaxial. To this end, we select the siltstone from the roof of a deep soft rock tunnel in central eastern China as the research object and use the creep disturbance dynamic impact loading system combined with acoustic emission monitoring to carry out experiments under different dynamic disturbance conditions and reveal the mechanical and acoustic emission characteristics of the soft rock. Our study aims to provide a theoretical basis for the control of the deformation and instability of coal and rocks in deep mines affected by mining activities.

The study of the effect of low-frequency dynamic disturbances on the mechanical properties of rocks is of greater engineering interest. In rock mechanics tests, the dynamic disturbance environment in the downhole mining process can be simulated by applying a certain frequency of cycles to the rock specimen. The magnitude of stress perturbation varies depending on the form of dynamic perturbation such as blasting and stress redistribution. In deep well mining, the rock is in a high stress state and a small stress perturbation may cause the rock to exceed its yield stress and cause damage to the rock. In summary, most of these experimental studies on the mechanical properties of coal rocks under combined dynamic and static loads have focused on single high-frequency impact loading with SHPB and drop hammers. By applying average values of stress values for different cycles of loading, the siltstone was damaged by dynamic disturbance, which inevitably has different deformation damage characteristics from that of uniaxial ones. To this end, we selected a siltstone in the roof of a deep soft rock tunnel in east-central China as the research object and used a creep perturbation dynamic impact loading system combined with acoustic emission monitoring to conduct experiments under different dynamic perturbation conditions to reveal the mechanical characteristics of the soft rock. Our study aims to provide a theoretical basis for controlling the deformation and instability of coal and rocks in deep mines affected by mining activities.

2. Experimental Scheme and Conditions

2.1. Experimental System

Experiments are conducted by using a creep disturbance dynamic impact loading system, which mainly consists of a loading system, a hydraulic pump station, a servo control system, and a data acquisition system. The maximum axial load of the static loading unit in the loading system is 800 kN, and the maximum axial load of the dynamic loading unit is 100 kN, as shown in Figure 1. The servo control and data acquisition system consists of a high-precision closed-loop servo control unit, pressure transducer, and magneto-displacement transducer to form a closed-loop control system to achieve accurate monitoring and acquisition of stress and displacement in a variety of stress paths. The acquisition system mainly includes indenter displacement and axial stress data acquisition, and servo control through the supporting software. Axial stress is monitored by pressure sensors and recorded with real-time feedback. Axial deformation and axial stress of the specimen are automatically collected, and acoustic emission monitoring is used to track the crack damage process. Multiple sets of cyclic loading experiments are applied using sine wave loading.

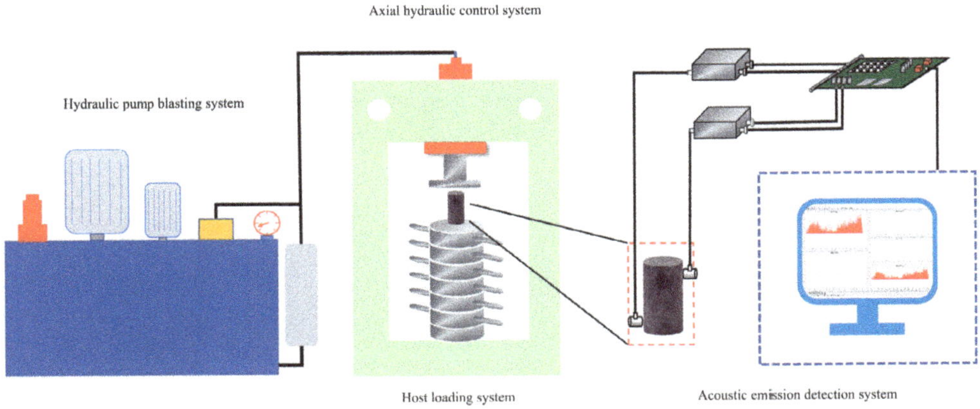

Figure 1. Dynamic impact loading simulation test system for creep disturbance.

2.2. Specimen Preparation

The siltstone used in this work was recovered from the roof of a mining roadway in Xinshanghai NO.1 Coal Mine in China, which is a typical Jurassic soft rock coal mine.

The samples were brownish-grey, with no visible structural defects on the surface. Specimen preparation procedure: (1), samples were taken from the same rock block to make standard cylindrical specimens of Φ50 mm × 100 mm. (2), the two ends of the specimens were polished by using a stone grinding machine so that the non-parallelism and non-perpendicularity were less than 0.2 mm, in accordance with the requirements of the International Society of Rock Mechanics (ISRM) and (3) finally, the specimens were divided into two groups, numbered RS-1~3 and RS-4~24, with a natural apparent density of 2325.4~2630.5 kg/m^3 and an average density of 2477.95 kg/m^3, as shown in Figure 2.

Figure 2. Part of siltstone sample.

2.3. Experimental Design

The siltstone was divided into two groups for uniaxial compression experiments and cyclic disturbance loading experiments. During the initial compression loading stage, displacement-controlled loading (4 μm/s) was applied to compact the microfractures inside the specimens. In the cyclic loading stage, specimens were first loaded to 5.48 MPa, 6.39 MPa, 7.31 MPa, 9.13 MPa, 10.05 MPa, 10.96 MPa, and 11.87 MPa (30%, 35%, 40%, 50%, 55%, 60%, and 65% of the mean uniaxial compressive strength). The sine wave cyclic loading method was then used to apply the dynamic loads. The uniaxial compressive strength of 18.264 MPa, obtained from the uniaxial compression tests, was used as the initial average of the cyclic load. The experimental sinusoidal loading waveform is shown in Figure 3. The specimen loading process was as follows: first, the specimen was loaded at a rate of 0.1 mm/min to the initial value of the cyclic stage, and then the sinusoidal waveform was cycled 1000 times until the siltstone specimen was damaged or the axial force was 0. A schematic diagram of the cyclic disturbance loading for the siltstone specimens is shown in Figure 3. By analyzing the relationship between different initial average stresses and axial strain and elastic modulus, the mechanical properties of siltstone under sinusoidal cyclic loading can be better understood, and the fitting trend of the maximum load with different initial average stresses can be seen more graphically.

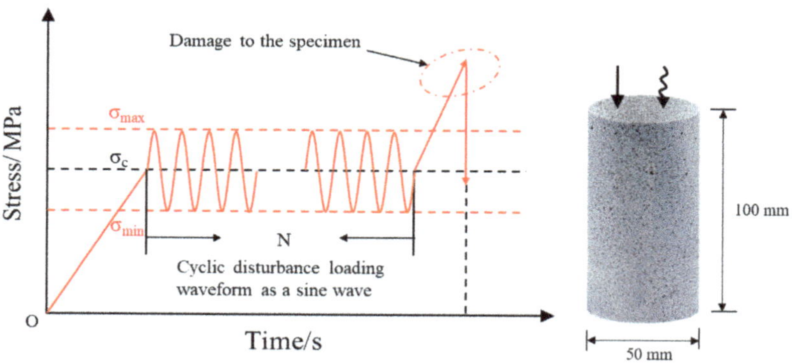

Figure 3. Sine waveform and force diagram of siltstone specimen during cyclic disturbance loading.

3. Analysis of Experimental Results

3.1. Uniaxial Compression Process and Experimental Results

The experimental design and physical parameters of the siltstone are provided in Table 1, where RS-1, RS-2 and RS-3 are uniaxial compression tests and the rest are cyclic disturbance loading tests. The stress-strain curves of uniaxial compression tests and the comparison of the specimens pre- and post-failure are shown in Figure 4, where the peak stress in the stress-strain curve is the compressive strength. The uniaxial compression stress-strain curve exhibits four stages: microfracture compaction stage (I), elastic deformation stage (II), yielding stage (III), and post-peak stage (IV).

Table 1. Testing program.

Specimen Number	Quality/g	Height/mm	Diameter/mm	Density/(kg·m^{-3})	Initial Average Stress Value/MPa	Compressive Strength/MPa
RS-1	467.49	100.18	50.20	2358.9	—	17.74
RS-2	452.52	99.86	49.53	2353.1	—	19.27
RS-3	464.57	100.16	50.07	2356.9	—	17.78
RS-4	458.63	100.02	50.12	2325.4	5.48	15.29
RS-5	467.58	100.11	50.08	2372.3	5.48	13.47
RS-6	463.86	100.13	50.01	2359.7	5.48	14.25
RS-7	460.28	100.01	50.09	2336.7	6.39	13.40
RS-8	462.49	100.20	50.03	2349.1	6.39	13.71
RS-9	464.82	100.18	50.05	2359.5	6.39	12.81
RS-10	465.25	100.28	50.05	2359.4	7.31	14.38
RS-11	481.52	100.15	50.12	2438.2	7.31	13.71
RS-12	470.29	100.18	50.13	2379.6	7.31	12.92
RS-13	466.76	99.98	50.03	2376.0	9.13	16.47
RS-14	462.09	100.20	50.12	2338.6	9.13	13.35
RS-15	463.73	100.21	50.03	2355.2	9.13	15.39
RS-16	482.86	100.35	50.03	2449.0	10.05	13.31
RS-17	474.18	100.02	50.06	2409.9	10.05	16.08
RS-18	475.16	100.10	50.12	2407.2	10.05	13.45
RS-19	471.15	100.21	50.11	2385.2	10.96	13.71
RS-20	460.00	100.23	50.10	2329.2	10.96	13.64
RS-21	520.18	100.32	50.11	2630.5	10.96	13.57
RS-22	467.45	100.14	50.02	2376.7	11.87	8.83
RS-23	484.84	100.10	50.01	2467.1	11.87	9.85
RS-24	473.13	100.08	50.10	2399.4	11.87	9.42

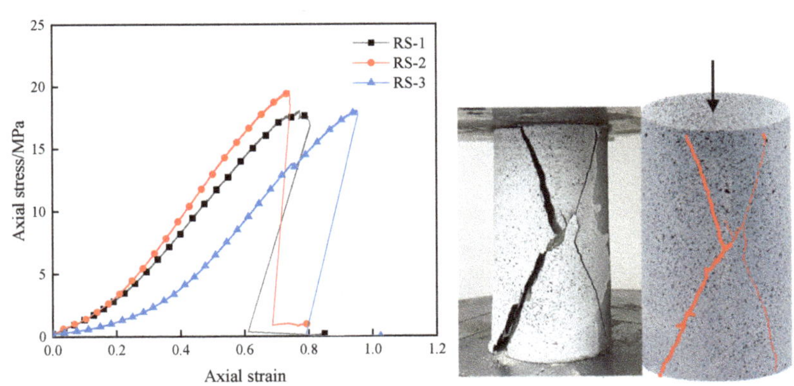

Figure 4. Axial stress-strain curves and specimen damage characteristics of siltstone specimens under uniaxial compression.

It can be seen from Figure 4 that during uniaxial compression, elastic modulus increases first and then decreases as the axial stress increases, indicating that when the axial stress is applied at the beginning of the experiment, the cracks inside the specimen are compacted, leading to an increase in elastic modulus. When the axial stress is applied to a certain extent, the cracks inside the specimen begin to propagate and then the main cracks coalesce, leading to a decrease in elastic modulus. Afterward, the axial stress continues to increase until it exceeds the bearing capacity of the specimen and then the specimen is ruptured.

3.2. Effect of Dynamic Disturbance on the Failure Pattern of Siltstone Specimens

Figure 5 shows the stress-strain curves of the specimens under various dynamic disturbances. When the initial average stresses of the specimens are similar, the stress-strain curves are also similar. However, there are differences in the stress-strain curves when the initial average stresses are different. It can be seen from the figure that the fracture compaction occurs mostly at the beginning of the loading, and the curve is upward convex in the compaction stage. After that, when the sine wave cyclic disturbance load σ_c for the siltstone specimens is greater than 60% of the uniaxial compressive strength, the specimens could be failed before the preset number of cycles (1000) is completed. During the cyclic dynamic disturbance loading, the hysteresis loop area indicating the energy loss of the specimen during the cyclic loading changes with σ_c. When the initial average stress value is 5.48 MPa (30% of σ_c), the hysteresis loop is "thin-lobed", and when the initial average stress value is 6.39 MPa (35% of σ_c), the hysteresis loop becomes "full" and its area increases. This phenomenon occurs due to the large deformation of the primary fractures in the specimen at the beginning of the cyclic dynamic disturbance stage. The hysteresis loops become "sharp-lobed" when the initial cyclic load is 7.31 MPa, 9.13 MPa (i.e., 40% and 50% of σ_c). When the initial cyclic load is 10.05 MPa (55% of σ_c), the hysteresis loop is "banded" and its area gradually decreases, indicating that the fractures inside the specimen are compacted and closed, and the energy dissipation is reduced. When the average stress is 10.96 MPa (60% of σ_c), the hysteresis loop becomes "full" and the specimen undergoes greater irreversible deformation, resulting in unstable failure. When the average stress is 11.87 MPa (65% of σ_c), the specimen is failed before the cycle starts and there is no hysteresis loop.

From the axial stress-strain curves of these seven specimens, it can be found that the deformation of the siltstone specimens increases as the initial average stress increases, indicating that the damage of the specimens during the cyclic disturbance also increases. When the axial stress exceeds the upper limit threshold of stress, the cracks continue to extend and intersect with each other, which in turn leads to the destabilization of the specimens.

Figure 6 shows a typical specimen failure plot under different initial average stresses. When the initial average stress value is 5.48 MPa, its impact on the damage of the specimen is low, and its damage characteristics are generally similar to that from uniaxial compression tests. Due to the low initial average stress, the specimen does not show obvious macroscopic damage after 1000 cycles of disturbance, and only new microfractures are created. The failure mode is shear, and there is partial damage in the middle of the specimen. When the initial average stress is 6.39 MPa, due to axial compressive stress, the shear stress on the fracture surface exceeds the shear strength and the tensile stress in the radial direction. The failure mode is mainly a shear failure, and there is a small part of splitting failure, and the specimen becomes cone-shaped inside after the failure. When the initial average stress is 7.31 MPa, the axial compressive stress causes the radial tensile stress of the specimen; once the radial tensile stress exceeds the tensile strength of the rock, split failure occurs; at this time, the extension direction of secondary cracks changes from longitudinal to dendritic dispersion. The initial average stress of 9.13 MPa has a greater impact on the overall failure mode of the specimen: a large number of secondary cracks are created in the specimen, and large pieces of rock fragments are peeled from the surface, and the failure mode is a combination of shear and splitting. When the initial average stress is 10.05 MPa, the failure mode is shear, with a clearly visible vertical main crack penetrating through the specimen. When the initial average stress is 10.96 MPa, the macroscopic crack of the specimen becomes the shear surface inclined at 60 degrees from the top to the bottom. When the initial average stress is 11.87 MPa, the specimen can no longer bear axial stress and cyclic disturbance, the failure mode is tensile/shear mixed, the integrity of the failed specimen is poor, the rupture surface is rough, showing the characteristics of brittle failure. When the specimen is ruptured, rock fragments burst with a loud sound.

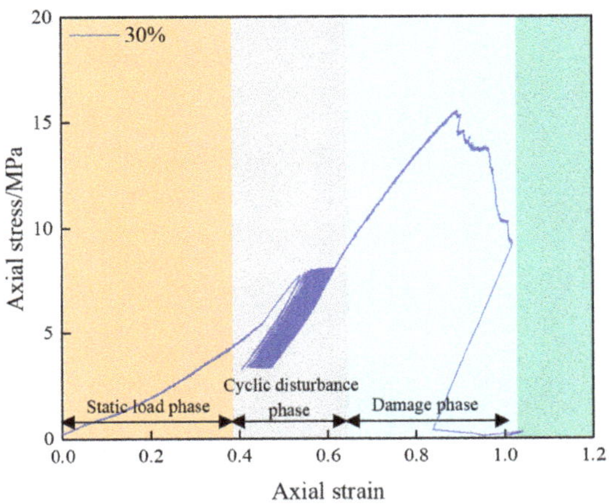

(**a**) The average stress is 5.48 MPa

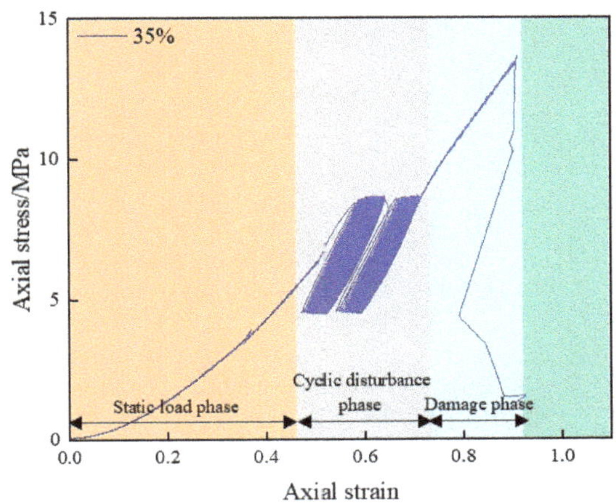

(**b**) The average stress is 6.39 MPa

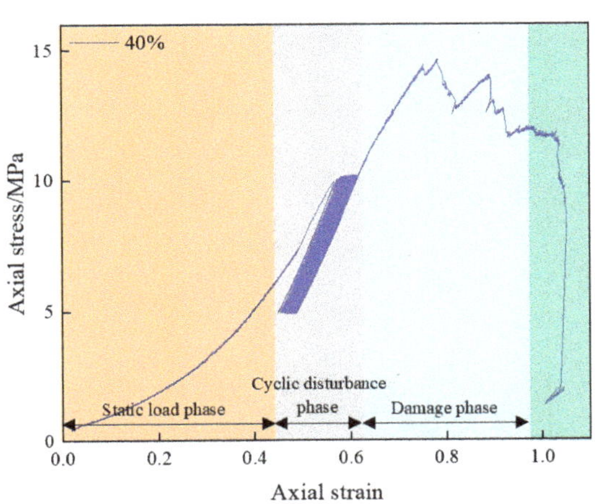

(**c**) The average stress is 7.31 MPa

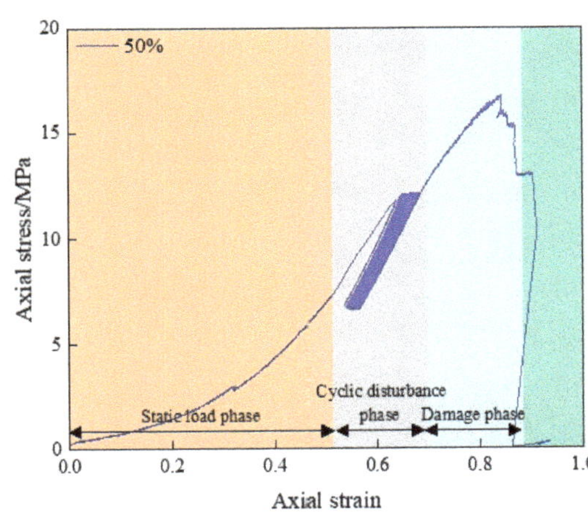

(**d**) The average stress is 9.13 MPa

Figure 5. *Cont.*

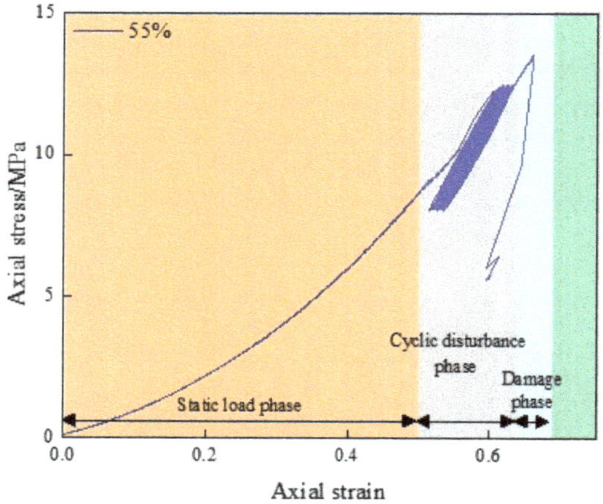

(e) The average stress is 10.05 MPa

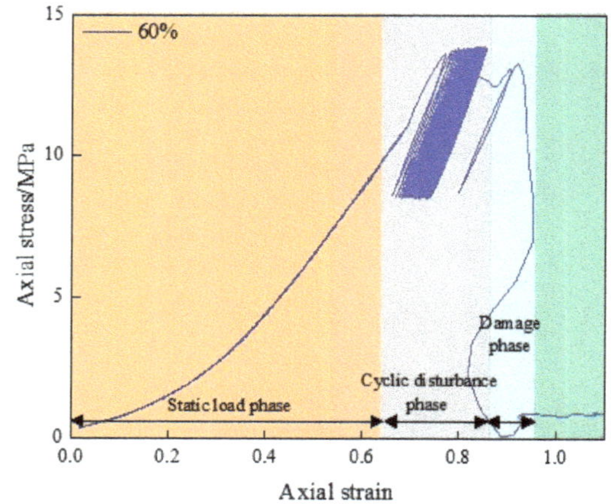

(f) The average stress is 10.96 MPa

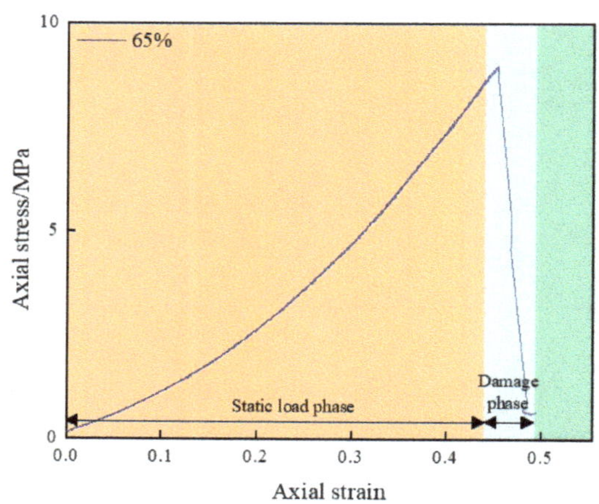

(g) The average stress is 11.87 MPa

Figure 5. Axial stress-axial strain curves of siltstone samples under different σ_c.

Therefore, it can be seen from the macroscopic failure of the specimen that under the cyclic dynamic disturbance, the failure mode gradually transitions from shear failure to shear-splitting mixed failure. Under large initial average stress (i.e., 18.264 MPa), a higher degree of scattering and fragmentation is observed, and the main crack extends to the bottom of the specimen, which shows that the failure mode of the specimen is closely related to the magnitude of the initial average stress.

(a) 5.48 MPa (b) 6.39 MPa (c) 7.31 MPa (d) 9.13 MPa (e) 10.05 MPa (f) 10.96 MPa (g) 11.87 MPa

Figure 6. Typical macroscopic damage characteristics of siltstone under different initial average stresses.

3.3. Numerical Model Development and Validation

The RFPA2D cyclic loading module was used for the numerical simulation. The RFPA software used in this paper, Rock Failure Process Analysis System (RFPA), is a tool based on an elastic damage model that takes into account the non-uniformity of rock materials and the distribution of defects using the intrinsic relationship of statistical damage randomness of defect distribution. We first established a two-dimensional plane strain model with a length × height of 50 mm × 100 mm with 50 elements in the X direction and 100 elements in the Y direction. The homogeneity of the model is 3, the density is 2477.95 kg/m^3, the damage criterion is the Mohr-ulomb criterion, the number of cyclic loading is 1000 times, the Poisson's ratio is 0.25, and the uniaxial compressive strength is 18.264 MPa. The model is numerically modeled according to the corresponding dimensions, and Figure 7 shows a cyclic disturbance model based on the Mohr-Coulomb failure criterion. In the sample model, a pre-pressure is applied to the upper boundary, followed by a sinusoidal cyclic loading with a setting of 1000 times, and no force is applied to the lower boundary. Taking the initial average stress of 10.96 MPa as an example, the elastic modulus of each loading step is shown in Figure 8, which can reproduce the macroscopic crack development of the specimen. The evolution of acoustic emission signals is shown in Figure 9.

As can be seen in Figure 8, at the beginning of the cyclic disturbance loading, the specimen is relatively intact, and there are no noticeable cracks on the surface. As the axial stress increases, shear cracks are created with an inclination of 60 degrees, which gradually penetrate through the specimen. The acoustic emission events of the siltstone specimen during the damage evolution are shown in Figure 9. The center of each circle in the figure represents the location of the acoustic emission event, and the size of the circle represents the amount of energy released by the event. Each circle represents an acoustic emission event, where a white circle represents an acoustic emission event resulting from compression-shear failure, and a red circle represents an acoustic emission event from tensile failure. At the beginning of the cyclic disturbance loading, sporadic acoustic emission events occur locally in the model, and with the increase in axial stress, acoustic emission events are concentrated, gradually forming macroscopic fractures, which in turn leads to instability failure of the siltstone specimen.

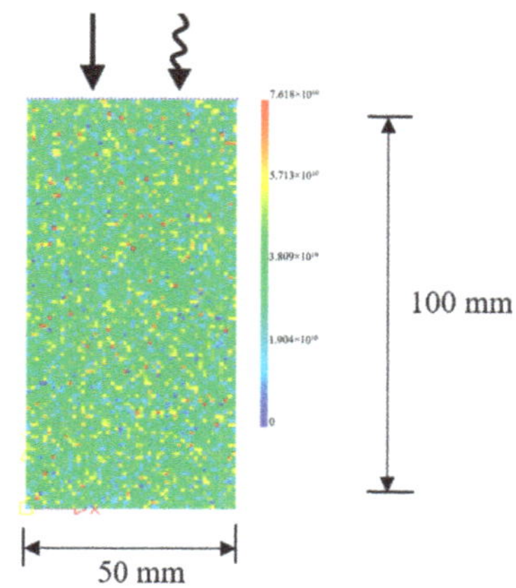

Figure 7. Numerical simulation model.

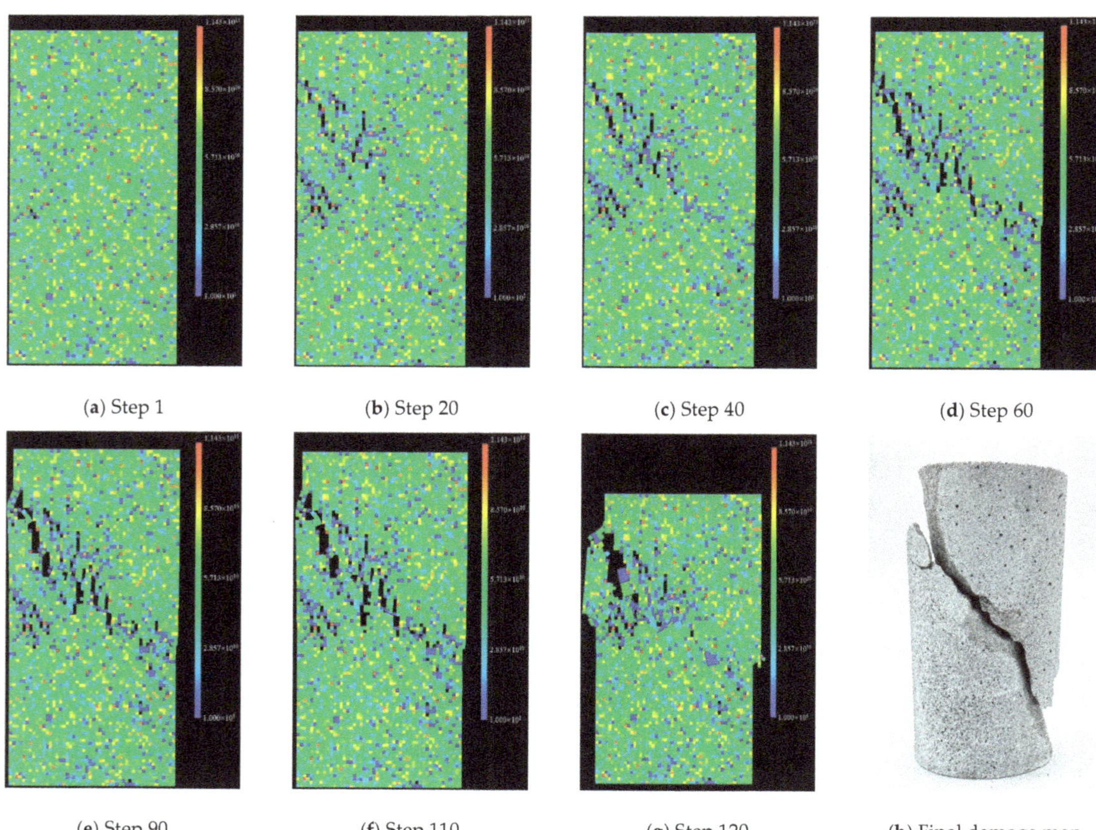

Figure 8. Modulus of elasticity.

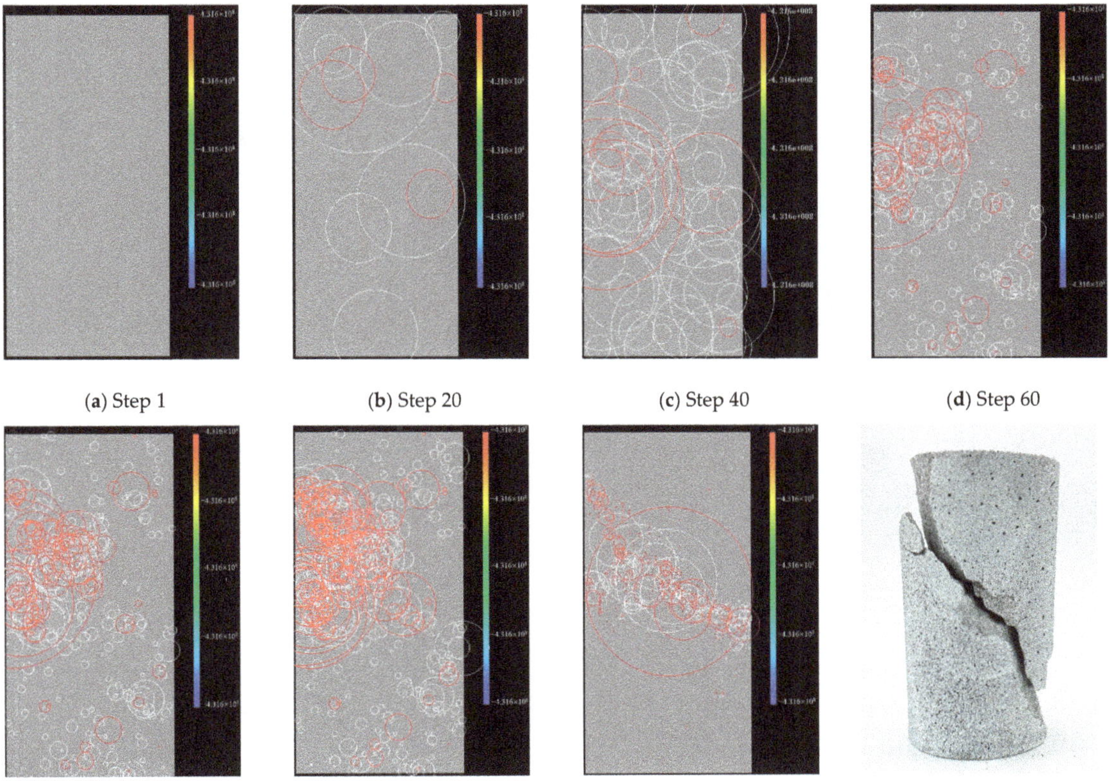

(a) Step 1 (b) Step 20 (c) Step 40 (d) Step 60

(e) Step 90 (f) Step 110 (g) Step 120 (h) Final damage map

Figure 9. Acoustic emission position.

3.4. Elastic Modulus Strengthening Effect at Different Initial Average Stresses

Figure 10 shows the number of cycles of siltstone specimens under different initial average stresses versus axial strain. The relationship between the cumulative axial strain and cyclic number can be divided into two cases: one is that when the initial average stress is small, no significant damage occurs in the siltstone specimen after completing 1000 cyclic disturbances and the other is that when the initial average stress is large, the siltstone specimen is damaged before completing 1000 cyclic disturbances or before the cycle begins. When the initial average stress was increased from 5.48 MPa to 10.05 MPa, the axial strain of the siltstone specimen increased slightly under cyclic disturbance, and then remained basically constant. When the initial average stress was 10.96 MPa, the axial strain increased with the increase in the number of cycles, and destabilization damage occurred at 219 cycles. When the initial average stress was 11.87 MPa, the specimen was damaged before the cycle started. In short, when the initial average stress does not reach a specific value, the specimen would not be damaged even if it is cyclically loaded multiple times, which indicates that there is an upper threshold for the damage of the siltstone specimen during cyclic loading.

Figure 11 shows the relationship between the elastic modulus of the siltstone specimen and the different initial average stresses, where A, B, C, D, E, Fand G represent the initial average stresses of 5.48 MPa, 6.39 MPa, 7.31 MPa, 9.13 MPa, 10.05 MPa, 10.96 MPa and 11.87 MPa, respectively. Taking the specimen with an initial average stress of 5.48 MPa as the reference, when the initial average stress was 6.39 MPa, 7.31 MPa, 9.13 MPa, 10.05 MPa,

10.96 MPa, and 11.87 MPa, the elastic modulus of the specimen increased by 15.5%, 19.6%, 49%, 59.2%, 64.5%, and 72.7%, respectively. The average of the elastic modulus of each group is used for data fitting, which shows that the elastic modulus increases as the initial average stress increases. Under the cyclic disturbance load, the pre-existing cracks in the siltstone specimen are closed, resulting in an increase in the stiffness and elastic modulus of the specimen.

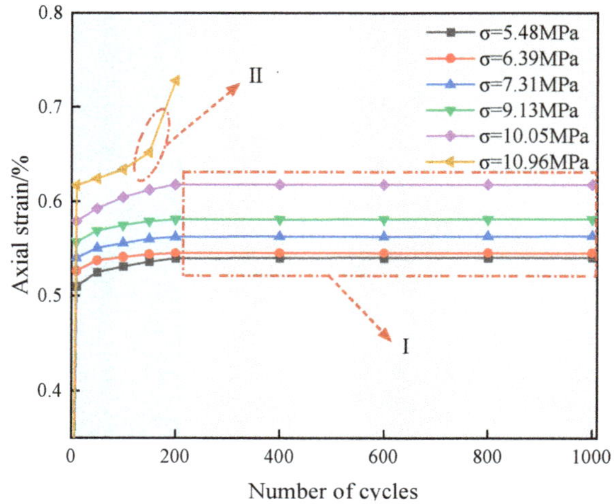

Figure 10. Number of cycles and axial strain curves in siltstone specimens at different initial average stresses.

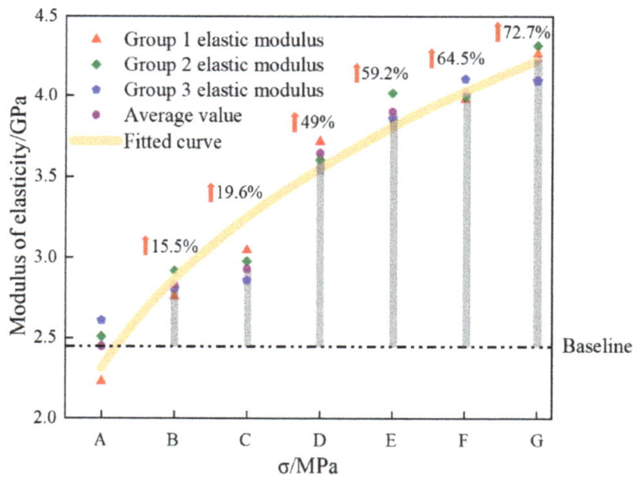

Figure 11. The relationship between siltstone sample and elastic modulus under different average stress.

The percentages corresponding to the different initial average stresses of the siltstone specimens are 30%, 35%, 40%, 50%, 55%, 60%, and 65%, and their corresponding maximum load relationship curves are shown in Figure 12. The testing process of the specimens can be divided into three stages, and the maximum load shows a trend of first decrease, then increase, and finally decrease, and the fitting curves are polynomial.

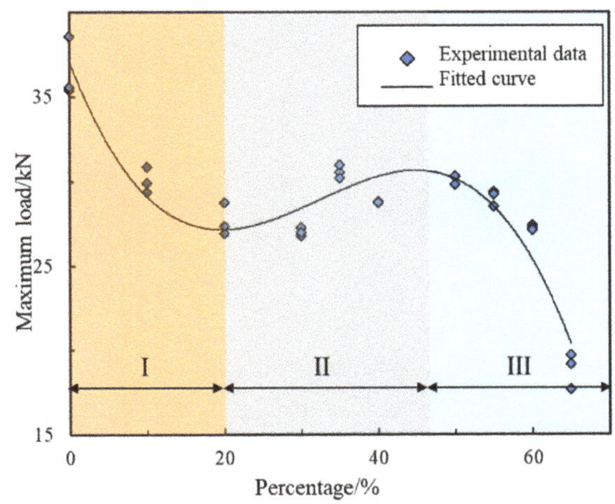

Figure 12. Diagram showing the relationship between different initial average stress percentages and maximum load.

3.5. Characterization of Acoustic Emission Waveforms

The acoustic emission system can monitor the elastic waves during the creation of fractures inside the specimen and the corresponding energy change during cyclic disturbance loading. The key points N1 and N2 are extracted according to the acoustic emission energy release characteristics. N1 is the initial period of specimen dynamic disturbance loading, and N2 is when the damage to the specimen occurs. The acoustic emission characteristics at key points N1 and N2 are extracted and the signals are transformed from the time domain to the frequency domain by using the fast Fourier transform to obtain the principal frequency eigenvalues. MATLAB is used to obtain a three-dimensional plot of amplitude-frequency-time, and then the effect of the dynamic disturbance on the acoustic emission characteristics of the siltstone is analyzed.

Figure 13 shows the three-dimensional plot of the principal frequency characteristics obtained by the Fourier transform of the acoustic emission key points N1 and N2 of the specimens under different stresses, and the sharp points corresponding to the amplitude are the principal frequency eigenvalues. In different sets of cyclic disturbance experiments, the specimen changes from the high principal frequency at key point N1 to the low principal frequency at key point N2, and the principal frequency eigenvalues are reduced to a large extent. The high-frequency eigenvalues of the waves indicate small-scale damage to the specimen, and the low-frequency eigenvalues of the waves indicate large-scale damage to the specimen. The principal frequency eigenvalues of key point N1 ranges from 172 KHz to 203 KHz. The initial average stresses are 10.96 MPa and 10.05 MPa, corresponding to the lowest and highest principal frequency eigenvalues, respectively. The principal frequency eigenvalues of key point N2 ranges from 73 KHz to 98 KHz. The initial average stresses are 9.13 MPa and 6.39 MPa, corresponding to the lowest and highest principal frequency eigenvalues, respectively. The principal frequency values of key points of the siltstone specimens under different average stresses are shown in Table 2.

Table 2. Principal frequency values of key points in siltstone specimens at different initial average stresses.

Key Point	5.48 MPa	6.39 MPa	7.31 MPa	9.13 MPa	10.05 MPa	10.96 MPa	11.87 MPa
N1	201 KHz	185 KHz	181 KHz	179 KHz	203 KHz	172 KHz	178 KHz
N2	87 KHz	98 KHz	85 KHz	73 KHz	82 KHz	79 KHz	87 KHz

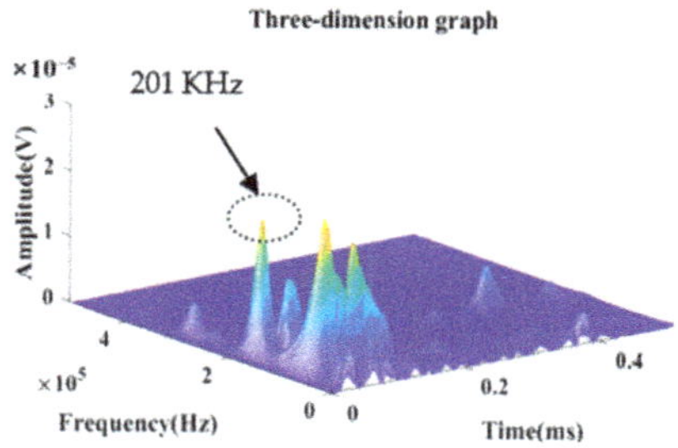

(1) Key point N1

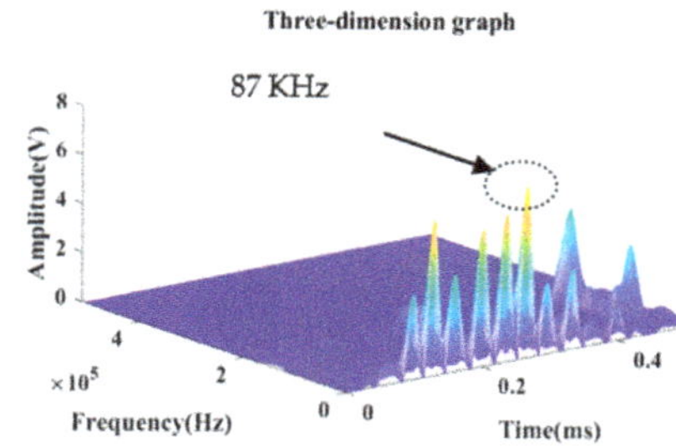

(2) Key point N2

(**a**) 5.48 MPa

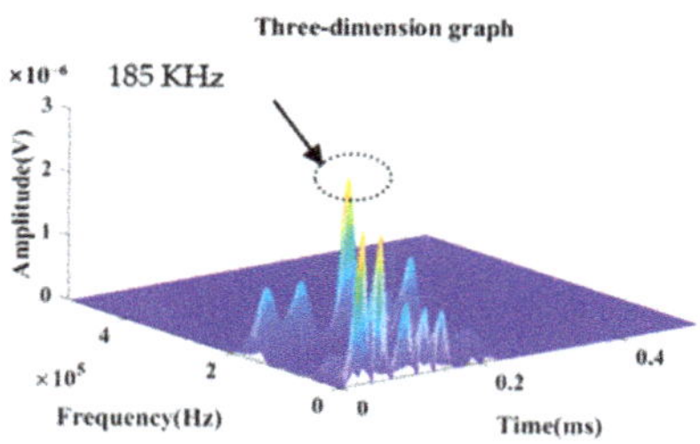

(1) Key point N1

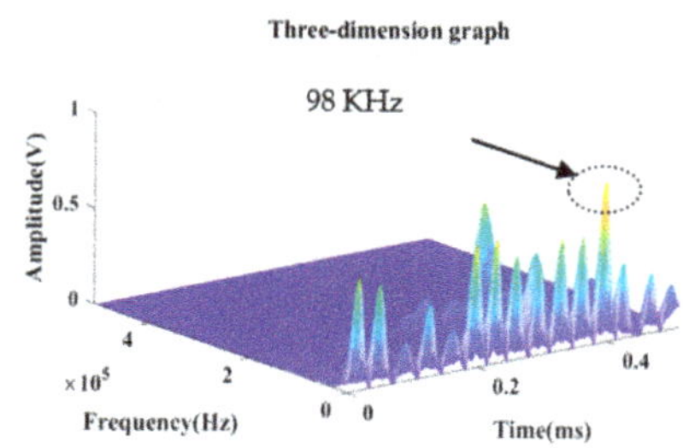

(2) Key point N2

(**b**) 6.39 MPa

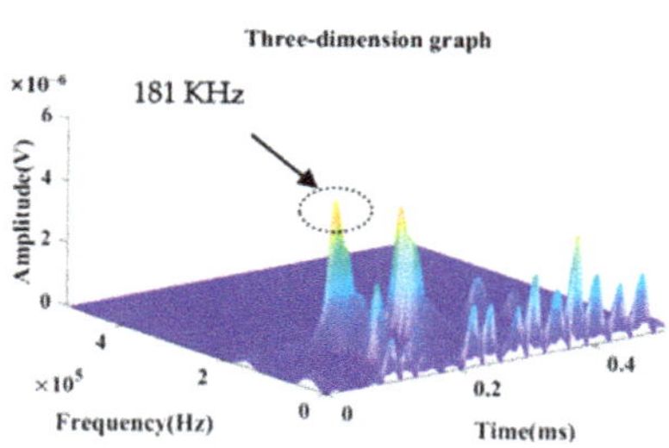

(1) Key point N1

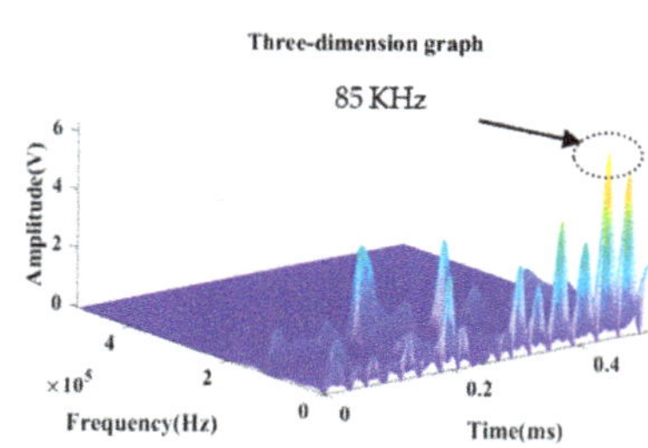

(2) Key point N2

(**c**) 7.31 MPa

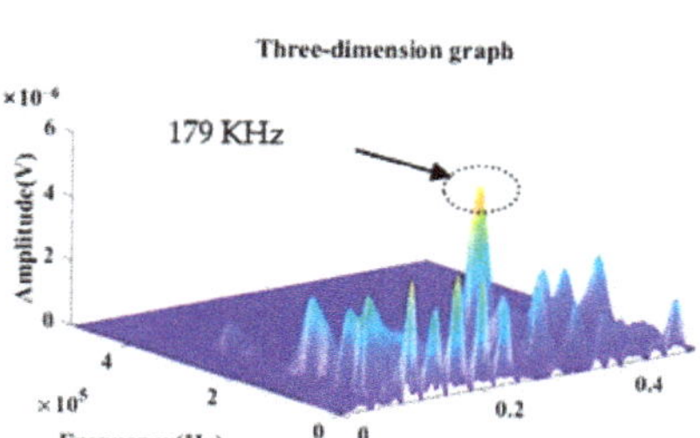

(1) Key point N1

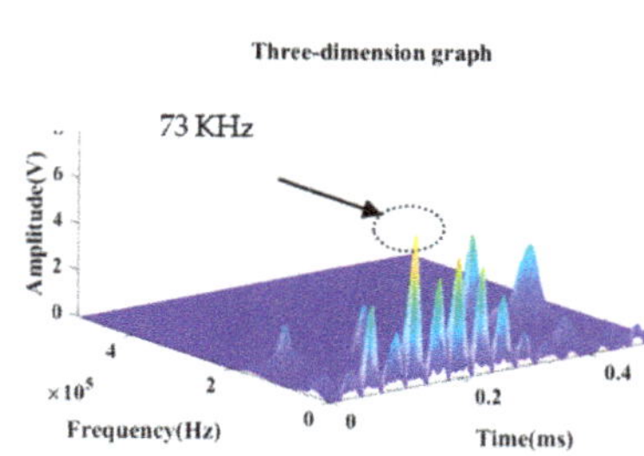

(2) Key point N2

(**d**) 9.13 MPa

Figure 13. *Cont.*

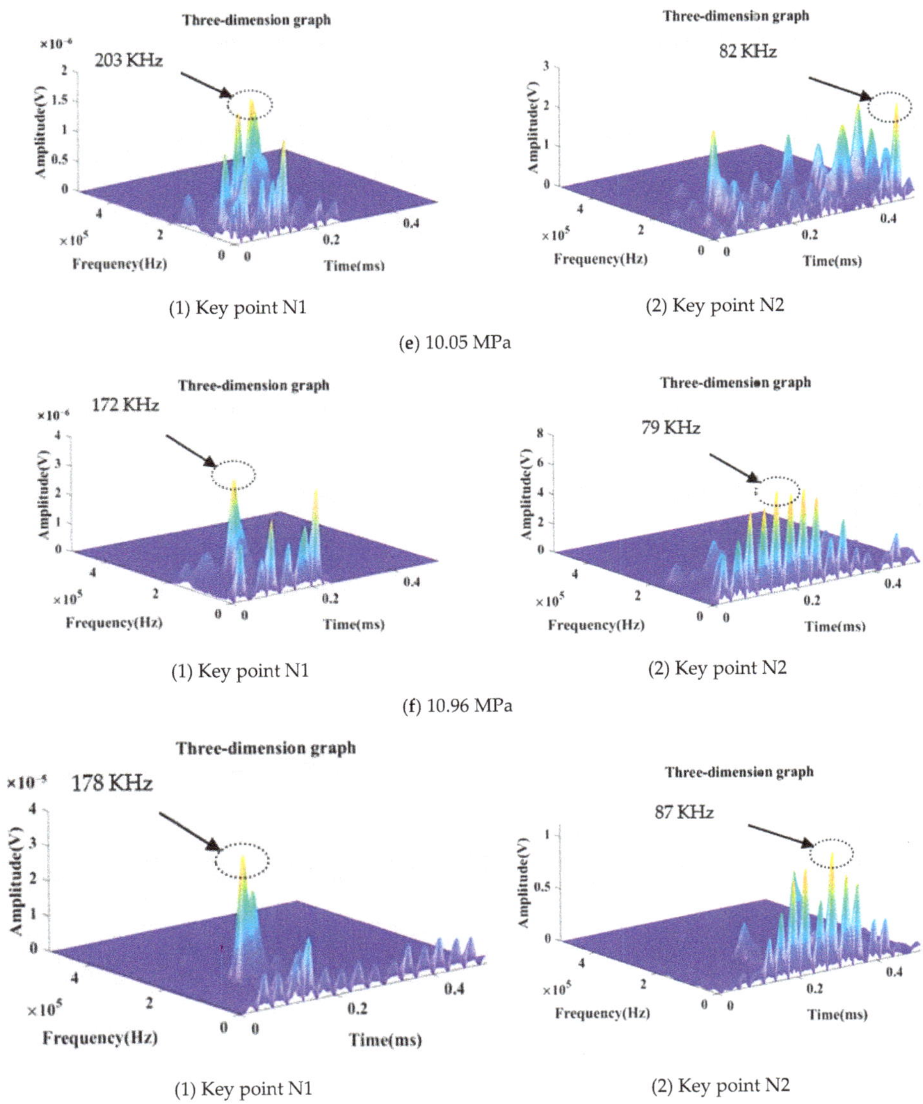

Figure 13. Acoustic emission main frequency characteristics of siltstone samples at key points under different average stresses.

4. Conclusions

(1) The deformation of the siltstone specimens increases as the initial average stress increases, indicating that the damage of the specimen during the cyclic disturbance process also increases. When the axial stress exceeds the threshold of the upper stress limit, the cracks continue to propagate and coalesce, which in turn leads to the destabilization of the specimen.

(2) The magnitude of the initial average stress is closely related to the degree of damage and failure mode of the siltstone. When the initial average stress is low, the failure model is mainly shear failure. With the increase in the initial average stress, the failure mode of the siltstone specimen gradually transitions from shear failure to

shear-splitting mixed failure. The larger the initial average stress, the higher the degree of fragmentation of the specimen. The primary crack is developed from the direction of the maximum principle stress to the bottom of the specimen.

(3) In the RFPA2D cyclic loading simulation, when the dynamic load starts, sporadic acoustic emission events are generated locally in the model. With the increase in axial stress, acoustic emission events become localized and macroscopic fractures are gradually formed, which in turn, leads to the destabilization of the siltstone specimen.

(4) As the initial average stress increases, the modulus of elasticity increases. The reason is that under the dynamic disturbance loading, the pre-existing cracks within the siltstone specimens are closed, which increases the stiffness and elastic modulus of the specimen.

Author Contributions: Conceptualization, Y.J. and D.W.; methodology, L.D. and B.L.; validation, H.L. and G.Z. (Guangchao Zhang); formal analysis, G.Z. (Guanglei Zhou); writing—original draft preparation, L.D. and D.W.; writing—review and editing, D.W. and X.S. All authors have read and agreed to the published version of the manuscript.

Funding: This research was supported by the National Natural Science Foundation of China (No. 52204101), the Natural Science Foundation of Shandong Province (No. ZR2022QE137), and the Open Project of State Key Laboratory for Geomechanics and Deep Underground Engineering in CUMTB (No. SKLGDUEK2023).

Institutional Review Board Statement: Not applicable.

Informed Consent Statement: Not applicable.

Data Availability Statement: The data is available upon reasonable request to the corresponding author.

Acknowledgments: We sincerely thank our study participants.

Conflicts of Interest: The authors declare no conflict of interest.

References

1. China National Bureau of Statistics. *Statistical Bulletin on National Economic and Social Development 2021*; China National Bureau of Statistics: Beijing, China, 2021.
2. NDRC. *The 14th Five-Year Plan for National Economic and Social Development of China and the Outline of Vision for 2035*; People's Press: Beijing, China, 2021.
3. *Progress Report on China's Energy Revolution*; Petroleum Industry Press: Beijing, China, 2021.
4. He, M.; Jing, H.; Sun, X. *Soft Rock Engineering Mechanics*; Science Press: Beijing, China, 2002.
5. He, M.; Sun, X. *Support Design and Construction Guide of Soft Rock Roadway Engineering in Chinese Coal Mines*; Science Press: Beijing, China, 2004.
6. He, M. Progress and challenges of soft rock engineering in depth. *J. Chin. Coal Soc.* **2014**, *39*, 1409–1417.
7. Chen, Y.W. Research on rock explosion phenomenon and exploration of countermeasures in deep shaft mining of Ashele copper mine. *Xinjiang Nonferrous Met.* **2018**, *41*, 80–82.
8. Li, J.; Liu, L.Q.; Miao, G.W.; He, H.X.; Li, M.H. Optimization and application of deep mining roadway support technology in Chen'er gold mine. *Gold* **2018**, *39*, 31–35.
9. Li, X.; Gong, F.; Wang, S.; Li, D.; Tao, M.; Zhou, J.; Huang, L.; Ma, C.; Du, K.; Feng, F. Coupled static-dynamic loading mechanical mechanism and dynamic criterion of rockburst in deep hard rock mines. *Chin. J. Rock Mech. Eng.* **2019**, *38*, 708–723.
10. Yin, T.; Li, X.; Ye, Z.; Gong, F.; Zhou, Z. Energy dissipation of rock fracture under thermo-mechanical coupling and dynamic disturbances. *Chin. J. Rock Mech. Eng.* **2013**, *32*, 1197–1202.
11. Gong, S.; Zhao, Y.; Zhou, L.; Wang, Z.; Guo, W.; Yao, S.; Yang, D. Study on dynamic fracture behavior of limestone specimens containing double holes and crack under impact loading. *J. China Coal Soc.* **2022**, 1–19. [CrossRef]
12. Yu, Y.; Liu, F.; Yue, H.L.; Zhang, W.; Zhang, S.P.; Wang, P.B. Mechanical properties of discontinuity in rock mass with different inclination angles under cyclic dynamic disturbance. *J. China Coal Soc.* **2020**, *45*, 3748–3758.
13. Li, C.; Zhang, Y.; Zhang, G.; Gao, S.; Wang, H. Crack propagation mechanisms and stress evolution of floor under dynamic disturbance in deep coal mining. *Chin. J. Geotech. Eng.* **2018**, *40*, 2031–2040.
14. Gong, F.; Zhang, L.; Li, X.; Luo, Y. Experimental study on fracture behaviors of hard rock under dynamic disturbance with different pre-static loads. *Chin. J. Rock Mech. Eng.* **2017**, *36*, 1841–1854.
15. Tang, L.; Cheng, L.; Wang, C.; Shu, J.; Wu, J.; Chen, Y. Dynamic characteristics of serpentinite under condition of high static load and frequent dynamic disturbance. *Geotech. Rock Soil Mech.* **2016**, *37*, 2737–2745.

16. Wasantha Lal, A.M. Determination of multiple aquifer parameters using generated water level disturbances. *Water Resour. Res.* **2006**, *42*, 3429.
17. Feng, F.; Chen, S.; Zhao, X.; Li, D.; Wang, X.; Cui, J. Effects of external dynamic disturbances and structural plane on rock fracturing around deep underground cavern. *Int. J. Coal Sci. Technol.* **2022**, *9*, 15. [CrossRef]
18. Willis, J.R.; Movchan, N.V. Second-order in-plane dynamic perturbation of a crack propagating under shear loading. *Math. Mech. Solids.* **2014**, *19*, 82–92. [CrossRef]
19. Gómez Gonzales Giancarlo, L.; Ortiz González Julián, A.; Antunes Fernando, V.; Neto Diogo, M.; Díaz Francisco, A. Experimental determination of the reversed plastic zone size around fatigue crack using digital image correlation. *Theor. Appl. Fract. Mech.* **2023**, *125*, 103901. [CrossRef]
20. Brackmann, L.; Röttger, A.; Treppmann, M.; Weber, S. The behavior of cutting discs for mechanized tunneling under cyclic loading conditions. *Tunn. Undergr. Space Technol.* **2023**, *137*, 105151. [CrossRef]
21. Liu, Z.Y.; Gan, D.E.Q.; Yu, Z.; Tian, X.; Wang, L. Experimental research on the dynamic mechanical properties and breakage behavior of magnetite under one-dimensional coupled dynamic and static loads. *Chin. J. Rock Mech. Eng.* **2022**, *41*, 2869–2880.
22. Chen, N.; Li, P.; Han, H.; Zeng, Z.; Qiu, S. Acoustic emission characteristics of deformation damage in soft rock with lamellar structure. *Eng. J. Wuhan Univ.* **2022**, *55*, 539–544.
23. Chen, Y.; Gao, T.; Gao, C.; Yin, Y.; Li, Y. Fracture propagation and acoustic emission energy damage law of fractured rock mass under hydraulic coupling action. *J. Cent. South Univ. (Sci. Technol.)* **2022**, *53*, 2325–2335.
24. Zhang, D.; Guo, W.; Zhao, T.; Gu, X.; Chen, Y. Experimental study on directional propagation of rock type-I crack. *Geotechnics* **2022**, *43*, 231–244.
25. Liu, J.; Li, J.; Liang, P.; Li, Z. Experimental study on characteristics of acoustic emission S-wave and P-wave of granite under shear failure process. *Chin. J. Rock Mech. Eng.* **2023**, *42*, 429–440.
26. Chen, L.; Guo, W.; Zhang, D.; Zhao, T. Experimental study on the influence of prefabricated fissure size on the directional propagation law of rock type-I crack. *Int. J. Rock Mech. Min. Sci.* **2022**, *160*, 105274. [CrossRef]
27. Song, Y.; Cheng, K.; Meng, F. Research on acoustic emission characteristics of fractured rock damage under freeze-thaw action. *J. Min. Saf. Eng.* **2023**, *40*, 408–419.
28. Zhao, Y.; Ran, H.; Feng, G.; Guo, Y.; Fan, Y. Study on damage evolution and damage characteristics of gangue colluvial fillings with different aspect ratios under uniaxial compression. *J. Min. Saf. Eng.* **2022**, *39*, 674–688.
29. Wang, Z.; Ma, K.; Tian, H.; Li, Q. Study on the element volume of permeability characteristics in fractured rock mass based on RFPA2D-flow. *J. China Coal Soc.* **2019**, *44*, 3012–3021.
30. Chen, J.; Liu, B.; Zhu, C.; Pu, Y.; Gao, J.; Cui, Y.; Zhang, C. Early-warning evaluation and warning of rock burst using acoustic emission characteristics of coal sample failure. *Coal Sci. Technol.* **2023**, *51*, 116–129.
31. Qasim, A.H.; Aryan, K.; Cagdas, A.; Adnan, K.; Mehmet, Y. Numerical and experimental methodologies to investigate the damage progression inside the axisymmetric composite cylinders with cutouts under torsion. *Compos. Struct.* **2023**, *315*, 116990.
32. Andronikos, L.; Ilias, S.; Dimos, T. Non-extensive statistical mechanics in acoustic emissions: Detection of upcoming fracture in rock materials. *Appl. Sci.* **2023**, *13*, 3249.
33. Shcherbakov, I.P.; Makhmudov Kh, F.; Chmel', A.E. Trigger effect in impact fracture of granite sample under uniaxial compression. *J. Min. Sci.* **2022**, *58*, 930–935. [CrossRef]
34. Barile, C.; Pappalettera, G.; Paramsamy, K.V.; Casavola, C. A neural network framework for validating information–theoretics parameters in the applications of acoustic emission technique for mechanical characterization of materials. *Materials* **2022**, *16*, 300. [CrossRef]
35. Skalskyi, V.R.; Makeiev, V.F.; Stankevych, O.M.; Huniovskyi Ya, R.; Huniovska, R.P.; Kyrmanov, O.S. Evaluation of the fracture resistance of removable denture base materials. *Mater. Sci.* **2022**, *58*, 229–236. [CrossRef]
36. Loukidis, A.; Tzagkarakis, D.; Kyriazopoulos, A.; Stavrakas, I.; Triantis, D. Correlation of acoustic emissions with electrical signals in the vicinity of fracture in cement mortars subjected to uniaxial compressive loading. *Appl. Sci.* **2022**, *13*, 365. [CrossRef]
37. Grabi, M.; Chellil, A.; Habibi, M.; Laperriere, L.; Grabi, H. Strain and damage assessment of treated and untreated luffa mat composite using acoustic emission and digital image correlation. *J. Nat. Fibers* **2022**, *19*, 12536–12547. [CrossRef]

applied sciences

MDPI

Article

Numerical Study on the Dynamic Propagation Model of Cracks from Different Angles under the Effect of Circular Hole Explosion

Junwei Zhang, Duanying Wan, Weiting Gao *, Lei Zhou and Meng Wang

Key Laboratory of Deep Underground Science and Engineering, Ministry of Education, College of Architecture and Environment, Sichuan University, Chengdu 610065, China; zjw@stu.scu.edu.cn (J.Z.); wanduanying1995@163.com (D.W.); zhouleittkx@126.com (L.Z.); wangmeng@scu.edu.cn (M.W.)

* Correspondence: gaoweitingscu@126.com

Abstract: A dynamic disturbance will induce cracks around the tunnel in tunnel blasting or shield construction. To investigate the overall stability of cracks with various angles during a fixed borehole (round hole explosion) blasting, models containing an individual crack with different angles were introduced for simulation research. The research set up a thin sheet model with a length of 350 mm and a width of 150 mm, with a 7 mm diameter hole and a pre-existing crack of 75 mm and 5 mm in the middle. The evolution of the stress wave propagation model and the crack propagation model were simulated using the AUTODYN software. And in this study, the theory of stress wave is used to creatively explain the dynamic load under the action of formation and reasons for the danger zone. The results indicate that pre-existing cracks from different angles will have an impact on the blast hole and the new cracks generated around itself. At 45–90°, pre-existing cracks will direct reflected stress waves to promote some cracks around the hole to have faster growth rates than others, and these special cracks with faster growths and longer lengths will more easily connect with the free surface or other cracks, resulting in overall instability. And these conditions are consistent with the prediction made by the stress wave propagation simulation study. The research results have certain guiding significance for the stability analysis and hazardous area prediction of tunnel blasting with existing cracks.

Keywords: P-wave; S-wave; Rayleigh wave; pre-cracks; crack propagation; dangerous area prediction

Citation: Zhang, J.; Wan, D.; Gao, W.; Zhou, L.; Wang, M. Numerical Study on the Dynamic Propagation Model of Cracks from Different Angles under the Effect of Circular Hole Explosion. *Appl. Sci.* **2023**, *13*, 7955. https://doi.org/10.3390/app13137955

Academic Editor: Tiago Miranda

Received: 13 June 2023
Revised: 3 July 2023
Accepted: 5 July 2023
Published: 7 July 2023

1. Introduction

With the increasing demand for mineral resources and the booming development of infrastructure projects for high-speed and convenient transportation, excavation projects such as tunnels [1], mineral mining [2], and pipelines [3] have become increasingly frequent. Such construction projects usually occur in complex rock environments, where large rock masses are the main blasting objects for easy excavation and mining [4]. In a wide range of construction projects, the drilling and blasting method has become the main construction method to effectively control post-blasting disasters and provide effective guarantees for construction safety and quality. However, due to geological activities and induced excavation, the presence of joint planes [5] and cracks in large rock masses is common [6], as shown in Figure 1. In the Figure, the red arrows represent the load generated by the impact of blasting, the black lines represent the naturally occurring joints and cracks in the rock structure, and the white round holes represent the gun holes. These pre-existing cracks have a significant impact on the stability of the overall structure. In addition, under external interference such as blasting, new cracks occur and expand around the blast holes and pre-existing cracks easily, ultimately leading to overall structural instability, seriously delaying work and production progress, threatening the safety of relevant personnel, and causing incalculable economic losses. Therefore, studying the impact of cracks in rocks on blasting has a profound significance on safe construction.

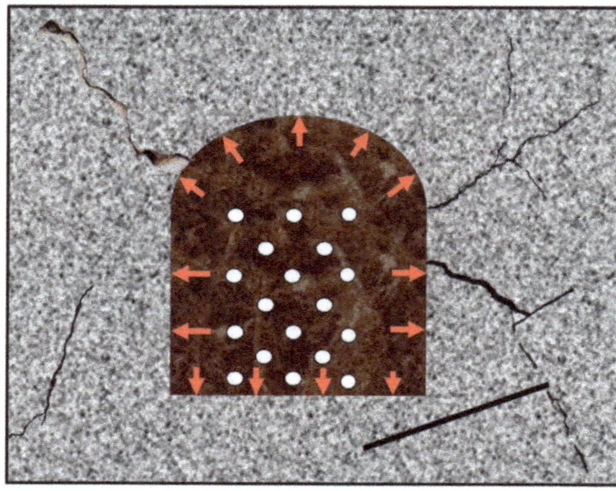

Figure 1. Schematic diagram of rock mass blasting with cracks.

In the past, most studies on excavation or mining have adopted the assumption of rock models without cracks. The advantage of this approach is that fewer factors affect the generation and propagation of cracks, making it easier to simulate. However, in recent years, more and more scholars have conducted research on this type of model with cracks. Suorineni et al. [7] stated that the presence of faults in rocks is an important factor that leads to overall instability. Freund [8–10] systematically studied the crack growth behavior under various dynamic loads and obtained some analytical results and conclusions, which confirmed that when subjected to blasting load, the cracks around the blast hole are more likely to generate new growth at the tip. However, because this process is affected by a complex environment, it is difficult to obtain a sufficiently accurate analytical expression. Moreover, under high-speed dynamic loads such as explosions, the interaction between the crack's tip and the dynamic stress field causes fluctuations in the crack propagation speed at the crack's tip [11]. Therefore, in response to the reasons that affect the initiation of tip cracks, some scholars [12] have studied the process of stress wave propagation and found that when the stress wave propagates to a static crack with two tips, it will scatter near the perturbed tip and reflect on the crack surface, forming a superimposed stress field, which is usually related to the incident angle. Additionally, based on such a model, Wheeler gives some analytical expressions for this field by using integral transform [13]. Regarding rock material models with cracks, some experimental explosion simulation studies [14] or theoretical summaries of rock materials [15] are based on the existence of a large number of natural structural joints or faults in the rock itself. Based on an experimental analysis and numerical simulation, Zhu [16] completed a study on the crack diffraction process using P-wave as an example and determined the influence of joint space and other factors on wave propagation. Uenishi [17] discussed the slope failure under the action of the Rayleigh wave, and the stress wave theory was adopted to analyze the model. Sun and Zhang used finite element models to study the effect of stress wave pairs on crack angles in the cited reference [18]. Zhou et al. [19] specifically studied the crack propagation mechanism of rock as an example material under different loading rates. Zhang et al. [20] specifically focused their research on actual coal mining and revealed the damage development process of this material using acoustic methods based on the spatial aggregation characteristics of cracks. They achieved good prediction and experimental fitting results. Zhou et al. [21] used a software simulation combined with physical experiments and found that compared to other fractured rock masses, the defect roadway with a single root crack and an inclination angle of 45° has the lowest static and dynamic stability. Li [22] also established a fractal damage joint model based on the fractal damage theory, studied the transmission and reflection of stress waves between joints, and derived the analytical expression of the transmission and reflection coefficient of stress waves across joints from the fractal damage joint model. In Lak's study [23], a general Green's function solution of elastic wave propagation due to rock blasting was analytically derived. Navier's equations of motion were assumed as governing equations, and a general two-dimensional elastodynamic Green's function was obtained in terms of displacement. The strain and stress fields related to the displacement of Green's function were also obtained using the theory of elasticity. In 2021, Zhu et al. [24] conducted a numerical study on the dynamic behavior of stress waves in cracks, indicating that Rayleigh waves have a significant impact on the propagation and variation of cracks in all directions.

The above research is based on practical engineering experiments; due to the short overall time, it is difficult to analyze the propagation of waves after explosions based on the stress wave theory, and the influencing factors cannot be explored by changing a single variable in the same situation. However, based on numerical model calculation research, although models containing existing cracks were set up for research, there is no in-depth and regular research on the different states of crack existence. The field to which this type of problem belongs is almost blank and quite valuable because the brittle characteristics of rock materials can cause the original cracks to rapidly expand and connect under high loading rates such as explosive loads, and even generate huge cracks. The repeated loading of conventional explosive stress waves will cause cracks to continuously expand in a very short period, even leading to hazards such as rock bursts, slope instability, and tunnel collapse. In recent years, some studies have found that interconnected cracks have a significant impact on the dynamic response of deep engineering projects such as tunnels under blasting stress waves. The stress distribution around tunnels changes significantly due to the presence of cracks, leading to the initiation and propagation of new cracks. More importantly, one study found that the dynamic crack propagation behavior is influenced by the direction of blasting stress wave incidence [25].

The main objective of this study is to confirm the feasibility of the stress wave theory for predicting the location of the dangerous area around the gun hole under dynamic load. In this paper, the dynamic propagation behavior of new cracks and existing cracks in a single crack model with different angles around the blasting hole is studied. This study focuses on stress waves as the main research object, but it is evident that the network structure formed by cracks has a significant guiding effect on the propagation of stress waves. Therefore, to eliminate the influence of fragmented regions on the propagation direction, in simulation, the model at the same angle is simulated twice, with the same peak load applied. Regarding the specific methods of this simulation, the experimental setup is divided into two groups, which, respectively, study the propagation of stress waves and crack fragmentation. The model adopts an infinite plane and refers to the most widely used cylindrical explosives. As stress waves play a major role in the propagation process, they can be used to replace the waves generated by explosions [26]. Relevant research has shown that multiple identical cylindrical explosives can be simplified into a single blast hole [27]. The material point method is firstly applied to treat this challenging task by Wan, et al. [28] In this way, some inherent weaknesses can be overcome by coupling the generalized interpolation material point (GIMP) and the convected particle domain interpolation technique (CPDI). The two-dimensional properties of the stress waves generated can be directly calculated using specific functions [13]. Shu et al [29] present a novel constitutive model with three smooth failure strength surfaces and three invariants. The new model can precisely and quantitatively capture the damage mechanisms of hydrostatic pressure based on test data and the interaction of the tensile and shear damages based on the Mohr-Coulomb criterion.

Therefore, in view of the research gaps of many scholars at the present stage, this paper, on the basis of a large number of numerical simulations, takes the angle as the influencing factor of blasting instability and studies the influence of the existing angle on the crack growth around the hole, and creatively uses the stress wave theory to effectively predict the dangerous area of the model. This study uses a numerical simulation to investigate the influence of the angle on the dynamic crack propagation behavior under stress wave action. The project uses AUTODYN's list of numerical analysis software, uses a triangle stress loading function to simulate blasting, establishes stress wave fields with cracks in different directions, and conducts detailed research on the dynamic propagation behavior of cracks and new cracks around the blast hole. Finally, in this numerical simulation, the stress wave theory is successfully used to analyze and predict the dangerous area after failure, proving that cracks ranging from 45° to 90° will induce characteristic cracks, which will accelerate the instability of the model.

2. Numerical Study

2.1. Determination of Sandstone Parameters

In this study, the numerical simulation is carried out based on the finite difference method, and the research object is divided into different sub-elements after grid division. The acceleration in the X and Y directions can be expressed as follows.

$$a_x = \frac{F_x}{m} \tag{1}$$

$$a_y = \frac{F_y}{m} \tag{2}$$

where F_x and F_y are the nodal forces in the x and y directions, respectively, and m is the mass.

The stress component can be expressed in the following form:

$$\sigma_i = P_i + S_i,\ (i = x, y) \tag{3}$$

$$\tau_{xy} = S_{xy} \tag{4}$$

where P is the external pressure, S_i is the deviatoric stress in the i direction, and there is a calculation relationship as follows

$$S_i^{n+1} = S_i^n + 2G\left(\dot{\varepsilon}_i - \frac{1}{3}\dot{e}\right)^{n+\frac{1}{2}} \Delta t,\ (i = x, y) \tag{5}$$

$$S_{xy}^{n+1} = S_{xy}^n + 2G\dot{\varepsilon}_{xy}^{n+\frac{1}{2}} \Delta t \tag{6}$$

$$\dot{e} = \dot{\varepsilon}_x + \dot{\varepsilon}_y \tag{7}$$

Here, G is the shear modulus of the material and $\dot{e}$ is the volume rate of the material.

For materials, the relationship between the pressure P and density described in the linear equation of state is

$$P = k \cdot (\frac{\rho}{\rho_0} - 1) \tag{8}$$

where P is pressure, k is the bulk modulus, k = 4.04 GPa, and ρ and ρ_0 are the initial and present densities. Here, as sandstone is a brittle rock material, the relationship between stress and strain during deformation can be studied using an elastic strength model.

At the same time, the experimental research is based on to the first strength theory and the third strength theory. If either the principal stress σ_1 reaches its maximum (σ), or the maximum shear stress τ_{max} is greater than the rock dynamic shear strength (τ), the material fails.

Here, a sound wave velocimeter is used to measure the dynamic material parameters of sandstone, as shown in Figure 2. It has two piezoelectric ceramic terminals, the transmitting end can emit waveform signals, and the receiving end can receive signals. According to the geometric configuration of the S-wave and P-wave, the relevant parameters of the material can be calculated by the time when the P-wave and S-wave pass through it.

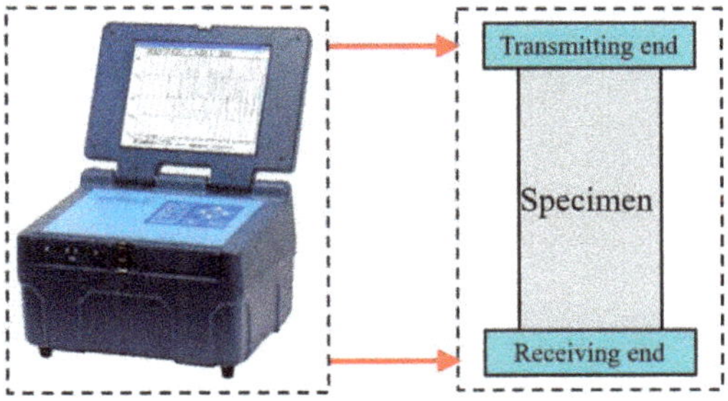

Figure 2. An acoustic velocimeter (Sonic Viewer-SX).

In this paper, the dynamic elastic modulus and dynamic Poisson's ratio of the rock materials are calculated by testing the expansion wave velocity V_p and shear wave velocity V_s of the material, as shown in Table 1. The calculation formula is shown in Equations (9) and (10). This paper adopts a rock sample ultrasonic tester to obtain the measurements of V_p and V_s, and the working principle of the instrument is shown in Figure 2, which consists of a high-voltage pulse transmitter, a receiver, and a converter. The test of the two wave speeds of the material is obtained in accordance with the provisions of the International Society of Rock Mechanics for 100 mm × 50 mm cylindrical specimens, and a total of five tests are prepared to obtain their average values.

Table 1. Parameters of sandstone used in this study.

Material	Density Kg/m^3	Bulk Modulus GPa	Elastic Modulus GPa	Poisson Ratio	V_s m/s	V_p m/s
Sandstone	2380	9.4624	17.60	0.19	2549.4	1459.2

If the velocities V_p and V_s of the P-wave and S-wave are taken into the following equation, the elastic modulus and Poisson's ratio of the material can be acquired.

$$V_p = \sqrt{\frac{E_d(1-\mu_d)}{(1+\mu_d)(1-2\mu_d)\rho}} \tag{9}$$

$$V_s = \sqrt{\frac{E_d}{2(1+\mu_d)\rho}} \tag{10}$$

$$E = \frac{\rho {V_p}^2(3{V_s}^2 - 4{V_p}^2)}{{V_s}^2 - {V_p}^2} \tag{11}$$

$$v = \frac{{V_s}^2 - 2{V_p}^2}{2{V_s}^2 - 2{V_p}^2} \tag{12}$$

where ρ is the density of sandstone, E is the dynamic elastic modulus, and μ is the Poisson ratio.

2.2. Establishment of Numerical Model

The numerical model is shown in Figure 3, which is a rectangular board with a length of 350 mm and a height of 150 mm. We set a hole with a radius of 7 mm to simulate a blast hole and generate cylindrical waves. The distance from the center of the hole to the left boundary is 55 mm and is the same from the center to the top edge. On the right side of the hole, there is a crack with an angle of θ with a length of L.

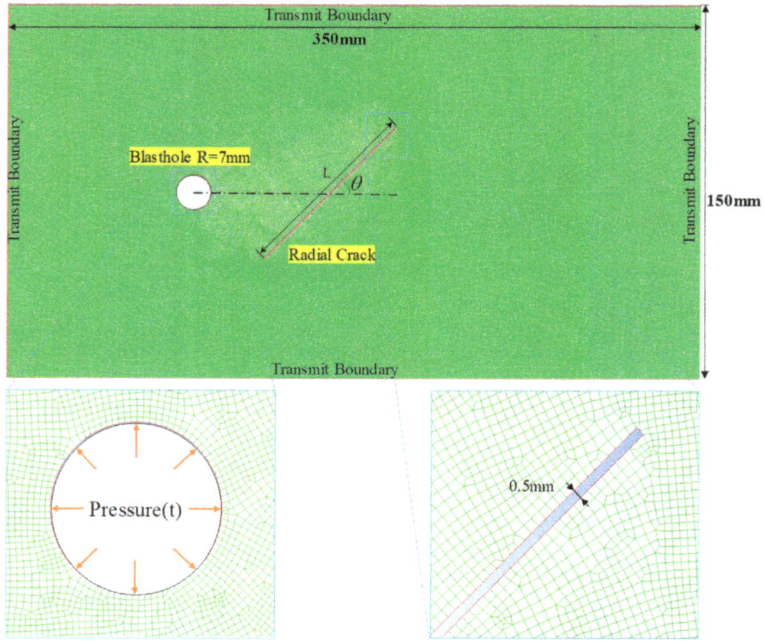

Figure 3. Numerical model.

Considering that the object of this study is deep rock, it can be approximated as an infinite rock plane, but the modeling scope is limited. So, when establishing the model, the boundary is set to transmit boundary, so that the stress waves generated by the blasthole at the boundary will not have any effect on the reflected waves in the subsequent process, greatly reducing the impact of tensile and compressive waves in the reflected P-wave on the model research. This can transform the research object from a finite model to an infinite region.

Taking the model with θ = 45° and L = 75 mm as the first exploratory research example, the velocity vector excited by stress waves was collected as shown in Figure 4. We set the loading condition to a peak of 50 Mpa with a loading time of 2 μs. The advantage of designing such a shorter wavelength is that it creates a good separation between different waves, making it easier to study the propagation and characteristics of different waves. By observing the images, it was found that both the P-wave and S-wave of the stress wave diffracted near the crack, especially when observed in the vector plots. And the P-wave and S-wave diffracted by excitation have different effects on the side of the crack near the blast hole and the side away from the blast hole. On the side that is the farther from the explosion hole, due to diffraction, the conventional diffraction P-wave and S-wave are generated, and the conventional lower Rayleigh wave propagates along the direction of the crack. On the side near the blast hole, due to the tensile wave inside the P-wave, its reflected wave propagates together with the diffracted P-wave, forming the first P-arc wave with stress concentration. The diffraction and reflection of the S-waves are mainly composed of shear waves that combine to form a second S-arc wave with stress concentration. It is worth mentioning that the two newly generated arc-shaped waves have the same properties as the diffraction wave at the tip that generates them, namely, the P-arc-shaped wave is also a tensile wave, and the S-arc-shaped wave is also a shear wave. However, the reflection wave still plays the main role in the two types of arc waves, so in future research on crack initiation, only the reflected wave can be considered as the main factor.

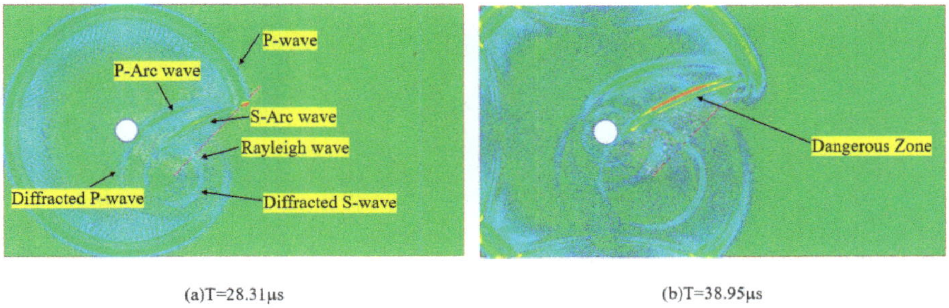

Figure 4. Stress-wave-induced particle velocity vectors.

After the above operations, in this study, the distance from the crack's center to the center of the circular hole is called the center distance, and the fixed center distance is 55 mm. The crack is rotated to change the angle for multiple sets of numerical model research, as shown in Figure 5.

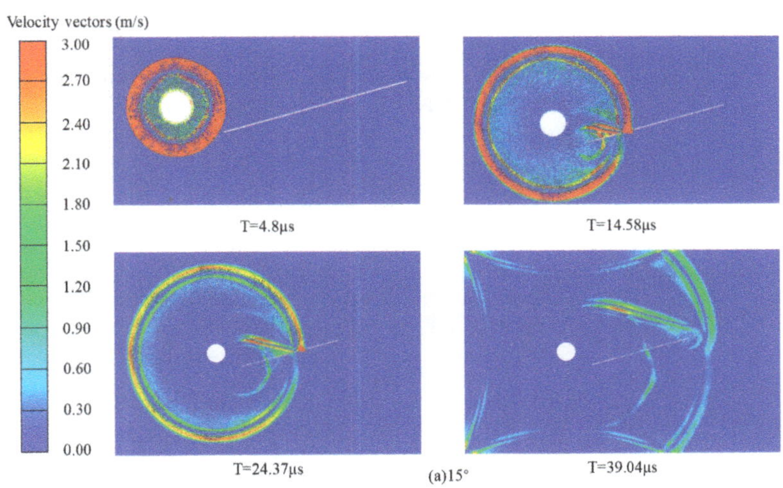

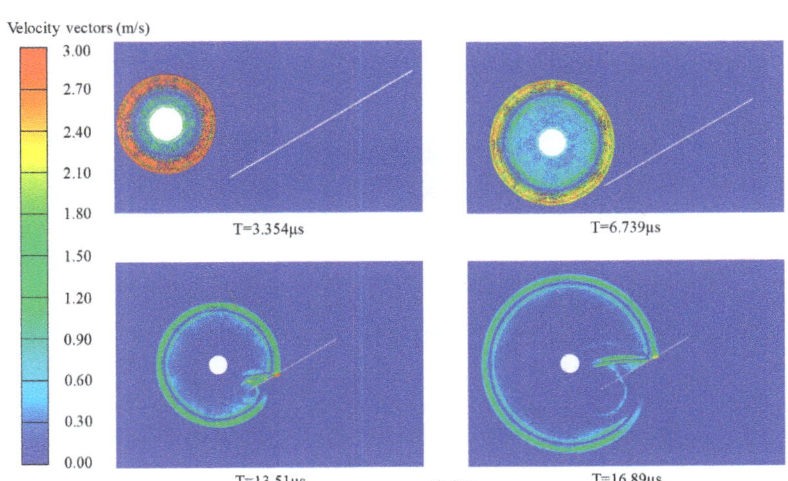

Figure 5. *Cont.*

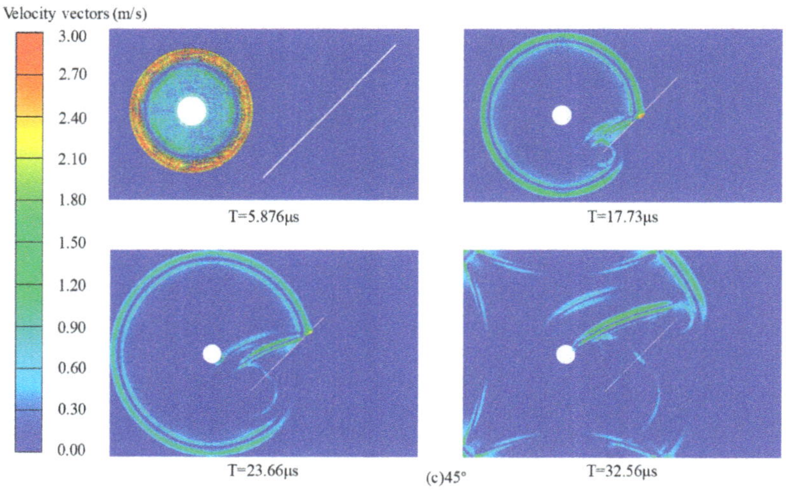

Figure 5. Reflection and propagation of stress waves at different angles.

As the angle increases, it can be observed that the Rayleigh waves on both sides of the crack, which play a dominant role in horizontal cracks, no longer become the main factor affecting the crack. At the same time, the arc wave formed by the combination of reflected and diffracted waves on the side of the crack passing through the blast hole becomes a high-stress concentration area. As the crack angle increases, the effect of the arc wave on the generation of the crack is enhanced.

For this phenomenon, we compare the contours of MIS.STRESS at the exact moment when the P-wave propagates to the far end of four different inclination cracks, as shown in Figure 6. It can be observed that in the first set of simulations at 0–30°, the main stress generated by the arc wave is significantly enhanced, but in the second set of simulations at 45–75°, the intensity of this effect decreases significantly.

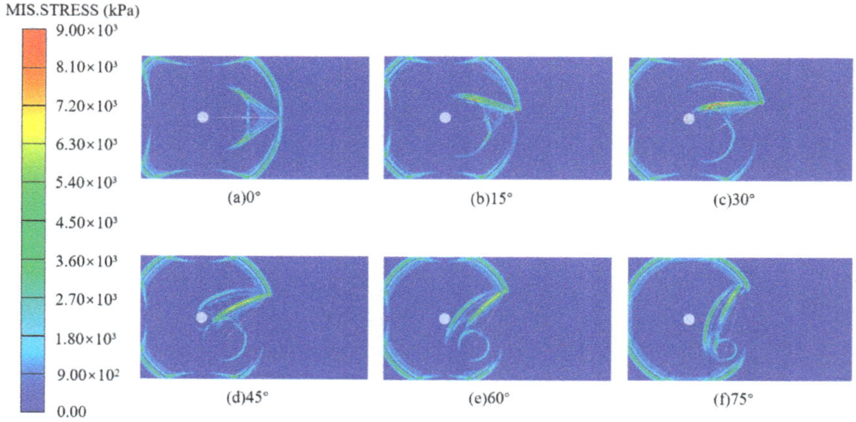

Figure 6. Contours of equivalent principal stress induced by stress waves.

For this phenomenon, the research subjects were divided into two groups for in-depth exploration. For the first group from 0° to 45°, as shown in Figure 6, as the crack inclination angle increases, the reflection wave of the composite-generated arc wave is significantly enhanced, resulting in a high-stress region. For the second set of simulations from 60° to 75°, after studying the velocity vector induced by stress waves at 60°, it was found that due to the large inclination angle, the reflected arc-shaped waves were reflected again at the borehole, reducing the effect of the arc-shaped waves located at the back, and ultimately directly reflecting the moderate effect of the main stress reduction in Figure 6. The schematic diagram of stress waves is shown in Figure 7.

Figure 7. Schematic diagram of stress wave propagation attenuation under inclination angle in the second simulation.

At the same time, by comparing the distance between the stress concentration area in the two stages and the near end of the crack when the P-wave just reaches the far end, it can be found that the distance simulated in the second group is greater than that in the first group. At the same time, because of the same loading conditions, compared with the first group, there is a longer near end distance, and the initial amplitude of the diffraction wave also decreases, which means that in the second group, the far end and the near end may not be easy to crack.

3. Dynamic Response Behavior of Cracks and Explosive Holes under Dynamic Impact

For the sandstone material used in this experiment, considering the occurrence and propagation of cracks in the material under conventional conditions, the maximum principal stress failure criterion is used to evaluate the material unit state, as shown below:

$$\sigma_1 < \sigma_T \tag{13}$$

σ_1 is the major principal stress in every element, and σ_T is the tensile strength of the sandstone material, with its value being 75 MPa. When σ_1 exceeds σ_T for an element, it will fail and cannot stand any tensile stress but can continue to be compressed.

For the dynamic load loading caused by a single blasting, a triangular curve function is used, with a total loading time of 20 μs. At 10 μs, the peak load reaches 50 MPa, as shown in Figure 8. The advantage of this is that it simplifies the calculation, focuses on the main role of the explosion load, and eliminates the influence of unimportant factors on the set conditions.

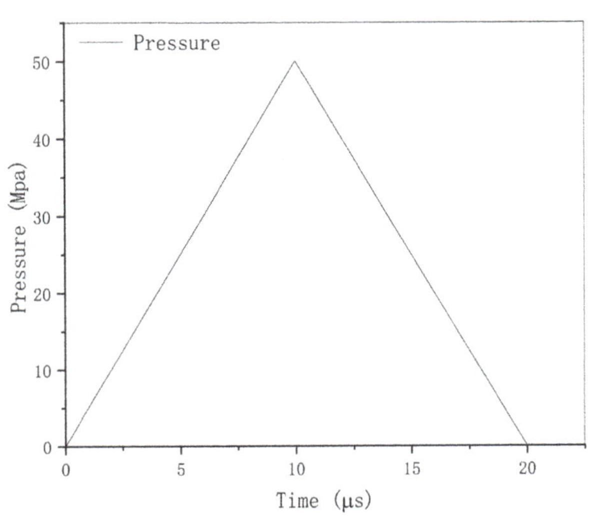

Figure 8. Pressure–time curve.

3.1. Packet Simulation

The loading curve of the simulation experiment is shown in Figure 9. The center distance of the fixed model is 35 mm, and the included angle with the horizontal plane starts from 0°. One study sample is taken every 15° until 90°. There are seven models in total, which are used to explore the impact of multi-angle problems on the crack growth behavior.

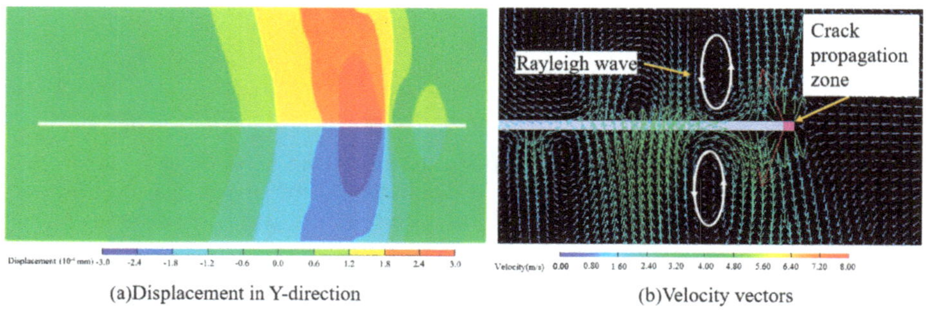

Figure 9. Cracking at the crack's tip under the action of Rayleigh waves.

3.1.1. Crack Inclination Angle = 0°

When observing the propagation characteristics of waves at 0°, due to the small geometric characteristics of cracks, there is almost no tensile or shear effect at the near end. At this point, the dominant factor causing the crack is reflected in the Rayleigh wave after the head wave. Due to the inherent nature of the Rayleigh wave, it will propagate on the surface of the crack, forming two vortex-shaped fields above and below, resulting in a displacement trend of separation between the upper and lower parts, as shown in Figure 9a. Ultimately, it will only propagate at the distal endpoint, as shown in Figure 9b.

3.1.2. Crack Inclination Angle = 15–45°

However, for 15°, 30°, and 45°, as shown in Figure 10, under similar conditions of blasting hole fragmentation, significant cracking occurs at the near end, and even connects with the cracks around the blasting hole, completely destabilizing the model. The cracking behavior at this point can be divided into two parts; the first part is the cracks around the blast hole, and the second part is due to the promoting effect of the existing cracks, resulting in the main crack, which is characterized by narrower and fewer branches compared to the surrounding cracks, and it may become the main cause of the overall instability. This part is further studied in the following text.

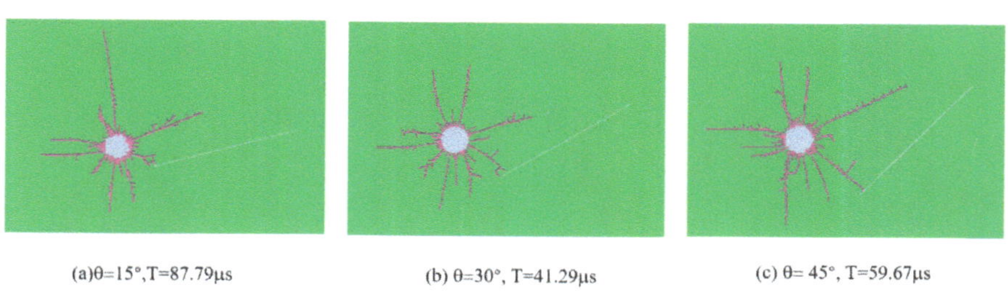

Figure 10. Crack propagation state generated at 15–45°.

The above phenomenon was explained in the first part, and the simulation results above well reflect the influence and results of the stress wave amplitude as the main factor for generating cracks on the cracking behavior of existing cracks, proving the reliability of the previous research and analysis results on waves.

3.1.3. Crack Inclination Angle = 60–90°

At this stage of inclination, the distance between the near end of the crack and the blast hole is relatively far, and the amplitude of the diffraction stress wave generated at the near end of the crack decreases. Moreover, due to the effect of the reflected wave, the effect of the Rayleigh wave is significantly reduced, so there is no cracking behavior at both ends of the crack. And due to the discovery of secondary reflection caused by blast holes in wave research, the reflection wave effect will decrease. At this time, the density of the cracks generated around the blast holes is significantly lower than the density of the cracks around the blast holes with crack angles ranging from 15° to 45°, as shown in Figure 11, which is also in line with the predicted effect.

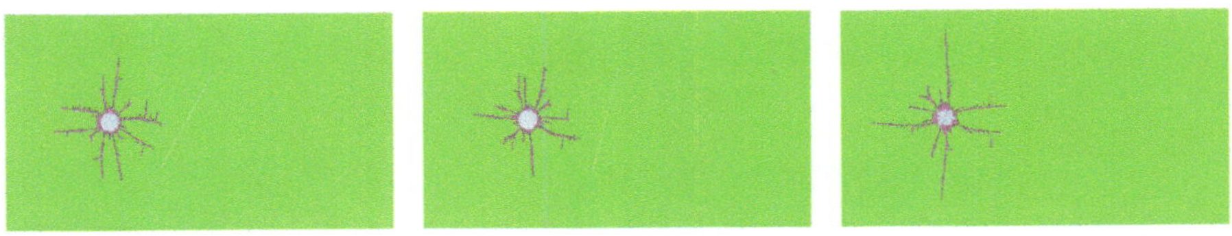

(**a**) θ = 60°, T = 50.87 μs (**b**) θ = 75°, T = 41.21 μs (**c**) θ = 90°, T = 99.12 μs

Figure 11. Crack propagation state generated at 60–90°.

At this point, it can be observed that the cracks generated by the blast hole on one side near the existing crack have more branches compared to the cracks on the other side, which is consistent with the possible effect of shear waves reflected between the blast hole and the crack as the main cause of cracking, which was analyzed earlier.

3.2. Research on the Influence of Waves Generated by Cracks on the Initiation of Explosive Holes

In the above research, it can be found that compared to existing cracks, the characteristics of the newly generated main cracks around the blast hole are quite obvious and have certain characteristics due to the reflection and diffraction effects of the corresponding stress waves of the pre-existing cracks.

Firstly, by simulating the crack-free model, it can be found that the crack propagation around the blast hole has a high degree of symmetry, and after extending to a certain extent, there will be many bifurcations at the end, as shown in Figure 12.

Figure 12. Model without cracks.

By comparing the simulation of the crack-containing model with the crack-free blast hole model, it can be found that in the crack-containing model, there is a narrow and less branched characteristic crack in the upper and lower positions of the blast hole that are approximately parallel to the crack, as shown in Figure 13. However, in the simulation, the generation and propagation behavior of new cracks are only affected by the diffraction of the tip and the reflection of the wave. In this process, the reflection wave plays a major role. Therefore, it can be considered that the interference of the reflection wave causes the generated cracks to become more narrow. In this study, the characteristic cracks with narrower and fewer branches are also known as the main cracks in the blast hole that are induced by the reflection wave (hereinafter referred to as the main cracks).

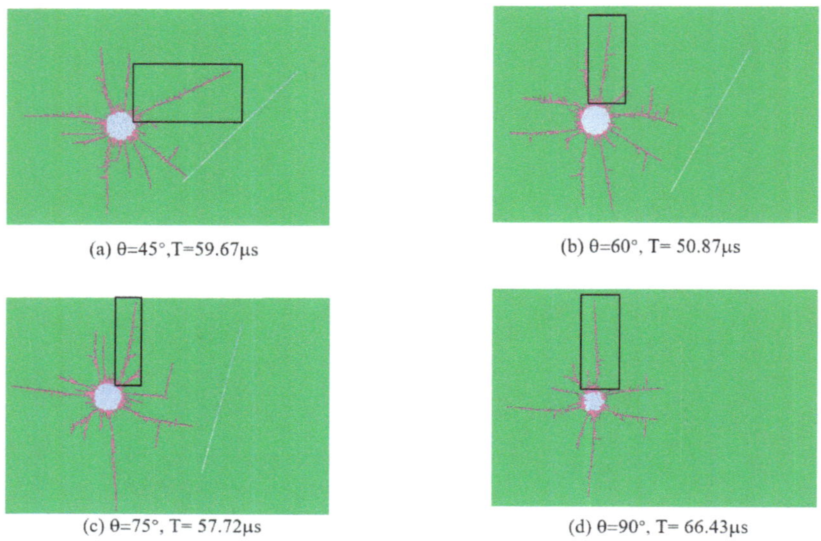

Figure 13. Main crack under different angle crack models.

By comparing the length of the cracks around the blast hole near the end of the crack propagation in the 45° to 90° simulation experiment, it can be observed that the main crack has a longer propagation distance compared to the other cracks, indicating that the main crack may be the main cause of instability. In conventional underground engineering, cracks with a longer or faster growth rate will obviously be more easily connected with the surrounding free surface or other cracks, which will lead to local crushing and instability.

At the same time, the data image of the main crack propagation length changing with time at the same angle is shown in Figure 14.

The study of the stress wave propagation in the first part found that at around 30 μs, the length of the main crack is similar to that of the surrounding crack, and even shows an approximate length in various angles of the model. Therefore, it can be considered that the process before 30 μs did not promote the propagation of cracks around the borehole, and the generation of cracks was caused by the stress wave of the initial load.

In the 45° model, the propagation speed of the main crack can be roughly calculated to be about 0.7 mm/μs from 0–30 μs, and the reflection wave begins to affect the main crack after 30 μs. At 30–40 μs, the average propagation speed of the crack is about 1.1 mm/μs, while at 40–50 μs, the propagation speed decreases to about 1 mm/μs, and even to about 0.89 mm/μs at 50–60 μs. This phenomenon indicates that in the 45° state, the reflected wave has a significant promoting effect on the main crack, resulting in a faster propagation rate compared to the initial stress wave load.

Moreover, by calculating the main crack propagation rates at different stages of 30–60 μs in Figure 14b–d, it can be found that after being affected by the reflected waves, the average propagation rate of the cracks is the highest during the period of 30–40 μs, and then gradually decreases at 40 μs. This phenomenon is common. This indicates that cracks with an inclination angle greater than 45° around them will significantly promote the propagation behavior of cracks that are approximately parallel to the blast hole and cracks.

This study also studied and calculated the 0–30° crack model and found that it also had a promoting effect on newly generated cracks at similar positions in the blast hole.

However, it may be due to the small angle, the formation of fragmented areas, and new free surfaces around the blast hole, which will change the propagation path of the wave. The promoting effect of the reflected waves is not significant, and it is difficult to form parallel or regular positional relationships with pre-existing cracks, which is why this is not included in the above discussion.

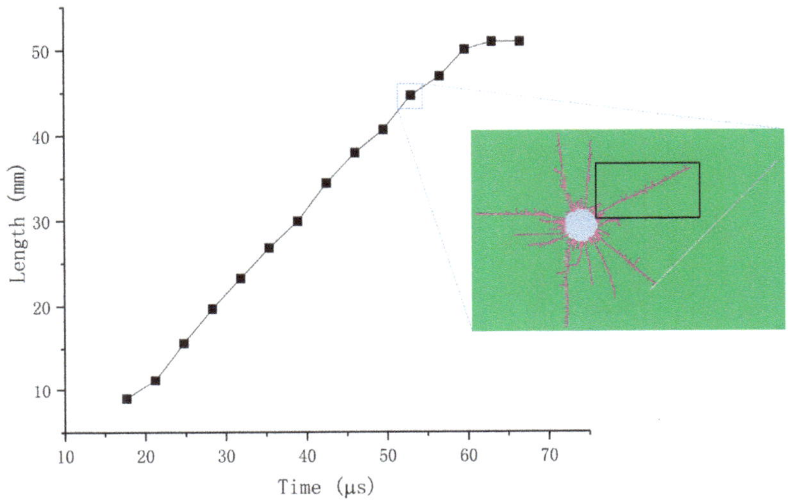

(a)

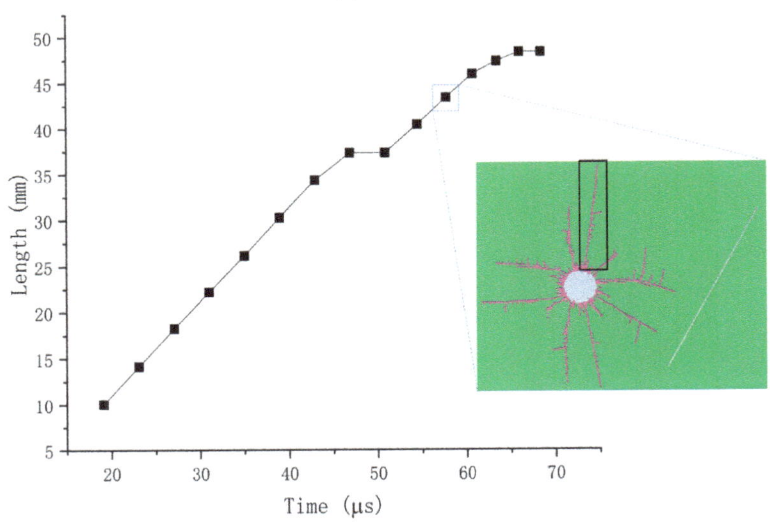

(b)

Figure 14. *Cont.*

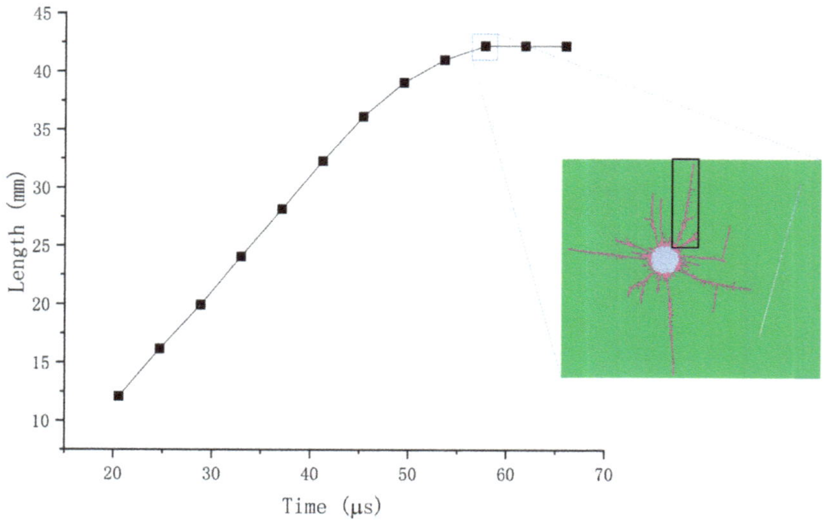

(c)

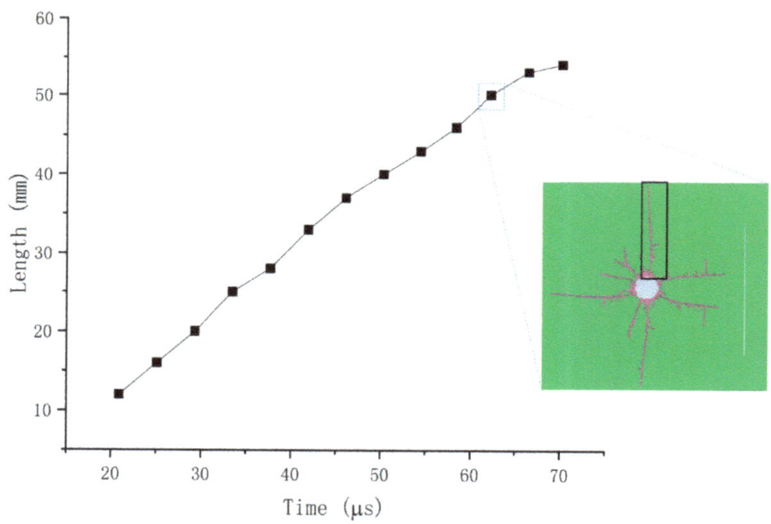

(d)

Figure 14. Models with main crack propagation and length–time curve; (**a**) 45°, (**b**) 60°, (**c**) 75°, (**d**) 90°.

4. Conclusions

The purpose of this paper is to use the stress wave theory to study the dynamic propagation mechanism of the crack and the influence law of the dangerous area from the perspective of the precast crack, and to predict the possible dangerous area in the model with the help of the wave propagation.

In terms of innovation, this study is based on the model of rock models with cracks in a large number of excavation and construction projects. On the one hand, the stress wave theory is considered to study the dynamic load conditions, and on the other hand, the crack propagation of prefabricated crack models with different inclination angles is studied in groups. Finally, the following conclusions are obtained:

(1) The action of the impact load causes the crack and the expansion of the crack behavior and hole around a pre-existing crack angle.
(2) The stress wave theory can predict the crack propagation behavior under the ideal condition of the example in this paper, and has a good fit with the final crack propagation results.
(3) When the inclination angle of the existing crack is 0–45°, the equivalent principal stress effect generated by the reflected wave is enhanced, and it is easier to form a dangerous area of high stress in the area near the crack hole. At 60–90°, the equivalent principal stress is reduced due to the secondary reflection of the reflected wave at the borehole position, and cracking behavior is less likely to occur at the far and near ends of the fracture. At this time, fracture behavior mainly occurs around the hole, and only a few cracks have a high propagation rate.
(4) The reflected wave is the main cause of the crack, followed by a shear wave; compared with the other side of the crack, the reflected wave may lead to more branches near the burst crack, forming a fracture area.
(5) The reflected waves generated by the presence of cracks at different angles will promote the propagation of cracks around the blasting hole. The crack with an angle of 45–90° has a particularly significant effect on the expansion rate of nearly parallel cracks (main cracks) around the blasting hole. Such rapidly expanding cracks will be more easily connected with the free surface or surrounding cracks, which may be one of the main reasons for regional instability.

Author Contributions: Conceptualization, J.Z. and W.G.; methodology, J.Z.; software, J.Z.; validation., W.G. and M.W.; formal analysis, J.Z.; investigation, J.Z.; resources, M.W.; data curation, D.W.; writing—original draft preparation, J.Z.; writing—review and editing, W.G.; visualization, W.G.; supervision, W.G. and M.W.; project administration, M.W. and L.Z.; funding acquisition, M.W. and L.Z. All authors have read and agreed to the published version of the manuscript.

Funding: This research was funded by the National Natural Science Foundation of China (12272247), the Natural Science Foundation of Sichuan, China (2022NSFSC1915), the National Key Project (GJXM92579), and the major research and development project of the Metallurgical Corporation of China LTD in the non-steel field (2021–5).

Institutional Review Board Statement: Not applicable.

Informed Consent Statement: Not applicable.

Data Availability Statement: The data are unavailable due to privacy.

Conflicts of Interest: The authors declare no conflict of interest.

List of Notations

a_x	acceleration in the x direction
a_y	acceleration in the y direction
F_x	force in the x direction
F_y	force in the y direction
σ_i	stress in the i direction
P_i	external pressure in the i
S_i	directiondeviatoric stress in the i direction
G	shear modulus of the material
$\dot{e}$	volume rate of the material
ε_i	strain in the i direction
k	bulk modulus of the material
ρ	initial density
ρ_0	present density
V_p	expansion wave velocity of the material
V_s	shear wave velocity of the material
E_d	dynamic elastic modulus
μ_d	dynamic Poisson ratio

References

1. Panji, M.; Ansari, B. Transient SH-wave scattering by the lined tunnels embedded in an elastic half-plane. *Eng. Anal. Bound. Elem.* **2017**, *84*, 220–230. [CrossRef]
2. Wang, M.; Ma, G.; Wang, F. Numerically investigation on blast-induced wave propagation in catastrophic large-scale bedding rockslide. *Landslides* **2021**, *18*, 785–797. [CrossRef]
3. Panji, M.; Ansari, B. Antiplane seismic ground motion above twin horseshoe-shaped lined tunnels. *Innov. Infrastruct. Solut.* **2020**, *5*, 1–16. [CrossRef]
4. Azimi, Y.; Osanloo, M.; Aakbarpour-Shirazi, M.; Bazzazi, A.A. Prediction of the blastability designation of rock masses using fuzzy sets. *Int. J. Rock Mech. Min. Sci.* **2010**, *47*, 1126–1140. [CrossRef]
5. Lei, Q.; Latham, J.P.; Xiang, J.; Tsang, C.F. Role of natural fractures in damage evolution around tunnel excavation in fractured rocks. *Eng. Geol.* **2017**, *231*, 100–113. [CrossRef]
6. Han, H.; Fukuda, D.; Liu, H.; Salmi, E.F.; Sellers, E.; Liu, T.; Chan, A. Combined finite-discrete element modelling of rock fracture and fragmentation induced by contour blasting during tunnelling with high horizontal in-situ stress. *Int. J. Rock Mech. Min. Sci.* **2020**, *127*, 104214. [CrossRef]
7. Suorineni, F.T.; Tannant, D.D.; Kaiser, P.K. Determination of fault-related sloughage in open stopes. *Int. J. Rock Mech. Min. Sci.* **1999**, *36*, 891–906. [CrossRef]
8. Freund, L.B. Crack propagation in an elastic solid subjected to general loading—III. Stress wave loading. *J. Mech. Phys. Solids* **1973**, *21*, 47–61. [CrossRef]
9. Freund, L.B. Crack propagation in an elastic solid subjected to general loading—I. Constant rate of extension. *J. Mech. Phys. Solids* **1972**, *20*, 129–140. [CrossRef]
10. Freund, L.B. Crack propagation in an elastic solid subjected to general loading—II. Non-uniform rate of extension. *J. Mech. Phys. Solids* **1972**, *20*, 141–152. [CrossRef]
11. Renshu, Y.; Yanbing, W.; Huajun, X.; Maoyuan, W. Dynamic Behavior Analysis of Perforated Crack Propagation in Two-Hole Blasting. *Procedia Earth Planet. Sci.* **2012**, *5*, 254–261. [CrossRef]
12. Wan, D.; Zhu, Z.; Liu, R.; Liu, B. Effect of two parallel cracks on main propagating cracks under blasting. *Baozha Yu Chongji/Explos. Shock. Waves* **2019**, *39*, 083105-1.
13. Wheeler, L.T. Wave Propagation in Elastic Solids. *SIAM Rev.* **1974**, *16*, 558. [CrossRef]
14. Li, J.; Zhu, Z.; Wan, D.; Liu, R. Dynamic propagation law and dynamic fracture parameters of mode I crack in sandstone under blasting. *Meitan Xuebao/J. China Coal Soc.* **2019**, *44*, 466–475.
15. Wang, F.; Wang, M.; Zhu, Z.M.; Hao, Q.; Peng, Y.; Xuya, W. Study on evolution law of rock crack dynamic propagation in complete process under impact loading. *Chin. J. Rock Mech. Eng.* **2019**, *38*, 1139–1148.
16. Zhu, J.; Ren, M.; Liao, Z. Wave propagation and diffraction through non-persistent rock joints: An analytical and numerical study. *Int. J. Rock Mech. Min. Sci.* **2020**, *132*, 104362. [CrossRef]
17. Uenishi, K. On a possible role of rayleigh surface waves in dynamic slope failures. *Int. J. Geomech.* **2010**, *10*, 153–160. [CrossRef]
18. Sun, H.; Zhang, S. Study on detection of angled surface cracks with laser-generated Rayleigh waves. *Shengxue Xuebao/Acta Acust.* **2013**, *38*, 405–412.
19. Zhou, M.; Li, Y.; Jiankui, W.; Yu, Y.; He, H. The characteristics of high speed crack propagation at ultra high loading rate. *Theor. Appl. Fract. Mech.* **2020**, *108*, 102650. [CrossRef]
20. Zhang, A.; Xie, H.; Zhang, R.; Gao, M.; Xie, J.; Jia, Z.; Ren, L.; Zhang, Z. Mechanical properties and energy characteristics of coal at different depths under cyclic triaxial loading and unloading. *Int. J. Rock Mech. Min. Sci.* **2023**, *161*, 105271. [CrossRef]
21. Zhou, L.; Chen, J.; Zhou, C.; Zhu, Z.; Dong, Y.; Wang, H. Study on failure behaviors of mixed-mode cracks under static and dynamic loads. *Geomech. Eng.* **2022**, *29*, 567–582.
22. Li, Y.; Zhu, Z.; Li, B.; Deng, J.; Xie, H. Study on the transmission and reflection of stress waves across joints. *Int. J. Rock Mech. Min. Sci.* **2011**, *48*, 364–371. [CrossRef]
23. Lak, M.; Marji, M.F.; Bafghi, A.Y.; Abdollahipour, A. Analytical and numerical modeling of rock blasting operations using a two-dimensional elasto-dynamic Green's function. *Int. J. Rock Mech. Min. Sci.* **2019**, *114*, 208–217. [CrossRef]
24. Zhu, Z.; Gao, W.; Wan, D.; Wang, M.; Shu, Y. Numerical Study of Fracture Characteristics of Deep Granite Induced by Blast Stress Wave. *Shock. Vib.* **2021**, *2021*, 2579569. [CrossRef]
25. Li, X.; Zhu, Z.; Wang, M.; Xiao, D.; Shu, Y.; Deng, S. Fracture mechanism of rock around a tunnel-shaped cavity with interconnected cracks under blasting stress waves. *Int. J. Impact Eng.* **2021**, *157*, 103999. [CrossRef]
26. Mellor, M. *The Dynamics of Explosion and Its Use. Developments in Civil Engineering, 1*; Henrych, J., Ed.; Elsevier: Amsterdam, The Netherlands, 1979; 558p.
27. Zuo, Y.J.; Tang, C.A.; Zhu, W.C.; Li, D.Y.; Li, S.C. Numerical analysis of tunnel reinforcing influences on failure process of surrounding rock under explosive stress waves. *J. Cent. South Univ. Technol.* **2008**, *15*, 632–638. [CrossRef]

28. Wan, D.; Wang, M.; Zhu, Z.; Wang, F.; Zhou, L.; Liu, R.; Gao, W.; Shu, Y.; Xiao, H. Coupled GIMP and CPDI material point method in modelling blast-induced three-dimensional rock fracture. *Int. J. Min. Sci. Technol.* **2022**, *32*, 1097–1114. [CrossRef]
29. Shu, Y.; Zhu, Z.; Wang, M.; Ying, P.; Wang, F.; Wan, D.; Li, X.; Gao, W. A plastic damage constitutive model for rock-like material focusing on the hydrostatic pressure induced damage and the interaction of tensile and shear damages under impact and blast loads. *Comput. Geotech.* **2022**, *150*, 104921. [CrossRef]

applied sciences

MDPI

Article

Inversion Analysis of the In Situ Stress Field around Underground Caverns Based on Particle Swarm Optimization Optimized Back Propagation Neural Network

Hong-Chuan Yan [1], Huai-Zhong Liu [1,*], Yao Li [2], Li Zhuo [1], Ming-Li Xiao [1], Ke-Pu Chen [2], Jia-Ming Wu [1] and Jian-Liang Pei [1]

[1] State Key Laboratory of Hydraulics and Mountain River Engineering, College of Water Resource and Hydropower, Sichuan University, Chengdu 610065, China
[2] Sinohydro Bureau 7 Co., Ltd., Chengdu 610213, China
* Correspondence: huaizhong.liu@scu.edu.cn

Abstract: The in situ stress distribution is one of the driving factors for the design and construction of underground engineering. Numerical analysis methods based on artificial neural networks are the most common and effective methods for in situ stress inversion. However, conventional algorithms often have some drawbacks, such as slow convergence, overfitting, and the local minimum problem, which will directly affect the inversion results. An intelligent inverse method optimizing the back-propagation (BP) neural network with the particle swarm optimization algorithm (PSO) is applied to the back analysis of in situ stress. The PSO algorithm is used to optimize the initial parameters of the BP neural network, improving the stability and accuracy of the inversion results. The numerical simulation is utilized to calculate the stress field and generate training samples. In the application of the Shuangjiangkou Hydropower Station underground powerhouse, the average relative error decreases by about 3.45% by using the proposed method compared with the BP method. Subsequently, the in situ stress distribution shows the significant tectonic movement of the surrounding rock, with the first principal stress value of 20 to 26 MPa. The fault and the lamprophyre significantly influence the in situ stress, with 15–30% localized stress reduction in the rock mass within 10 m. The research results demonstrate the reliability and improvement of the proposed method and provide a reference for similar underground engineering.

Keywords: in situ stress; underground engineering; inversion method; numerical simulation; neural network; particle swarm optimization algorithm

Citation: Yan, H.-C.; Liu, H.-Z.; Li, Y.; Zhuo, L.; Xiao, M.-L.; Chen, K.-P.; Wu, J.-M.; Pei, J.-L. Inversion Analysis of the In Situ Stress Field around Underground Caverns Based on Particle Swarm Optimization Optimized Back Propagation Neural Network. *Appl. Sci.* **2023**, *13*, 4697. https://doi.org/10.3390/app13084697

Academic Editors: Bei Jiang, Nuwen Xu and Zhengzhao Liang

Received: 20 February 2023
Revised: 6 April 2023
Accepted: 6 April 2023
Published: 7 April 2023

1. Introduction

With the larger scale, deeper burial, and more complex geological conditions of underground engineering, accidents during construction occur frequently. Such as large deformations in soft rocks, rockbursts in hard rocks, etc., were related to their regional in situ stress [1–4]. Hence, an accurate in situ stress field is not only the basis for the design and construction of underground engineering but also for the deformation and failure research. In situ tests provide an effective way to obtain in situ stress directly. Currently, the primary measurement methods are the hydraulic fracturing method, the stress relief method, and the acoustic emission method [5–9]. However, the investigation of in situ stress through boreholes is too expensive to set up sufficient boreholes to cover the entire underground cavern.

To overcome the limitation of on-site test data, researchers generally adopt three-dimensional numerical simulation methods to conduct the back analysis of the in situ stress field in the engineering field [10–13]. The main inversion methods are the boundary load adjustment method, the multiple regression fitting method, and the artificial neural network (ANN) algorithm. The boundary load adjustment method involves repeatedly adjusting the numerical model's boundary loads to approximate the calculated stress from the test data. This method can effectively and directly obtain a reasonable in situ stress field through iterative calculation. However, in practice, it may not always provide a unique and optimal solution, and the iterative process can be irregular and time-consuming [14]. Multiple regression is a method that takes many factors into account to ensure the uniqueness and rationality of the results. For example, Chen et al. [15], Yu et al. [16], and Li et al. [17] utilized the least squares regression method, the lasso regression method, and the partial least squares regression method, respectively, to conduct multiple linear regression analysis of the in situ stress field. In this method, the regression relationship between model conditions and stress state has been established to reduce the computational cost. Due to the basic assumptions of linearity and continuity, this method may be less applicable for underground engineering with complex geological conditions, where the in situ stress field is nonlinear with burial depth of rock. Hence, the application of artificial neural networks to in situ stress inversion has become a popular methodology in the 21st century. Zhang et al. [18], Li et al. [19], and Li et al. [20] utilized the genetical algorithm (GA), the back propagation (BP) neural network, and the GA-BP artificial neural network, respectively, to optimally solve the model boundary loads. They inversely obtained a more accurate in situ stress field due to the effectiveness of ANN in solving nonlinear and complex problems.

However, traditional neural network algorithms often have several drawbacks in practice. These include overfitting, slow convergence, and sensitivity to initial conditions, which may make them less suitable for complex underground problems. Particle Swarm Optimization (PSO), first proposed by Eberhart and Kennedy [21], is a widely used optimization algorithm. Due to its capability for global optimization, PSO can serve as a pretraining step to optimize the initial parameters before actual training with the BP algorithm. As a result, this approach may enhance the performance of network training and improve the accuracy of prediction. To the best knowledge of the authors, the PSO-BP algorithm was seldom applied to the back analysis of in situ stress.

In this paper, an intelligent in situ stress inversion method combining the BP algorithm and the PSO algorithm is proposed. To improve the accuracy of the inversion results, this method employs the global search capability of the PSO algorithm to optimize the BP initial parameters. The training network was then utilized to search for optimal boundary parameters of the numerical model based on the measured data. The next section outlines the theory of the inversion method and the implementation procedure of the PSO-BP algorithm. Subsequently, the proposed method was applied to an engineering project, the underground powerhouse of Shuangjiangkou Hydropower Station, to demonstrate its practicability.

2. Intelligent Inversion Method of the In Situ Stress Field

2.1. Determination of Boundary Conditions

In the analysis of the in situ stress field in the underground caverns, it is widely acknowledged that the gravity stress and the geotectonic movement are the primary contributing factors, while the influence of temperature and groundwater effects is generally considered to be less significant and, hence, often ignored [22,23]. This paper adopts the fast stress boundary method to conduct the in situ stress inversion by using the FLAC^{3D} 7.0 software. This method assumes that the in situ stress field results from a combination of self-weight and tectonic stresses. By adjusting the boundary conditions, a stress field that best matches the test data can be determined, which is known as the inverse in situ stress.

In the fast Lagrangian method, the model's horizontal stress state can be adjusted by specifying the coefficient k (the ratio of horizontal stress to vertical stress), and the shear stress state is simulated by applying the corresponding shear stress τ. In this paper, six boundary coefficients are used to correspond to six different stress boundary conditions, as shown in Figure 1. These coefficients include the self-weight coefficient (k_g), the horizontal stress coefficients (k_x and k_y), and the shear stress coefficients in three directions (τ_{xz} ,τ_{yz}, and τ_{xy}). In this way, the challenging problem of in situ stress inversion can be transformed into finding an optimal solution for the six boundary conditions.

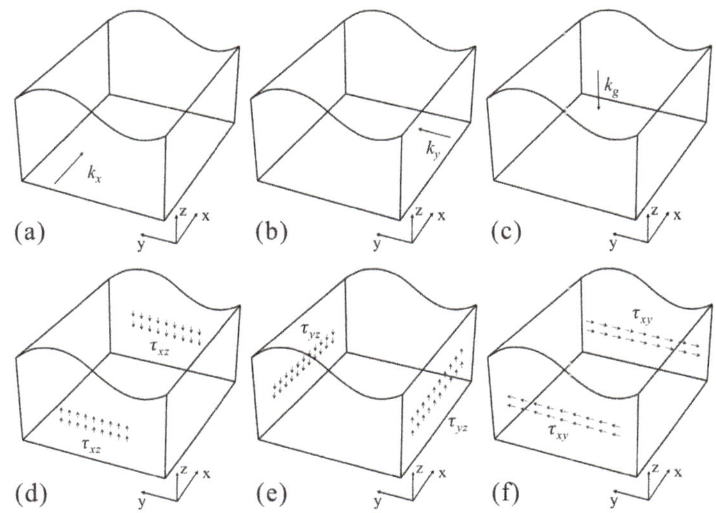

Figure 1. Different boundary conditions. (**a**) horizontal stress in the X-direction; (**b**) horizontal stress in the Y-direction; (**c**) gravity stress; (**d**) shear stress in the XZ-plane; (**e**) shear stress in the YZ-plane; and (**f**) shear stress in the XY-plane.

2.2. Optimization Solution Based on PSO-BP

2.2.1. Coordinate Transformation and Orthogonal Design

The in situ stress measurements are typically presented in the geodetic coordinate system. However, the inversion analysis for the in situ stress distribution in the underground cavern area is generally carried out in the computational coordinate system. Hence, before performing the inversion analysis, the stress state at the test points needs to be transformed from the geodetic coordinate system to the computational coordinate system, as demonstrated in Equation (1) [24].

$$\left.\begin{array}{l}\sigma_x = L_1^2\sigma_1 + L_2^2\sigma_2 + L_3^2\sigma_3 \\ \sigma_y = M_1^2\sigma_1 + M_2^2\sigma_2 + M_3^2\sigma_3 \\ \sigma_z = N_1^2\sigma_1 + N_2^2\sigma_2 + N_3^2\sigma_3 \\ \tau_{xy} = L_1M_1\sigma_1 + L_2M_2\sigma_2 + L_3M_3\sigma_3 \\ \tau_{yz} = N_1M_1\sigma_1 + N_2M_2\sigma_2 + N_3M_3\sigma_3 \\ \tau_{xz} = L_1N_1\sigma_1 + L_2N_2\sigma_2 + L_3N_3\sigma_3\end{array}\right\} \quad (1)$$

where, $L_i = \cos\beta_i\cos\alpha_{ix}$, $M_i = \cos\beta_i\sin\alpha_{ix}$, $N_i = \sin\beta_i$, β_i is the angle between σ_i and the horizontal plane, and α_{ix} is the angle between σ_i and the direction of the x-axis.

In order to conduct in situ stress inversion using neural network intelligence algorithms, it is necessary to generate sufficient numerical data as input and output samples. The orthogonal design method is used to design the training sample scheme and thus calculate more comprehensive and representative combinations of boundary conditions at a lower computational cost. This method utilizes orthogonal design tables to effectively address the challenges posed by multi-factor design and optimal level determination. Table 1 shows a typical four-factor and three-level L9(3^4) orthogonal design. The designed boundary condition scheme is then input into FLAC^{3D} software, respectively. The simulated stresses at the in situ stress test points, together with the boundary conditions, serve as the training samples for the network.

Table 1. Typical orthogonal design table L9(3^4).

No.	A	B	C	D
1	A_1	B_1	C_1	D_1
2	A_1	B_2	C_2	D_2
3	A_1	B_3	C_3	D_3
4	A_2	B_1	C_2	D_3
5	A_2	B_2	C_3	D_1
6	A_2	B_3	C_1	D_2
7	A_3	B_1	C_3	D_2
8	A_3	B_2	C_1	D_3
9	A_3	B_3	C_2	D_1

2.2.2. PSO Optimized BP Network Algorithm

The fundamental principle of in situ stress inversion involves the establishment of a nonlinear relationship between the stress state and the model boundary conditions in order to identify the optimal boundary conditions that correspond with the measurement data. The BP neural network is widely used in underground engineering due to its capacity to deal with nonlinear problems [25,26]. However, the BP algorithm has some unavoidable limitations in addressing complex nonlinear problems. For example, many parameters should be initialized in the BP algorithm, such as weights, thresholds, error tolerance, etc. Some of these parameters that are determined empirically and roughly are sometimes not appropriate for a specific problem, and this limitation makes the BP algorithm quite unstable. Thus, in this paper, the PSO has been used to optimize the BP initial parameters to improve the stability and accuracy of the final result. The specific implementation steps of the PSO-BP intelligent optimization algorithm are as follows [21,27]:

(1) Determine the structure of the BP neural network.

(2) Encode the BP connection weights and thresholds; the particle dimension D is the sum of all weights in neural networks and is expressed as:

$$D = S_1 \times R + S_1 \times S_2 + S_1 + S_2 \tag{2}$$

where, R, S_1, and S_2 are the numbers of neurons in the input, hidden, and output layers, respectively.

(3) Initialize the basic parameters of the particle swarm.

(4) Define the fitness function: the mean square error of BP is used as the function of fitness value, and the function is expressed as:

$$Fitness = \frac{1}{N}\sum_{i=1}^{N}\sum_{o=1}^{S_2}(d_{io}(x_i) - y_{io})^2 \tag{3}$$

where N represents the total number of input samples, $d_{io}(x_i)$ is the o-th output result for the i-th group of input samples, and y_{io} is the corresponding actual value.

(5) Based on the input-output sample pair, each particle's fitness value is calculated using the forward algorithm and the fitness function of the BP network.

(6) For each particle, if $Fitness[i]$ is less than the P_{best}, let P_{best} equal $Fitness[i]$; if $Fitness[i]$ is less than the G_{best}, let G_{best} equal $Fitness[i]$;

(7) Update the velocity and position of the particles by Equations (4) and (5) and ensure that the particle velocity does not exceed the velocity limit ($v_{\max}$).

$$v_{ij}(t+1) = v_{ij}(t) + c_1 r_1(t)\left[p_{ij}(t) - x_{ij}(t)\right] + c_2 r_2(t)\left[p_{gj}(t) - x_{ij}(t)\right] \tag{4}$$

$$x_{ij}(t+1) = x_{ij}(t) + v_{ij}(t+1) \tag{5}$$

where x_{ij} and v_{ij} are the position and velocity of particles i in the j-th dimensional component, respectively; P_{ij} and P_{gj} are the individual extremum and global extremum of particles i in the j-th dimensional component, respectively; c_1 and c_1 are acceleration constants, which normally take the same value between 1.5 and 2.5; r_1 and r_2 are random numbers ranging from 0 to 1.

(8) Calculate the error of the PSO-BP algorithm as:

$$E = \frac{1}{t}\sum_{i=1}^{t} J({G_{best}}^{(i)}) \tag{6}$$

where $J({G_{best}}^{(i)})$ is the fitness of the global optimal value in the i-th iteration and t is the current iteration time.

(9) If the error (E) is within the preset tolerance or the number of iterations reaches its maximum, the final solution is the vector of the global optimal position (G_{best}) in the last iteration. Otherwise, return to step 5.

2.2.3. PSO-BP Inversion Method

To enhance the accuracy of in situ stress inversion results, a nonlinear numerical analysis approach was proposed that combines the model boundary conditions with the PSO-BP algorithm. In this method, the complex geological movements are resolved into six basic boundary conditions, and then, through FLAC^{3D} software, the stress state of each boundary is calculated to form the network samples. The PSO-BP neural network is used to establish a nonlinear relationship between the boundary conditions and the stress state at specific testing locations. Finally, the optimal boundary conditions that coincide with the measurement results are obtained through iterative searching. The primary implementation steps are shown in Figure 2:

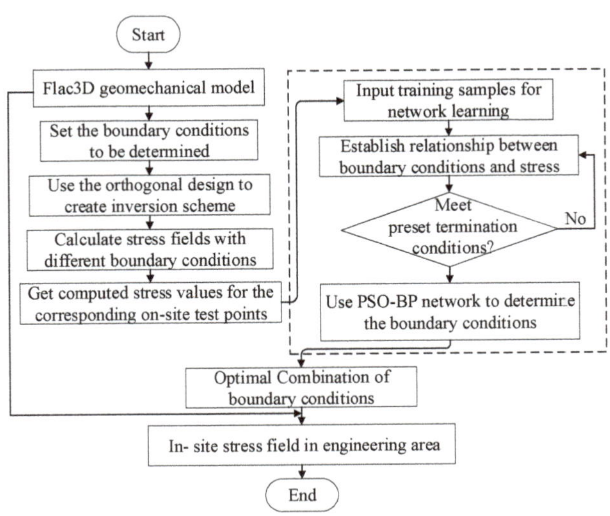

Figure 2. Flow chart of the PSO-BP in situ stress inversion method.

3. Application to the Shuangjiangkou Hydropower Station

3.1. Measured Results of In Situ Stress in Underground Cavern Zones

3.1.1. General Situation of Engineering Geology

The Shuangjiangkou hydropower station in Jinchuan County, Sichuan Province, is the fifth power station in the Dadu River basin. Figure 3 shows the site of the Shuangjiangkou hydropower station, which has the highest dam in the world. The underground cavern is deeply buried on the left bank with a horizontal depth of 300–500 m and a vertical depth of 42 m, and its axial azimuth is N10°W. F1 is the main fault distributed around the underground cavern area, which has a width of 0.5 to 0.6 m, and the occurrence is about

N79°W/SW∠48°. A lamprophyre also occurred about N35°–50°W/SW∠72°–75° at the cavern area, with a width of 0.8–1 m. In general, the Shuangjiangkou underground cavern area mainly comprises porphyritic granite with different degrees of weathering. Currently, empirical systems, such as rock-mass rating (RMR), tunneling quality index (Q) system, and geological strength index (GSI), are mainly utilized in international rock engineering projects [28–30]. In the geological survey report of Shuangjiangkou Hydropower Station, the (basic quality index of rock mass) BQ method [31] is employed to classify the rock mass into four types of materials, namely, class II (fresh granite), class III (moderately weathered granite), class IV (strongly weathered granite), and class V (geological structures), as listed in Table 2.

Figure 3. Dam site of Shuangjiangkou Hydropower Station.

Table 2. The mechanical parameters.

Type of Rock Mass	Rock Classification	Density (kg/m^3)	Young's Modulus E (GPa)	Poisson's Ratio μ	Cohesion c (MPa)	Friction Coefficient f
Fresh granite	II	2600	32.9	0.25	1.5	1.3
Moderately weathered granite	III	2550	25.0	0.30	1.0	1.0
Strongly weathered granite	IV	2350	20.0	0.32	0.5	0.8
Geological structures	V	2300	6.0	0.35	0.1	0.35

3.1.2. Analysis of the Field Test

In situ stress measurements are a direct method for identifying the magnitude, direction, and distribution of in situ stress in the engineering field. Figure 4 displays the layout of in situ stress measurement points around the underground caverns on the left bank. The five measurement points are situated parallel to the main powerhouse's axis and located inside the SPD9# structure at an elevation of 2268 m. The measured data is obtained in the geodetic coordinate system, and thus it is necessary to transform them into the model coordinate system using Equation (1). The transformed results are listed in Table 3.

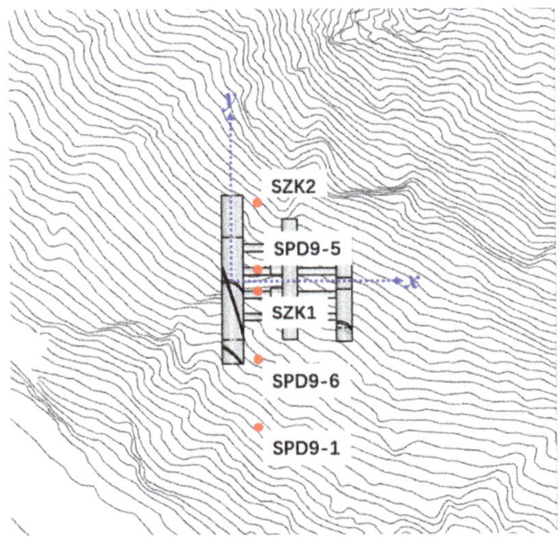

Figure 4. Distribution of in situ stress measured points.

Table 3. Stress components from in situ stress measurements around the underground caverns.

Test No.	Test Point	Burial Depth	σ_x/MPa	σ_y/MPa	σ_z/MPa	τ_{xz}/MPa	τ_{yz}/MPa	τ_{xy}/MPa
1#	SZK2	549 m	16.92	23.50	15.04	4.44	−3.19	1.54
2#	SPD9-5	470 m	19.26	22.74	16.66	0.33	−7.11	4.71
3#	SZK1	431 m	10.21	15.95	9.10	0.86	−2.60	−0.52
4#	SPD9-6	357 m	18.64	21.86	13.56	4.29	−2.47	6.81
5#	SPD9-1	308 m	15.84	22.18	24.01	1.64	−3.61	1.37

As shown in Table 3, the measured stress in the x-axis direction in the cavern area is 10.21–19.26 MPa, which is approximately 1.1 to 1.4 times greater than that in the vertical direction. Similarly, the stress in the horizontal y-axis direction ranges from 15.95 to 23.50 MPa, which is around 1.4 to 1.8 times greater than the vertical stress. The measured stress values are ranked as σ_y, σ_x, and σ_z in descending order, with horizontal stress σ_y and σ_x being higher than the vertical stress σ_z. This suggests that the tectonic movements have a significant impact on the stress distribution in the Shuangjiangkou underground cavern area. The shear stress values in the cavern area are generally low, with an average of 2.3 MPa for τ_{xz}, −3.7 MPa for τ_{yz}, and 2.8 MPa for τ_{xy} in the three directions.

The monitoring data used in the back analysis may directly affect the accuracy of the inversed in situ stress distribution, as it is the main input for the back analysis algorithm. Therefore, filtering the illogical measured points is necessary for the inversion process. As shown in Table 3, the stress values at the 3# measured point are smaller than the others, and considering that the 3# point is located near the F1 fault, the stress decreases in this region may be reasonable. Notably, despite the fact that the 5# point is located at the lowest burial depth of just 308 m compared to the highest 1# point of 549 m, the σ_z value at this point is about 24 MPa, ranking first among these five measurement points (the average value being 15.6 MPa). This means that there probably exist stress fluctuations near the 5# point area, and the measured data at this point may not properly reflect the overall in situ stress distribution in the engineering area. However, to consider the comprehensiveness and diversity of the input sample, all five measurement points were included in the inversion calculation.

3.2. Inversion Analysis of the In Situ Stress Distribution

3.2.1. Numerical Model

The inversion model should provide a comprehensive analysis of the impact of topography, geomorphology, lithology, and geological structures on the engineering field. After considering the precise locations of the testing points, a numerical model centered on the main powerhouse was established. As shown in Figure 5a, the 3D model can be mainly divided into three layers. The upper layer, represented by the color red, consists of strongly weathered granite. The middle layer, depicted in blue, is composed of moderately weathered granite. Finally, the lower layer, shown in green, comprises fresh rock. The geological structures, namely the F1 fault and lamprophyre, are simulated in the model through the fine mesh with specific thicknesses, as shown in Figure 5b. Besides, the spatial relationships between the F1 fault, lamprophyre, and underground caverns were determined based on a geological survey report. The underground cavern group of the Shuangjiangkou hydropower station consists of three parallel caverns (the main powerhouse, transformer chamber, and tailrace surge chamber) and some auxiliary caverns such as the penstock and generatrix tunnels. The range of the model requires an adequate length to minimize boundary effects and ensure an appropriate distance between the underground caverns and boundaries. The Flac^{3D} model has dimensions of 700 m in the x-axis (vertical axial direction), 700 m in the y-axis (axial direction), and 1100 m in the z-axis (elevation direction). The model has been extended to include the ground surface above the underground caverns, consisting of 477,685 nodes and 753,151 brick elements. The mesh is primarily composed of eight-node hexahedral elements, which exhibit uniform size and excellent morphology. Besides, to improve the computational accuracy, the mesh in the underground cavern area was further refined, with a minimum mesh density of 0.5m. In FLAC^{3D} software, the Mohr-Coulomb criterion is commonly used to model the behavior of rock under different loading conditions, which is the criterion utilized in this back analysis. The required mechanical parameters of all materials for this criterion, such as Young's modulus, Poisson's ratio, and strength parameters, are listed in Table 2, according to the geological survey report of Shuangjiangkou Hydropower Station.

3.2.2. Orthogonal Design Experiment

Obtaining accurate in situ stress distribution through neural networks requires a sufficient number of comprehensive training samples. Therefore, the in situ stress at the measurement points under different boundary conditions was generated through FLAC^{3D} software. The training sample scheme was established using an orthogonal design. Firstly, the six boundary condition coefficients of in situ stress inversion in FLAC^{3D} were determined according to the inversion theory in Section 2.1. Then, the test data in Table 3 was used as the base values for each parameter. The weight coefficient k_g was 1.3, the horizontal stress coefficients k_x and k_y were 1.3 and 1.6, respectively, and the shear stress coefficients were 2.3 MPa, −3.7 MPa, and 2.8 MPa, respectively. The weight coefficient, the horizontal stress structure coefficients, and the shear stress coefficients were varied within a range of 10%, 20%, and 30%, respectively, based on the base values. As shown in Table 4, an L49 (7^6) orthogonal scheme was designed. The 49 sets of boundary condition schemes were then input into FLAC^{3D} software. The boundary conditions and the simulated in situ stress at the testing points constitute the training samples for the network.

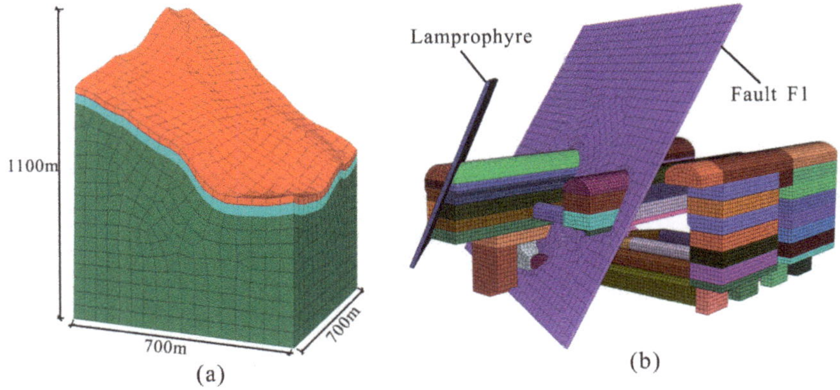

Figure 5. Numerical model for the in situ stress inversion. Mesh model of the surrounding mountain (**a**) and underground cavern and main geological structures (**b**).

Table 4. Boundary condition scheme $L_{49}(7^6)$.

Test No.	k_g	k_x	k_y	τ_{xz}/MPa	τ_{yz}/MPa	τ_{xy}/MPa
1	0.91	0.91	1.12	1.61	−2.59	1.96
2	1.04	1.17	1.60	2.53	−4.44	3.64
3	1.17	1.43	2.08	1.84	−3.70	3.36
4	1.30	1.69	1.44	2.76	−2.96	3.08
5	1.43	1.04	1.92	2.07	−4.81	2.80
6	1.56	1.30	1.28	2.99	−4.07	2.52
7	1.69	1.56	1.76	2.30	−3.33	2.24
8	0.91	1.69	1.92	2.53	−3.70	2.52
9	1.04	1.04	1.28	1.84	−2.96	2.24
10	1.17	1.30	1.76	2.76	−4.81	1.96
11	1.30	1.56	1.12	2.07	−4.07	3.64
12	1.43	0.91	1.60	2.99	−3.33	3.36
13	1.56	1.17	2.08	2.30	−2.59	3.08
14	1.69	1.43	1.44	1.61	−4.44	2.80
15	0.91	1.56	1.60	1.84	−4.81	3.08
16	1.04	0.91	2.08	2.76	−4.07	2.80
17	1.17	1.17	1.44	2.07	−3.33	2.52
18	1.30	1.43	1.92	2.99	−2.59	2.24
19	1.43	1.69	1.28	2.30	−4.44	1.96
20	1.56	1.04	1.76	1.61	−3.70	3.64
21	1.69	1.30	1.12	2.53	−2.96	3.36
22	0.91	1.43	1.28	2.76	−3.33	3.64
23	1.04	1.69	1.76	2.07	−2.59	3.36
24	1.17	1.04	1.12	2.99	−4.44	3.08
25	1.30	1.30	1.60	2.30	−3.70	2.80
26	1.43	1.56	2.08	1.61	−2.96	2.52
27	1.56	0.91	1.44	2.53	−4.81	2.24
28	1.69	1.17	1.92	1.84	−4.07	1.96
29	0.91	1.30	2.08	2.07	−4.44	2.24
30	1.04	1.56	1.44	2.99	−3.70	1.96
31	1.17	0.91	1.92	2.30	−2.96	3.64
32	1.30	1.17	1.28	1.61	−4.81	3.36
33	1.43	1.43	1.76	2.53	−4.07	3.08
34	1.56	1.69	1.12	1.84	−3.33	2.80
35	1.69	1.04	1.60	2.76	−2.59	2.52
36	0.91	1.17	1.76	2.99	−2.96	2.80
37	1.04	1.43	1.12	2.30	−4.81	2.52

Table 4. *Cont.*

Test No.	k_g	k_x	k_y	τ_{xz}/MPa	τ_{yz}/MPa	τ_{xy}/MPa
38	1.17	1.69	1.60	1.61	−4.07	2.24
39	1.30	1.04	2.08	2.53	−3.33	1.96
40	1.43	1.30	1.44	1.84	−2.59	3.64
41	1.56	1.56	1.92	2.76	−4.44	3.36
42	1.69	0.91	1.28	2.07	−3.70	3.08
43	0.91	1.04	1.44	2.30	−4.07	3.36
44	1.04	1.30	1.92	1.61	−3.33	3.08
45	1.17	1.56	1.28	2.53	−2.59	2.80
46	1.30	0.91	1.76	1.84	−4.44	2.52
47	1.43	1.17	1.12	2.76	−3.70	2.24
48	1.56	1.43	1.60	2.07	−2.96	1.96
49	1.69	1.69	2.08	2.99	−4.81	3.64

3.2.3. Identification of Boundary Conditions

A reliable nonlinear relationship between boundary conditions and the stress at the measured points is crucial for the accuracy of the inversed in situ stress distribution. However, the traditional method may encounter some issues such as 'over-fitting' or 'under-fitting' due to the uncertainty in the training sample size. The limitations restrict the ability of nonlinear mapping to approximate and subsequently affect the accuracy and reliability of the algorithm results. To address this issue, the PSO optimization method was used to optimize the BP neural network and establish a more accurate and reliable nonlinear model.

The PSO-BP neural network was trained using the boundary conditions listed in Table 4 as input samples and the corresponding calculated stresses at the measured points as sample outputs. The Shuangjiangkou hydropower station has five measurement points around the left underground caverns, each with six stress values; thus, a 30-16-6-6 topological network was constructed. The mean square error of the network model was set to 10^{-8}, which was used to judge the training accuracy of iterations. The PSO-BP network was stopped after 3367 iterations when the error reached the preset value. Subsequently, the actual stresses at each measured point are input into the trained PSO-BP model to search for the optimal combination of boundary conditions. The inversed coefficients obtained by the PSO-BP neutral network are k_g = 1.43, k_x = 1.26, k_y = 1.60, τ_{xz} = 1.61 MPa, τ_{yz} = −2.51 MPa, and τ_x = 1.44 MPa.

To validate the accuracy of the determined boundary conditions, the obtained coefficients are utilized in the FLAC^{3D} software for additional forward analysis. Table 5 gives the stress results for each measurement point. It shows that the analysis results are generally acceptable for most points, except for the resulted σ_z of point 5#, where the analysis error is 20.33%. The average analysis error for the remaining points is 8.09%. Compared with Table 3, the measured σ_x and σ_y of point 5# are almost equal to those of other points, but the measured σ_z of point 5# is 7–15 MPa greater than that of other points, so it may not be accurately measured. As a result, the back-analysis σ_z of point 5# does not match very well with the measured data. Figure 6 compares the inversed and tested stresses. The dashed line indicates where the two stresses are equal. It can be observed that there is excellent agreement between the PSO-BP results and measured data, with a correlation coefficient of about 0.95.

Table 5. Computational stress components at test points.

Test No.	σ_x/MPa	σ_y/MPa	σ_z/MPa	τ_{xz}/MPa	τ_{yz}/MPa	τ_{xy}/MPa
1	18.33	20.07	14.83	1.02	−2.63	1.22
2	18.08	20.91	14.45	1.12	−2.18	1.27
3	11.98	18.10	9.41	1.63	−0.80	0.37
4	17.72	22.48	14.02	1.69	−2.34	0.73
5	17.96	23.49	14.00	2.65	−2.45	0.79

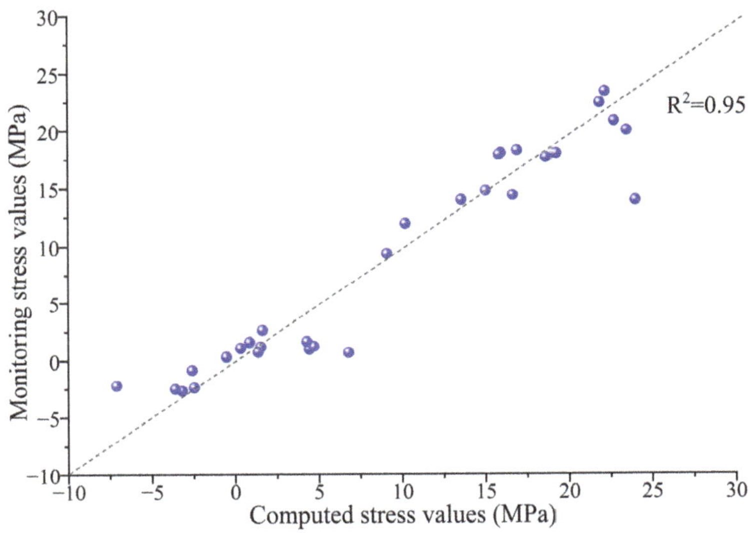

Figure 6. Comparison of inverted results with measured results.

3.3. Characteristics of In Situ Stress around Underground Caverns

The purpose of the in situ stress inversion is to investigate the distribution of in situ stress in the engineering area. By studying its distribution characteristics, it is possible to provide some references for the engineering's design, construction, and long-term stability. Hence, based on the boundary conditions obtained from the PSO-BP inversion, the distribution characteristics of the in situ stress in the Shuangjiangkou underground cavern area were computed.

Figure 7 shows the stress distribution in the cross-section of the main powerhouse, while Figure 8 plots that in the longitudinal section direction. As can be seen, there is an increasing trend in the distribution of in situ stresses with the increase in rock burial depth. The shallow surface presents local stress relaxation with relatively low stress values in this area. This is followed by a stress transition zone, where the principal stress distribution gradually becomes horizontal with a progressive increase in stress. The deeper part of the rock has a higher stress level and a more stable stress state. The first principal stress ranges from 20 to 26 MPa, with some stress concentration occurring around the river valley. The third principal stress ranges from 14 to 18 MPa and is predominantly oriented in the vertical direction. Fault F1 and the lamprophyre notably influence the stress distribution in the main powerhouse area, resulting in a significant reduction in the stresses within 10 m.

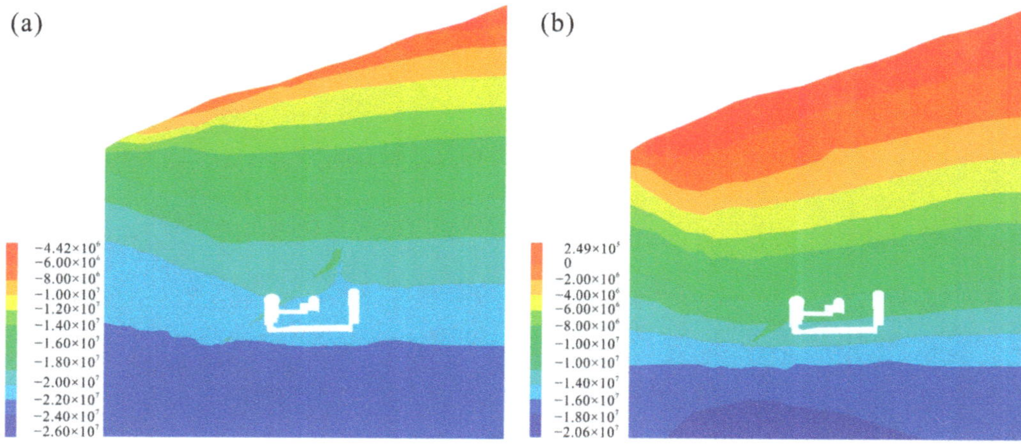

Figure 7. Stress distribution in the cross-section of the main powerhouse. (**a**) the first principal stress, (**b**) the third principal stress.

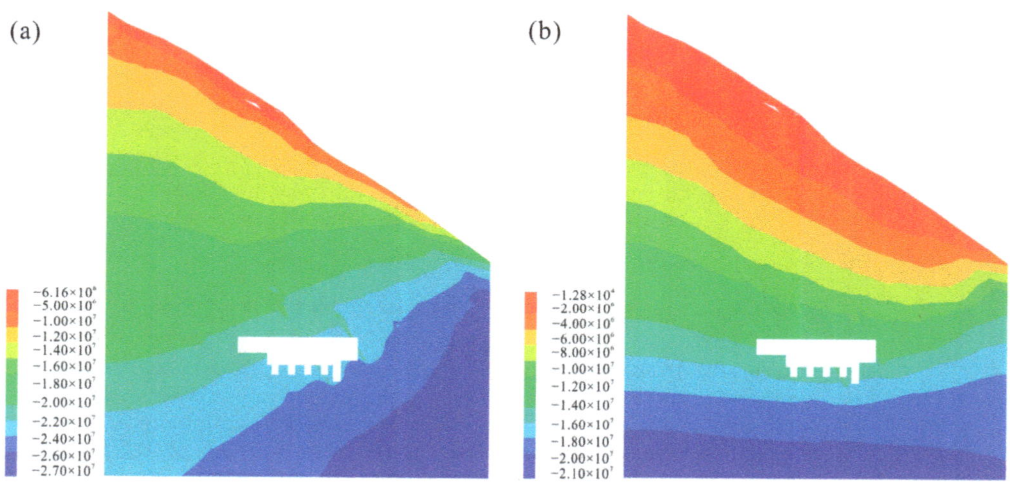

Figure 8. Stress distribution in the longitudinal section of the main powerhouse. (**a**) the first principal stress; and (**b**) the third principal stress.

The results of the in situ stress inversion indicate that weak structural planes may have the potential to significantly influence the stress state of an engineering area. As shown in Figure 9, all three stress components increase with the buried depth while remaining relatively consistent within the main powerhouse area at the same elevation. There is a sudden drop in stress values at the fault area, with the σ_x falling to around 10–12 MPa, σ_y about 17–20 MPa, and σ_z about 8–10 MPa. However, the stress state around the tailrace surge chambers does not fluctuate since the fault and the lamprophyre do not intersect with the tailrace surge chambers. Thus, faults and lamprophyre could notably influence the in situ stress distribution, leading to a reduction in the in situ stress value of the surrounding rock.

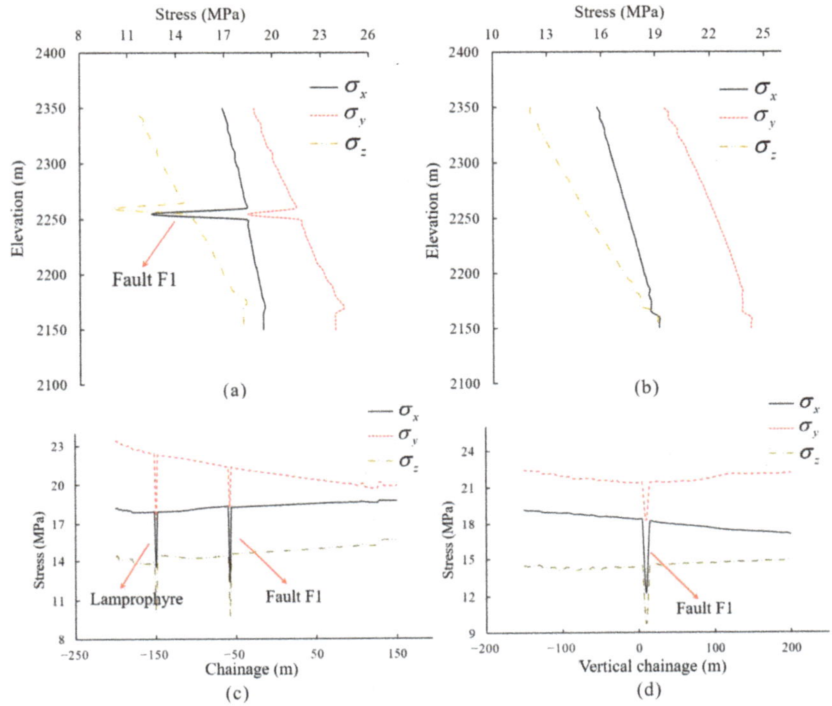

Figure 9. Stress variations along the depth (**a**) in the main powerhouse and (**b**) in the tailrace surge chambers; Stress variations in the plane of 2260 m elevation along the axis of the main powerhouse (**c**) and along the vertical axis of the main powerhouse (**d**).

4. Discussion

To validate the improvement of the proposed method, a comparison between the PSO-BP in situ stress numerical analysis method and the conventional method is conducted in Shuangjiangkou Hydropower Station engineering.

For comparison, the BP network was constructed the same as the PSO-BP network in this study, with a topology of 30-16-6-6 and the mean square error of the network model set to 10^{-8}. The BP neural network was also trained using the boundary conditions listed in Table 4 as input samples and the calculated stresses at the measured points as sample outputs. After the iterative search, the optimal combination of boundary conditions from the BP network method can be obtained. The inversed coefficients are k_g = 1.35, k_x = 1.18, k_y = 1.72, τ_{xz} = 0.73 MPa, τ_{yz} = −2.29 MPa, and τ_x = 1.75 MPa. Then, the obtained coefficients are utilized in the FLAC^{3D} software for additional forward analysis, which generates the computational results from the conventional method, as shown in Table 6. For comparison, the measured data and the results from the PSO-BP method are also listed in this table.

As can be seen, the BP network method also inversed a relatively reasonable combination of boundary conditions, while the average relative error of the PSO-BP results is 3.45% lower than that of the conventional method, indicating a closer approximation to the actual in situ stress. Therefore, for in situ stress inversion using the same engineering, the PSO-BP method may evidently improve the accuracy of the inversion results compared with the conventional BP network method. It should be noted that the PSO-BP method has some limitations, one of them being the requirement of sufficient input samples. While the huge training samples result in a significant increase in computational cost, the accuracy of the results may improve as the number of samples increases. Hence, it is necessary to select a reasonable and sufficient number of training samples for different engineering problems.

Table 6. Computational stress components at test points. A comparison of the computational stress of the two back analysis methods.

Test No.	Algorithm	σ_x/MPa	σ_y/MPa	σ_z/MPa	τ_{xz}/MPa	τ_{yz}/MPa	τ_{xy}/MPa
1	Measured	16.92	23.50	15.04	4.44	−3.19	1.54
	BP	14.50	21.07	13.47	0.69	−2.18	1.29
	PSO-BP	18.33	20.07	14.83	1.02	−2.63	1.22
2	Measured	19.26	22.74	16.66	0.33	−7.11	4.71
	BP	14.36	21.87	13.31	0.47	−1.88	1.45
	PSO-BP	18.08	20.91	14.45	1.12	−2.18	1.27
3	Measured	10.21	15.95	9.10	0.86	−2.60	-0.52
	BP	10.52	19.21	8.75	1.21	−0.21	0.71
	PSO-BP	11.98	18.10	9.41	1.63	−0.80	0.37
4	Measured	18.64	21.86	13.56	4.29	−2.47	6.81
	BP	14.22	23.49	13.21	1.30	−1.92	1.82
	PSO-BP	17.72	22.48	14.02	1.69	−2.34	0.73
5	Measured	15.84	22.18	24.01	1.64	−3.61	1.37
	BP	14.27	24.38	13.33	1.61	−2.07	0.95
	PSO-BP	17.96	23.49	14.00	2.65	−2.45	0.79

5. Conclusions

To improve the accuracy of the in situ stress inversion, a PSO-BP in situ stress numerical analysis method was proposed and applied to the in situ stress inversion of the Shuangjiangkou Hydropower Station underground engineering. Based on the results of the numerical simulation, the feasibility of the method was verified, and the characteristics of the in situ stress distribution around the underground caverns were obtained. The main conclusions are drawn as follows:

(1) The complex geological movements can be decomposed into six basic boundary conditions for numerical simulation, and the PSO-BP neural network can then identify the optimal boundary conditions that best represent the measured in situ stresses.

(2) The average relative error of the inversion results obtained by the proposed method is 3.45% lower than those obtained by the traditional method; thus, the proposed method is more effective and the inversed in situ stress distribution is more reasonable.

(3) The distribution of in situ stresses increases as the buried depth of rock increases. The first principal stress ranges from 20 to 26 MPa, with stress concentration around the river valley. The third principal stress ranges from 14 to 18 MPa and is oriented approximately vertically.

(4) Fault F1 and the lamprophyre notably influence the stress distribution in the main powerhouse area, with 15–30% localized stress reduction in the rock mass within 10m, which may provide a reference for similar engineering with weak structural planes.

Author Contributions: Conceptualization, H.-C.Y., H.-Z.L. and L.Z.; Data curation, Y.L. and K.-P.C.; Formal analysis, H.-C.Y.; Funding acquisition, H.-Z.L.; Investigation, H.-C.Y.; Methodology, H.-C.Y., H.-Z.L. and L.Z.; Project administration, M.-L.X.; Resources, H.-Z.L. and Y.L.; Software, H.-C.Y. and J.-M.W.; Validation, H.-C.Y.; Writing—original draft, H.-C.Y.; Writing—review and editing, H.-C.Y., H.-Z.L., L.Z., M.-L.X., K.-P.C., J.-M.W. and J.-L.P. All authors have read and agreed to the published version of the manuscript.

Funding: This research was funded by the National Natural Science Foundation of China (No. 52109135), and the Postdoctoral Research Foundation of China (2019M653402).

Institutional Review Board Statement: Not applicable.

Informed Consent Statement: Not applicable.

Data Availability Statement: Data sharing is not applicable.

Conflicts of Interest: The authors declare that they have no known competing financial interests or personal relationships that could have appeared to influence the work reported in this paper.

References

1. Song, Z.F.; Sun, Y.J.; Lin, X. Research on In Situ Stress Measurement and Inversion, and its Influence on Roadway Layout in Coal Mine with Thick Coal Seam and Large Mining Height. *Geotech. Geol. Eng.* **2017**, *36*, 1907–1917. [CrossRef]
2. Li, S.; Feng, X.T.; Li, Z.H.; Chen, B.; Zhang, C.; Zhou, H. In situ monitoring of rockburst nucleation and evolution in the deeply buried tunnels of Jinping II hydropower station. *Eng. Geol.* **2012**, *137–138*, 85–96. [CrossRef]
3. Zhang, G.; Li, Y.; Meng, X.; Tao, G.; Wang, L.; Guo, H.; Zhu, C.; Zuo, H.; Qu, Z. Distribution Law of In Situ Stress and Its Engineering Application in Rock Burst Control in Juye Mining Area. *Energies* **2022**, *15*, 1267. [CrossRef]
4. Gong, M.; Qi, S.; Liu, J. Engineering geological problems related to high geo-stresses at the Jinping I Hydropower Station, Southwest China. *Bull. Eng. Geol. Environ.* **2010**, *69*, 373–380. [CrossRef]
5. Haimson, B.C. Deep in-situ stress measurements by hydrofracturing. *Tectonophysics* **1975**, *29*, 41–47. [CrossRef]
6. Aadnoy, B.S. Inversion technique to determine the in-situ stress field from fracturing data. *J. Pet. Sci. Eng.* **1990**, *4*, 127–141. [CrossRef]
7. Ljunggren, C.; Chang, Y.; Janson, T.; Christiansson, R. An overview of rock stress measurement methods. *Int. J. Rock Mech. Min. Sci.* **2003**, *40*, 975–989. [CrossRef]
8. Bao, T.; Burghardt, J. A Bayesian Approach for In-Situ Stress Prediction and Uncertainty Quantification for Subsurface Engineering. *Rock Mech. Rock Eng.* **2022**, *55*, 4531–4548. [CrossRef]
9. Meng, Q.; Chen, Y.; Zhang, M.; Han, L.; Pu, H.; Liu, J. On the Kaiser Effect of Rock under Cyclic Loading and Unloading Conditions: Insights from Acoustic Emission Monitoring. *Energies* **2019**, *12*, 3255. [CrossRef]
10. Xu, D.; Huang, X.; Jiang, Q.; Li, S.; Zheng, H.; Qiu, S.; Xu, H.; Li, Y.; Li, Z.; Ma, X. Estimation of the three-dimensional in situ stress field around a large deep underground cavern group near a valley. *J. Rock Mech. Geotech. Eng.* **2021**, *13*, 529–544. [CrossRef]
11. Wang, B.L.; Ma, Q.C. Boundary element analysis methods for ground stress field of rock masses. *Comput. Geotech.* **1986**, *2*, 261–274. [CrossRef]
12. McKinnon, S.D. Analysis of stress measurements using a numerical model methodology. *Int. J. Rock Mech. Min. Sci.* **2001**, *38*, 699–709. [CrossRef]
13. Meng, W.; He, C. Back Analysis of the Initial Geo-Stress Field of Rock Masses in High Geo-Temperature and High Geo-Stress. *Energies* **2020**, *13*, 363. [CrossRef]
14. Zhang, Z.; Gong, R.; Zhang, H.; Lan, Q.; Tang, X. Initial ground stress field regression analysis and application in an extra-long tunnel in the western mountainous area of China. *Bull. Eng. Geol. Environ.* **2021**, *80*, 4603–4619. [CrossRef]
15. Chen, B.; Ren, Q.; Wang, F.; Zhao, Y.; Liu, C. Inversion Analysis of In-situ Stress Field in Tunnel Fault Zone Considering High Geothermal. *Geotech. Geol. Eng.* **2021**, *39*, 5007–5019. [CrossRef]
16. Yu, R.; Tan, Z.; Gao, J.; Wang, X.; Zhao, J. Inversion and Analysis of the Initial Ground Stress Field of the Deep-Buried Tunnel Area. *Appl. Sci.* **2022**, *12*, 8986. [CrossRef]
17. Yong, L.; Yunhua, G.; Weishen, Z.; Shucai, L.; Hao, Z. A modified initial in-situ Stress Inversion Method based onFLAC3D with an engineering application. *Open Geosci.* **2015**, *7*, 824–835. [CrossRef]
18. Zhang, S.; Yin, S. Determination of in situ stresses and elastic parameters from hydraulic fracturing tests by geomechanics modeling and soft computing. *J. Pet. Sci. Eng.* **2014**, *124*, 484–492. [CrossRef]
19. Li, X.; Zhou, X.; Xu, Z.; Feng, T.; Wang, D.; Deng, J.; Zhang, G.; Li, C.; Feng, G.; Zhang, R.; et al. Inversion Method of Initial In Situ Stress Field Based on BP Neural Network and Applying Loads to Unit Body. *Adv. Civ. Eng.* **2020**, *2020*, 15. [CrossRef]
20. Li, G.; Hu, Y.; Li, Q.B.; Yin, T.; Miao, J.X.; Yao, M. Inversion Method of In-situ Stress and Rock Damage Characteristics in Dam Site Using Neural Network and Numerical Simulation—A Case Study. *IEEE Access* **2020**, *8*, 46701–46712. [CrossRef]
21. Eberhart, R.; Kennedy, J. A new optimizer using particle swarm theory. In Proceedings of the Sixth International Symposium on Micro Machine and Human Science (Cat. No.95TH8079), MHS'95, Nagoya, Japan, 4–6 October 1995; pp. 39–43. [CrossRef]
22. Guo, M.W.; Yin, S.D.; Li, C.G.; Wang, S.L. 3D In Situ Stress Estimation by Inverse Analysis of Tectonic Strains. *Appl. Sci.* **2021**, *11*, 5284. [CrossRef]
23. Matsuki, K.; Nakama, S.; Sato, T. Estimation of regional stress by FEM for a heterogeneous rock mass with a large fault. *Int. J. Rock Mech. Min. Sci.* **2009**, *46*, 31–50. [CrossRef]
24. Pielke, R.A. Chapter 6—Coordinate Transformations. In *International Geophysics*; Pielke, R.A., Ed.; Academic Press: Cambridge, MA, USA, 2013; Volume 98, pp. 111–141.
25. Baghbani, A.; Choudhury, T.; Costa, S.; Reiner, J. Application of artificial intelligence in geotechnical engineering: A state-of-the-art review. *Earth-Sci. Rev.* **2022**, *228*, 26. [CrossRef]
26. Jiang, Q.; Feng, X.T. Intelligent Stability Design of Large Underground Hydraulic Caverns: Chinese Method and Practice. *Energies* **2011**, *4*, 1542–1562. [CrossRef]
27. Yuhui, S.; Eberhart, R.C. Parameter selection in particle swarm optimization. In Proceedings of the 7th International Conference, EP98, San Diego, CA, USA, 25–27 March 1998; pp. 591–600. [CrossRef]
28. Bieniawski, Z.T. *Engineering Rock Mass Classifications: A Complete Manual for Engineers and Geologists in Mining, Civil, and Petroleum Engineering*; Wiley: Hoboken, NJ, USA, 1989.

29. Rehman, H.; Ali, W.; Naji, A.; Kim, J.-j.; Abdullah, R.; Yoo, H.-K. Review of Rock-Mass Rating and Tunneling Quality Index Systems for Tunnel Design: Development, Refinement, Application and Limitation. *Appl. Sci.* **2018**, *8*, 1250. [CrossRef]
30. Cai, M.; Kaiser, P.K.; Uno, H.; Tasaka, Y.; Minami, M. Estimation of rock mass deformation modulus and strength of jointed hard rock masses using the GSI system. *Int. J. Rock Mech. Min. Sci.* **2004**, *41*, 3–19. [CrossRef]
31. The Ministry of Water Resources of China. *Chinese National Standard GB/T 50218-2014: Standard for Engineering Classi-fication of Rock Mass*; China Planning Press: Beijing, China, 2014.

applied sciences

MDPI

Article

Formation Mechanism and Control Technology of an Excavation Damage Zone in Tunnel-Surrounding Rock

Hongxian Fu [1,2], Xiaoming Guan [3], Chun Chen [4], Jianchun Wu [4], Qiqiang Nie [4], Ning Yang [3,*], Yanchun Liu [3] and Junwei Liu [3]

1 Key Laboratory of Urban Underground Engineering of Ministry of Education, Beijing Jiaotong University, Beijing 100044, China
2 College of Civil Engineering, Beijing Jiaotong University, Beijing 100044, China
3 College of Civil Engineering, Qingdao University of Technology, Qingdao 266520, China
4 China Railway No. 10 Engineering Group No. 3 Construction Co., Ltd., Hefei 230031, China
* Correspondence: nee_young@163.com

Abstract: Loosened rock circle is formed around the tunnel when the tunnel is constructed by the drilling and blasting method. The size of the loosened rock circle around the tunnel and the degree of internal rock fragmentation has an important influence on the support parameters, durability, and safety of the tunnel. Firstly, referencing an existing tunnel project, blasting tests using nonelectronic and electronic detonators were carried out to determine the influence of blasting construction on the scope of the rock loose circle and the degree of rock fragmentation. Then, a numerical simulation was used to study the contribution of the blasting impact and surrounding rock stress redistribution on the loosened rock circle around the tunnel. The results showed that the range of the loosened rock circle around the tunnel generated by the normal blasting of nonelectronic detonators was 1.5~2.3 m, and the wave velocity of the rock mass in the loosened rock circle around the tunnel decreased to 23~36%. The size of the loosened rock circle around the tunnel generated by the blasting impact was 0.66 m, accounting for 33% of the range of the loosened rock circle around the tunnel. The range of the loosened rock circle around the tunnel produced by electronic detonator blasting was 0~1.4 m. The wave velocity of the rock mass in the loosened rock circle around the tunnel decreased to 12~17%. The range of the loosened rock circle around the tunnel was approximately 60~76% of that of detonator blasting, and the broken degree of the surrounding rock in the loosened rock circle around the tunnel was small. The research results can provide a reference for the optimization design of preliminary support parameters of tunnels, such as anchors and steel arches in blasting construction.

Keywords: tunnel blasting; loosened rock circle; acoustic method; nonelectronic detonators; electronic detonator; control technology

Citation: Fu, H.; Guan, X.; Chen, C.; Wu, J.; Nie, Q.; Yang, N.; Liu, Y.; Liu, J. Formation Mechanism and Control Technology of an Excavation Damage Zone in Tunnel-Surrounding Rock. *Appl. Sci.* **2023**, *13*, 1006. https://doi.org/10.3390/app13021006

Academic Editors: Zhengzhao Liang, Bei Jiang and Nuwen Xu

Received: 28 November 2022
Revised: 7 January 2023
Accepted: 9 January 2023
Published: 11 January 2023

1. Introduction

The drilling and blasting methods are the most commonly used in the construction of rock tunnels. The protected surrounding rock is damaged, and its stress is redistributed after the rock within a tunnel outline is thrown out by the blasting load during excavation. The strength of the protected surrounding rock is reduced, and the original cracks of it are further widened to form a circle of loosened rock around the tunnel. This has a significant influence on the supporting parameters, durability, and structural safety of the tunnel.

The range, fragmentation degree, and permeability of loosened rock circles around the tunnel were extensively studied in terms of hydropower station chamber excavation, tunnel engineering, and waste storage cavern engineering [1]. Failure zone, fracture zone, and disturbance zone are formed in the surrounding rock after tunnel excavation. A large number of rocks are separated from the surrounding rock and become unstable bodies in the failure zone, and the surrounding rock generates microcracks or fissures in a fractured zone. In the disturbance area, the stress of the surrounding rock and groundwater pressure has changed, yet other physical and mechanical properties of the surrounding rock have changed little [2].

At present, the range of loosened rock circles around the tunnel formed and their physical and mechanical properties after tunnel excavation have been studied by many scholars through theoretical analysis, field tests, and numerical simulation. A Russian scholar named Protodyakonov proposed the natural equilibrium arch theory and initiated a theoretical study of the loosened rock circle around a tunnel in the early 20th century. Subsequently, scholars established wedge theory, elastic medium theory, flexible medium theory, loose fracture theory, elastic–plastic medium theory, fracture zone diagram theory, and discontinuous zone theory [3]. Zhang et al. [4] and Jiang et al. [5] derived an analytical solution of the radius and stress of the plastic zone of a circular roadway based on elastic–plastic theory, the Drucker–Prager criterion, and the double-shear unified strength criterion. Wang et al. [6] and Wang et al. [7] derived an approximate solution to the boundary equation of the plastic zone of the surrounding rock of an unequal-pressure circular roadway based on the Hoek–Brown strength criterion and the Mohr–Coulomb strength criterion. Meng et al. [8] and Guan et al. [9,10] verified the elastic–plastic damage mechanical model of the surrounding rock and computed the plastic zone range. Based on elastic–plastic theory, many scholars have used different strength criteria and approved analytical methods to calculate the loosened rock circles around standard circular tunnels. Nevertheless, these computational results are problematic when meeting engineering needs. In order to obtain more practical theoretical analysis results, Ma et al. [11] deduced an analytical solution in which the plastic zone of a tunnel is controlled by the yield criterion during the failure process, and stress field changes were obtained in the plastic zone of the tunnel.

At the same time, the development of science and technology has brought a variety of advanced monitoring instruments, which have made outstanding contributions to the monitoring test of surrounding rock loose circles. Many scholars adopted multipoint displacement measurements, the seismic wave method, the geological radar method, and the borehole imaging method to study loose circles in tunnels. At present, the single-hole acoustic monitoring method [12] and the borehole acoustic monitoring method [13,14] are the most widely used methods to study the range of tunnels surrounding rock loose circles.

With improvements in computing ability and numerical algorithms, researchers have begun to use numerical simulations to calculate the range of loose circles. This method can fully combine the results of field monitoring and theoretical deduction, which is a hot topic in the research of the measuring range of loose circles [15]. Wei et al. [16], Han et al. [17], and Pan et al. [18,19] studied the deformation and failure characteristics of the surrounding rock by using particle flow and discrete element simulation methods that referenced a number of microfeature indicators, such as the microcrack field, crack propagation direction and final trend, and they provided a scientific explanation for the nature of many physical phenomena. In addition to blasting factors, the existence of in situ stress can also lead to stress fracture cycles in a tunnel excavation section. Siren et al. [20] conducted research to distinguish between blasting and stress-induced fracture zones.

Blasting technology has a great influence on the range and mechanical properties of tunnels surrounding rock loose circles. The tunnel blasting method can be categorized as either ordinary blasting, smooth blasting, or micro-vibration blasting. Through field tests and research, the loose circle formed by ordinary blasting has been deemed the largest, followed by smooth blasting, and the smallest circle is created by micro-vibration blasting [21]. However, there are few studies on the loose ring under the combined use of various blasting methods. Zhou et al. [22] believed that multiple blasting methods only produced cumulative damage at the cracks or fissures and increased the displacement at the joint surface, and the loose ring size depended on the maximum burst. Liu et al. [14] and Li et al. [23] found that multiple explosions expanded the loose circle from 2.8 m to 3.2 m through the acoustic wave method and borehole peeping method and increased the degree of rock breakage, and the damage degrees of sidewalls and vaults can increase by 36~50% and 47~55% in loose circles, respectively. Ji et al. [24] used field tests and finite element simulations to study the changes in the cumulative damage zones of tunnel-surrounding rock under multiple excavation and blasting conditions, and the damage degrees of the surrounding rock were sorted, divided, and summarized. Yang et al. [25,26] studied the difference between the blasting vibration on a tunnel surface and the blasting vibration inside the surrounding rock through field tests and the differences between the vibration peaks and dominant frequencies on the tunnel surface and the interior of the surrounding rock.

At present, there is no unified and clear definition of a loose tunnel circle, and the disturbance zone, plastic zone, and loose circle are often muddled. The technical construction specification for rock foundation excavation in hydraulic structures stipulates that the change rate of wave velocity greater than 10% after blasting is the criterion of rock mass failure. This study used this standard to determine the loose circle of a tunnel. The loosening circle is formed under the dual actions of blasting and stress redistribution in the surrounding rock. However, the contribution of the two to the final loose circle remains unclear. In this study, referencing the construction of a tunnel, blasting construction tests of nonelectric detonators and electronic detonators are carried out. The vibration velocity and the range of the loose circle produced by the two blasting methods were tested. The field test data and numerical simulation method were used to explore the influence of the surrounding rock stress redistribution and blasting load on the range of the loose circle. The results show that the use of electronic detonators is more conducive to the protection of tunnel surrounding rock than non-electric detonators and can significantly reduce the rock damage range.

2. General Information of the Project

The research background project is Renhechang Tunnel, which is an important project on the Chongqing-Huaihua Railway and the total length is 4734 m. The rock layer is mainly mudstone with sandstone with some developed joints, and water seepage often occurs during excavation.

In order to study the formation mechanism of the loose circle produced by tunnel blasting, tunnel nonelectric detonator blasting, and electronic detonator blasting tests are carried out in a section. The new railway tunnel passes underneath an existing highway tunnel in this section. The distance between the new tunnel and the existing tunnel is 30~31 m, and the length of the underpass is 70 m. The tunnel of this section passes through a geological section of mudstone clamp sandstone, and the surrounding rock grade is IV. The physical and mechanical properties of the surrounding rock of the tunnel are shown in Table 1.

Table 1. Material property parameters of main stratum types of excavated tunnels.

Geotechnical Name	Unit Weight γ (kN/m^3)	Cohesion C (kPa)	Angle of Internal Friction φ (°)	Basic Bearing Capacity σ_0 (kPa)
Gravel soil	21		20	100
Mollison	18.5	12	7	100
Silty clay	19	20	15	150
Mudstone clamp	22		35	300
Sandstone	23		45	450

3. Blasting and Vibration Tests

Firstly, the first blasting damage test of the surrounding rock was carried out with nonelectric detonators. The hole arrangement and initiation sequence are shown in Figure 1, and the blasting parameters are shown in Table 2. Because the section passes through an operating highway tunnel, the blasting vibration test points are arranged in the operating highway tunnel and above the excavation face. The specific monitoring points are arranged in Figure 2. The y-direction blasting vibration velocity received by the three-direction velocity sensor is the largest according to the spatial position of the sensor and the explosion source, so these tests focus on the y-direction vibration velocity. The blasting vibration velocity waveform is shown in Figure 3.

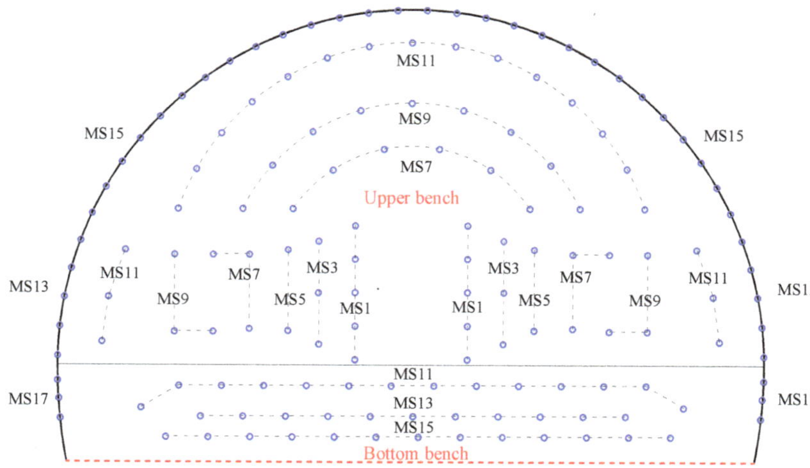

Figure 1. The hole arrangement and initiation sequence of nonelectric detonator blasting in the test. Note: The marked MS represents the detonator delay time series. The holes connected by dotted lines are in the same delay sequence and represent simultaneous initiation.

Table 2. Details of blasting holes and explosive parameters.

Bench	Name of the Boreholes	Number of Holes	Borehole Depth (m)	Single Hole Charge (kg)	Detonator Segment	Subsection Charge (kg)
The arch part of upper bench	Auxiliary hole	6	3	1.0	MS7	6.0
	Auxiliary hole	9	3	1.0	MS9	9.0
	Auxiliary hole	15	3	1.0	MS11	15.0
	Peripheral hole	28	3	0.6	MS15	16.8
The lower part of upper bench	Cutting hole	10	4.5	2.0	MS1	20.0
	Auxiliary hole	6	4	1.8	MS3	10.4
	Auxiliary hole	4	4	1.6	MS5	6.4
	Auxiliary hole	12	3	1.2	MS7	7.2
	Auxiliary hole	6	3	1.0	MS9	6.0
	Auxiliary hole	6	3	1.0	MS11	6.0
	Peripheral hole	10	3	0.6	MS13	6.0
Bottom bench	Auxiliary hole	12	3	1.8	MS11	21.6
	Auxiliary hole	13	3	1.8	MS13	23.4
	Auxiliary hole	14	3	1.8	MS15	25.2
	Peripheral hole	6	3	0.6	MS17	3.6
Total		157				182.6

Then, the blasting test of surrounding rock damage was carried out by using electronic detonators. In order to ensure that the two tests have high contrast, the same hole arrangement of the two tests is ensured, as shown in Figure 4. The hole charge weight and blasting vibration test point arrangements are the same as those for nonelectric detonator blasting. The vibration velocity waveform of the electronic detonators is shown in Figure 5.

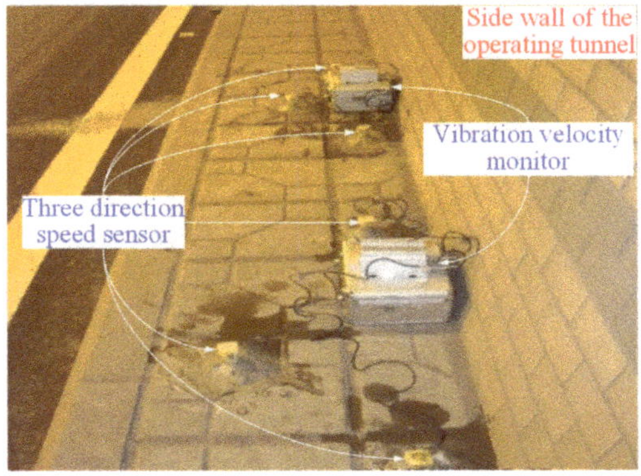

Figure 2. Vibration velocity monitoring points in operating tunnel above excavation face.

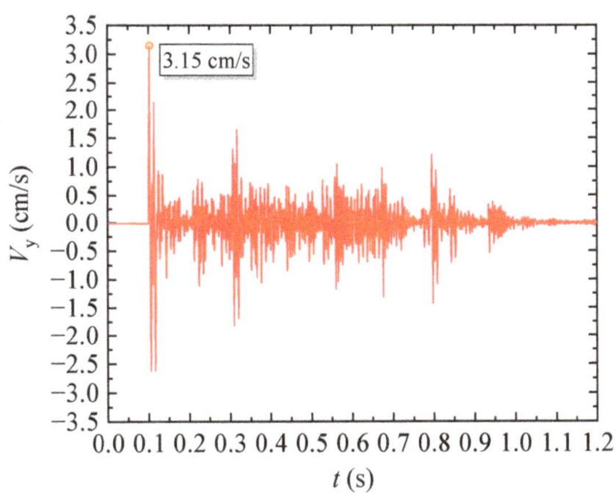

Figure 3. Vibration waveform of tunnel nonelectric detonator blasting.

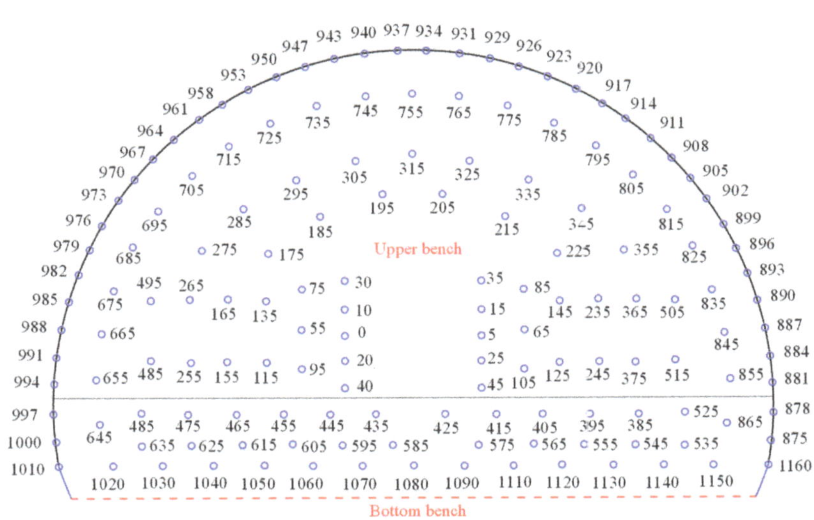

Figure 4. The blasting hole arrangement and initiation sequence of tunnel electronic detonators.

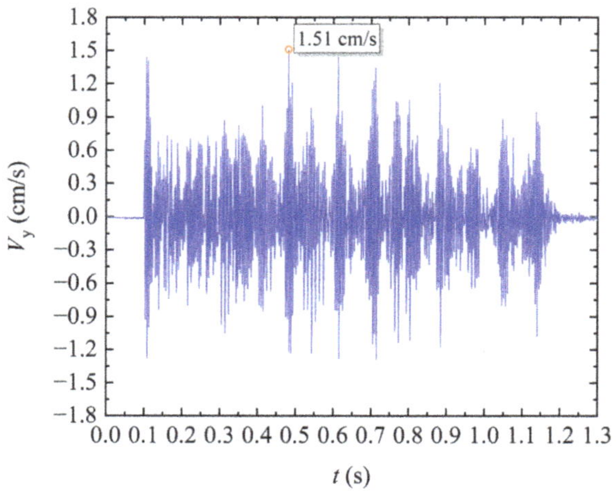

Figure 5. Blasting vibration velocity waveform of tunnel electronic detonators.

The maximum vibration velocity is 3.15 cm/s when the nonelectric detonators are used, and the maximum vibration velocity is reduced to 1.51 cm/s after the electronic detonator test, as shown in Figures 3 and 5, respectively. The maximum vibration velocity caused by electronic detonators is about 52.1% lower than that caused by nonelectric detonators, and the decrease is obvious. An important factor affecting the damage of surrounding rock is the maximum vibration velocity generated by cutting blasting. However, the maximum vibration velocity is randomly generated in a blast hole when using electronic detonators. Therefore, vibration reduction measures such as drilling empty holes to reduce cut blasting do not have advantages in the use of electronic detonator initiation. Another important reason is that the nonelectric detonator initiation delay accuracy is not easy to control. The delay parameter error of the design based on the burning velocity of the explosive is large, which makes it necessary to detonate multiple holes at the same time. On the contrary, the delay accuracy of electronic detonators has been significantly improved. It can realize the sequential initiation of single hole without considering the influence of cumulative error. In this way, the amount of explosive detonated at the same time is significantly less than that of nonelectric detonator. The technical improvement has led to the control of the possible higher vibration velocity from the source.

The vibration velocity monitoring results of the two tests are analyzed. It is found that the peak particle velocity produced by cutting blasting is the largest in the nonelectric detonator test. The vibration velocities generated by subsequent blastholes are significantly reduced. The continuous high vibration velocity of electronic detonators may have greater adverse effects on the surrounding rock. Therefore, the method of monitoring the loose range of surrounding rock on site is adopted to explore this hypothesis.

4. Monitoring the Loose Circle after Two Blasting Tests

The longitudinal wave velocity of the surrounding rock was monitored by the conventional single-hole test method using an RSM-SY5 acoustic detector after tunnel blasting excavation. The compressional wave velocity of the tunnel-surrounding rock had a good correlation with the degree of surrounding rock breakage. The longitudinal wave velocity of surrounding rock decreases with the increase in its fracture development degree. The conventional single-hole acoustic wave testing method is shown in Figure 6, and the layout of the test holes is shown in Figure 7. A simple trolley and pneumatic drill were used to drill the holes. According to field test experience, the hole depths were 4.0~4.4 m, and the hole diameters were 40 mm. The blasthole was set perpendicular to the tunnel wall as much as possible to meet the loose ring test requirements [27,28]. The drilling and testing process is shown in Figure 8. Due to the relatively developed cracks in the surrounding rock, the test holes in the vault could not be filled with water, resulting in the test holes not being tested.

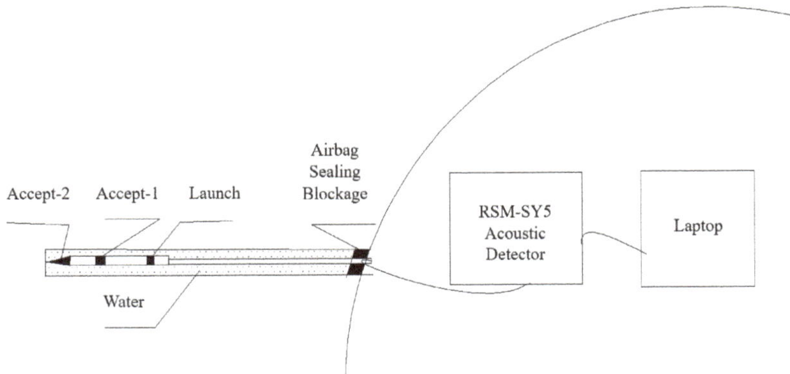

Figure 6. Schematic diagram of single-hole test by acoustic method.

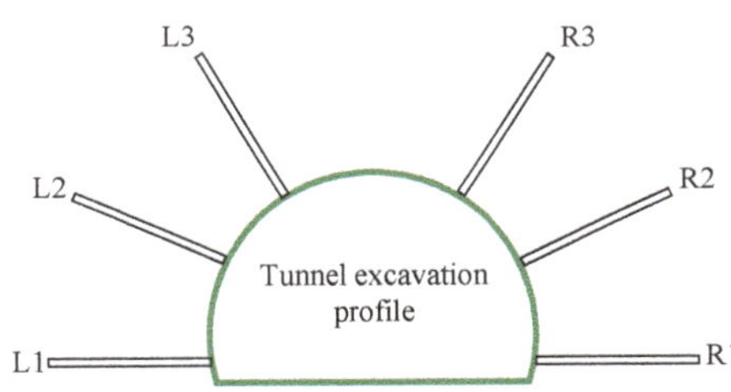

Figure 7. Schematic diagram of test hole layout.

Figure 8. Field test of the loose circle. (**a**) Use support trolley to assist drilling; (**b**) install monitoring sensors; (**c**) observe the monitoring results.

At present, there is no unified and clear definition of a loose tunnel circle. The technical code for the construction of rock foundation excavation engineering of hydraulic structures [29] notes that the change rate of wave velocity greater than 10% after blasting is the criterion of rock mass failure. Firstly, two footage sections with similar surrounding rock properties were first selected. In order to ensure the accuracy of the monitoring results, priority will be given to test work after blasting. Each monitoring test lasted nearly 12 h. The test analysis of the tunnel loose circle range excavated by nonelectric and electronic detonator blasting is shown in Figures 9 and 10. The position where the first rapid increase of wave velocity was monitored can be considered as the boundary position of the loose circle, and the blue dotted line is used as the auxiliary line to mark the boundary position of the loose circle. The shape of the loose circle during nonelectric and electronic detonator blasting is shown in Figure 11. The properties of the loose circle are shown in Table 3.

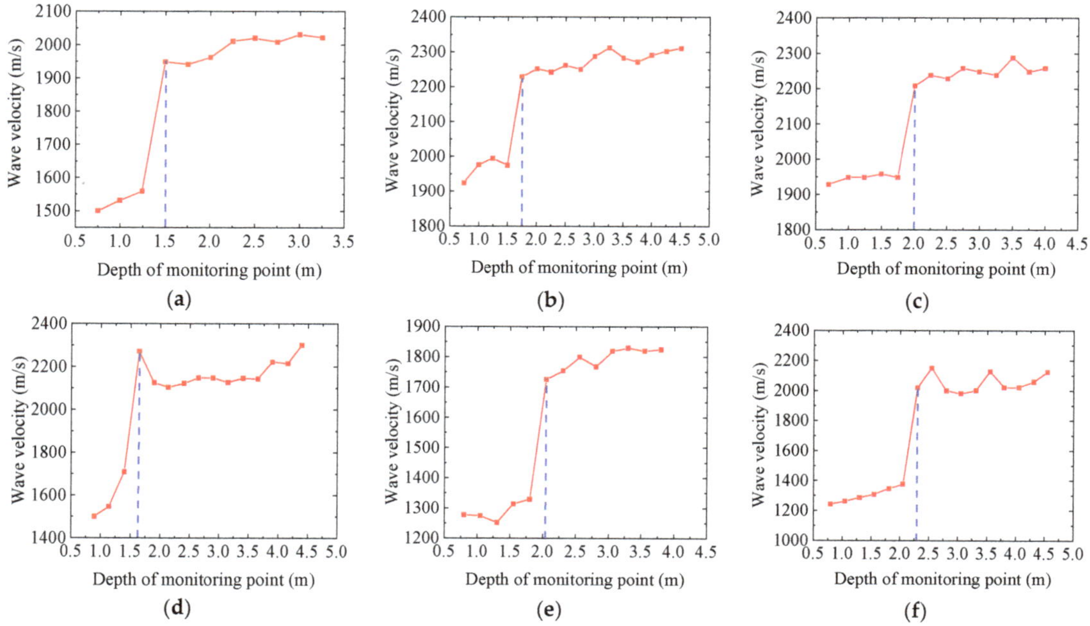

Figure 9. Variation in the longitudinal wave velocity of the surrounding rock with the test hole depth during tunnel nonelectric detonator blasting excavation. (**a**) R1, (**b**) R2, (**c**) R3, (**d**) L1, (**e**) L2, and (**f**) L3.

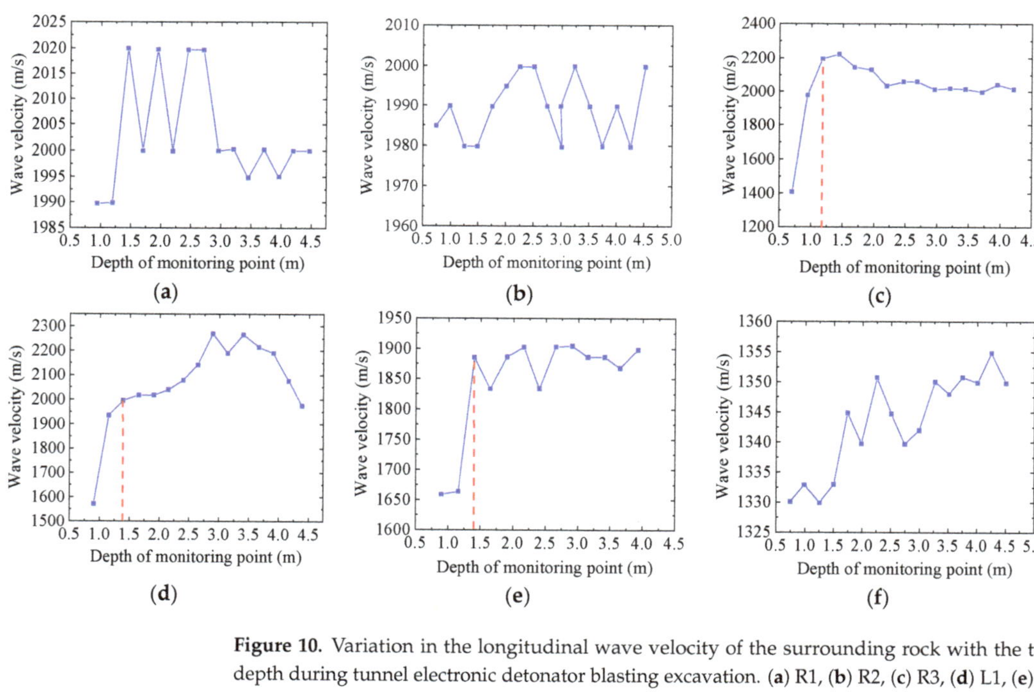

Figure 10. Variation in the longitudinal wave velocity of the surrounding rock with the test hole depth during tunnel electronic detonator blasting excavation. (**a**) R1, (**b**) R2, (**c**) R3, (**d**) L1, (**e**) L2, and (**f**) L3.

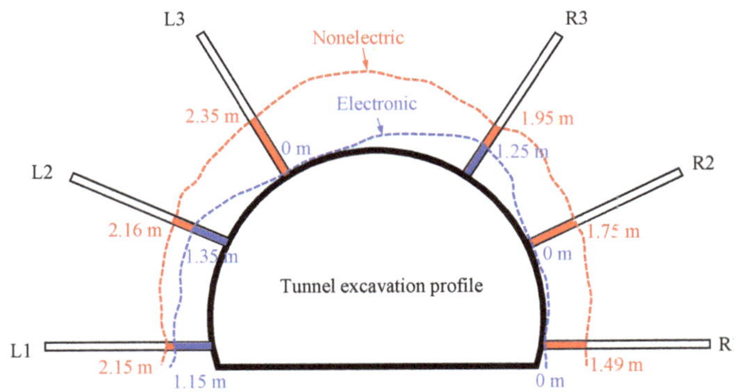

Figure 11. Schematic diagram of a loose circle during tunneling by nonelectric and electronic detonator blasting.

Table 3. Property parameters of the tunnel nonelectric and electronic detonator blasting loose circle.

Test Holes	Loosening Range (m)		Reduction Degree of Longitudinal Wave Velocity of Surrounding Rock	
	Nonelectric	Electronic	Nonelectric	Electronic
R1	1.49	0	22%	2%
R2	1.75	0	14%	2%
R3	1.95	1.25	13%	16%
L1	2.15	1.15	33%	17%
L2	2.16	1.35	28%	20%
L3	2.35	0	40%	2%

Contrastive analysis of loose circles produced by nonelectric detonator blasting and electronic detonator blasting in tunnels. Nonelectric detonators and electronic detonators were used in field tests, while ensuring the similarity of surrounding rock properties at the two blasting footages as much as possible.

(1) Fragmentation degree of the rock mass in a loose circle. In terms of loose range of surrounding rock, the depth of the loose circle caused by the nonelectric detonator was 1.49~2.35 m, and the depth caused by the electronic detonator was 0~1.35 m. The loose range caused by nonelectric detonators is significantly larger than that of electronic detonators in each radial direction, as shown in Figure 11. The serious damage depth of 2.35 m is avoided in the L3 direction, and the reduction in other directions is basically above 1 m. Two sections with the same surrounding rock properties cannot be found in the field test, and the random distribution of joint fissures cannot be avoided. There are a large number of joint fissures in the left tunnel wall and the right tunnel spandrel of the test section using electronic detonators after field investigation. The rock is mixed with a weak sand layer, and the water seepage phenomenon is obvious. This leads to the phenomenon of large damage depth at the L1, L2, and R3 positions in the electronic detonator test results. The electronic detonators can reduce the damage depth of more than 1 m, which can significantly reduce the use of supporting bolts, resulting in greater economic benefits, according to the construction experience. At the same time, it also avoids the construction process of post grouting reinforcement. More importantly, it can ensure the safety and stability of the tunnel after blasting.

(2) In addition, the use of electronic detonators not only reduces the depth of the damage range but also has a significant effect on retaining the integrity of the surrounding rock. The wave velocity of the rock mass in the loose circle caused by the nonelectric detonator decreases to 13~40%, and that of the rock mass in the loose circle caused by

the electronic detonator decreases to 16~20%, as shown in Table 3. The use of electronic detonators can significantly weaken the reduction of longitudinal waves in the surrounding rock. It is conducive to giving full play to the supporting role of the original surrounding rock and ensuring the stability of the excavated tunnel.

The field tests consume a lot of human and material resources and delay the construction period. It cannot be carried out on a large scale. Therefore, in order to further study the differences between electronic and nonelectric detonators in tunnels surrounding rock damage, it is necessary to use finite element software for further research and analysis.

5. Finite Element Analysis and Removal of Stress Redistribution Factors

Large finite element software FLAC3D has been widely used in geotechnical and underground engineering. It can be analyzed by the "mixed discrete method" when simulating plastic failure. It can provide more accurate and reasonable results than the ordinary discrete integration method. Therefore, FLAC3D software was used for numerical simulation in this study.

The tunnel excavation section area was approximately 150 m^2 with a span of 14.7 m and a height of 12.5 m. To avoid the influence of the model boundary effect, in the numerical model, the distance between the tunnel center axis and the left and right boundaries was greater than four times the tunnel span, the distance between the tunnel center axis and the upper and lower boundaries was also greater than three times the tunnel span. The excavation footage was 3 m. The whole calculation model was 120 m wide and 100 m high, with 49,152 units in total. The models are shown in Figures 12 and 13.

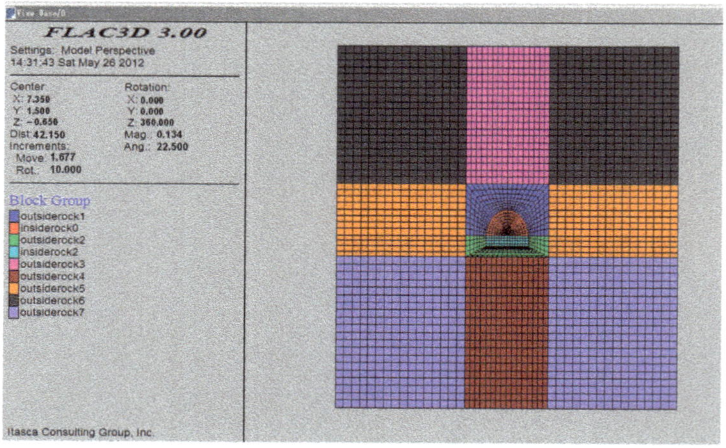

Figure 12. Plane view of finite element model.

Based on the physical and mechanical parameters of the surrounding rock, the final parameter values were as follows after trial calculation by referring to relevant manuals [29]: Elasticity modulus $E = 10$ GPa. Poisson's ratio $u = 0.3$. Frictional angle $\varphi = 35^{\circ}$. The cohesive force of the surrounding rock was 1×10^5 Pa. The tensile strength of the surrounding rock was 3×10^5 Pa.

In the FLAC3D calculation, the rock mass deformation parameters were the volume modulus (K) and shear modulus (G). Therefore, the elastic modulus or deformation modulus (E) and Poisson's ratio (μ) must be transformed into volume modulus (K) and shear modulus (G), and the transformation formula is as Equations (1) and (2).

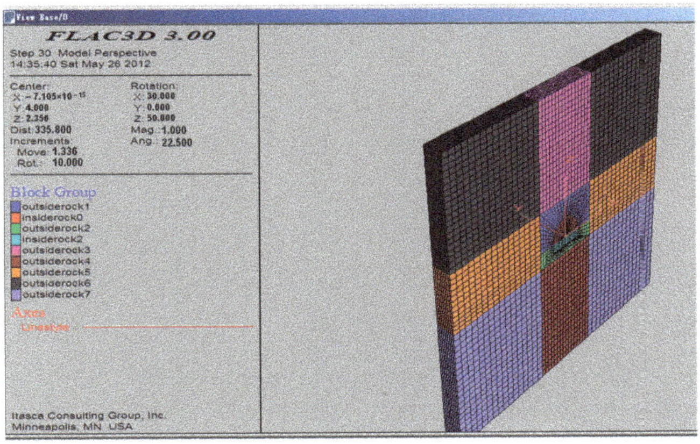

Figure 13. Three-dimensional view of finite element model.

$$K = \frac{E}{3(1-2\mu)} \quad (1)$$

$$G = \frac{E}{2(1+\mu)} \quad (2)$$

The bulk modulus K and shear modulus G of the model can be calculated by Equations (3) and (4).

Bulk modulus:

$$K = \frac{E}{3(1-2\mu)} = 8.33 \times 10^9 \text{ (Pa)} \quad (3)$$

Shear modulus:

$$G = \frac{E}{2(1+\mu)} = 3.85 \times 10^9 \text{ (Pa)} \quad (4)$$

In the FLAC3D dynamic calculation, four kinds of dynamic loads can be input: acceleration time history, velocity time history, displacement time history, and stress time history. The dynamic load input of this simulation adopted the monitored blasting vibration velocity time–history curve since the actual particle vibration velocity data were measured by field test. This effectively ensures that the numerical simulation calculations are as consistent as possible with the field blasting tests.

The monitored vibration velocity in Figures 3 and 5 was transformed into the vibration velocity time–history curve required by FLAC3D, as shown in Figures 14 and 15.

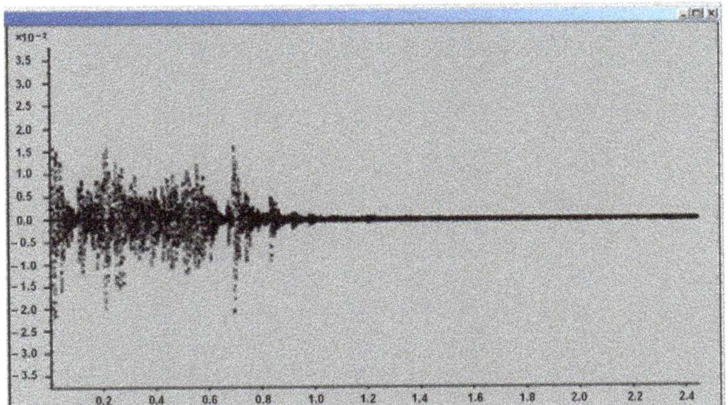

Figure 14. Data after nonelectric detonator vibration test and transition.

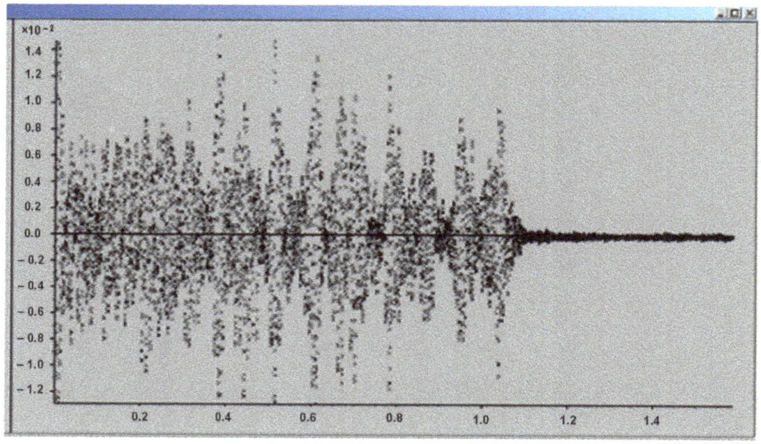

Figure 15. Data after electronic detonator vibration test and transition.

The loose circles of the tunnel blasted by nonelectric detonators and electronic detonators were obtained by numerical calculation, as shown in Figures 16 and 17. The blue elements in the figures represent the undamaged surrounding rock. The green, brown and red elements represent the damage of surrounding rock, and the damage degree decreases in turn.

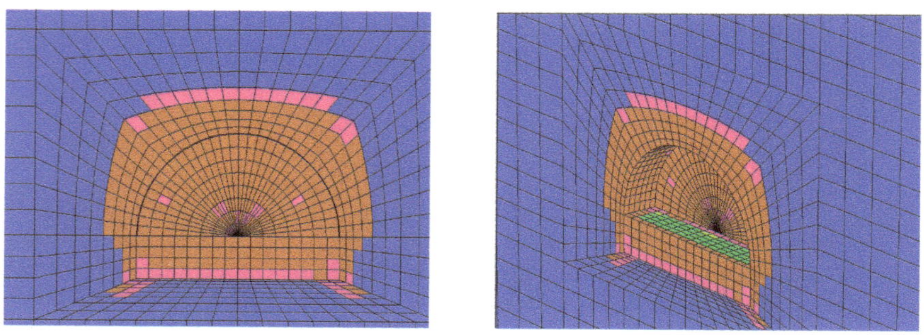

(a) (b)

Figure 16. Numerical calculation of loose circle for nonelectronic detonator blasting excavation. (**a**) Plane view; (**b**) 3D view.

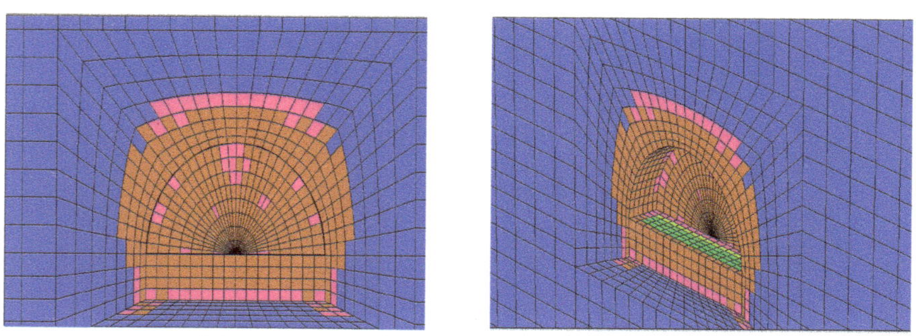

(a) (b)

Figure 17. Numerical calculation of loose circle for electronic detonator blasting excavation. (**a**) Plane view; (**b**) 3D view.

In order to combine the field tests, the numerical simulation is carried out according to the actual blasting footage of 3 m. The nonelectric detonator test caused 137.33 m^3 surrounding rock unit failure in the footage through the statistical data in Table 4, and the average thickness of the surrounding-rock loose circle is shown in Table 5. The average damage depth of the entire excavation section caused by explosive initiation can be obtained by dividing the surface area of the excavation profile within the footage length. The average failure depth of surrounding rock damage caused by the nonelectric detonator test is close to 2 m. This is basically consistent with the results monitored in the field test, which verifies the correctness of the numerical model. Similarly, 104.86 m^3 surrounding rock units failed in the electronic detonator results, and the average failure depth was 1.514 m. The calculated average depth of electronic detonators is obviously smaller than that of nonelectric detonators. The finite element simulation can ensure that the material properties of surrounding rock are the same in two times of blasting. The surrounding rock is assumed to be a homogeneous material without joint fissure. Therefore, the calculated results of nonelectric and electronic detonators are smaller than those of field tests, which is reasonable. This further validates the irregularity of the damage profile in Figure 11 in Section 4. It shows that in the area where the electronic detonator test is used, the joint fissures at the left wall and right spandrel positions of the tunnel have a great influence on the test results. It significantly reduces the blast resistance of surrounding rock.

Table 4. The average thickness of the loose ring calculated by simulation.

Computational Model	Excavation Footage (m)	Yield Volume (m^3)	Average Thickness (m)
Nonelectronic detonator	3	137.33	1.980
Electronic detonator	3	104.86	1.514

Table 5. The average thickness of the loose ring in the comparative test.

Computational Model	Loading Situation	Yield Unit Volume (m^3)	Loose Ring Average Thickness (m)	Thickness of Loose Ring Due to Blasting Load (m)	Influence of Blasting Action on Loose Ring
Nonelectronic detonator	Yes	137.33	1.98	0.66	33%
	No	91.34	1.32		
Electronic detonator	Yes	104.86	1.51	0.20	13%
	No	91.34	1.32		

The field blasting tests can only monitor the final damage range of surrounding rock because of the instantaneity of explosion. In fact, the loose circle of surrounding rock is generated by the double effects of stratum stress redistribution caused by excavation and blasting load impact. Here, the finite element simulation method is used to analyze these two factors in depth. The numerical simulation consists of two cases. One was to simulate the blasting effect without applying a dynamic load, and the other was to simulate the blasting effect while applying a dynamic load. The two simulation results are shown in Table 5.

From Table 5, the following conclusions can be drawn:

(1) The thickness of the loose ring was approximately 1.32 m without considering the blasting load when the tunnel was excavated by conventional nonelectric detonator blasting. The thickness of the loose ring generated by the blasting load was approximately 0.66 m. The final thickness of the loose ring was 1.98 m. It can be considered that the extent of the loose circle generated by the redistribution of surrounding rock stress accounts for 67% of the final loose circle. The range of the loose circle produced by the blasting load accounts for 33% of the final loose circle range. Although the thickness of the rock loose

circle caused by surrounding rock excavation accounts for the main part, the influence of the blasting load is also significant.

(2) The thickness of the loose ring was approximately 1.32 m without considering the blasting load when the tunnel was excavated by electronic detonator blasting. The thickness of the loose ring produced by the blasting load was approximately 0.20 m. The final thickness of the loose circle was 1.51 m. It could be considered that the range of the looseness circle generated by the stress redistribution of the surrounding rock accounts for 87% of the final loose circle. The range of the loose circle produced by the blasting load accounts for 13% of the final loose range. The damage proportion of electronic detonator blasting load to surrounding rock loose circles is much lower than that of nonelectric detonators. Therefore, it can be considered that the damage degree of explosives controlled by electronic detonators to surrounding rock is much smaller than that of nonelectric detonators because of their better delay and smaller explosive detonating quality each time.

6. Recommendations and Limitations

This study explored the damaging effect of the use of electronic detonators and nonelectric detonators on the protected surrounding rock. The monitoring and calculation results of damage range and rock crushing degree show that a nonelectric detonator has a positive effect on the protection of surrounding rock. Its application in practical engineering can be embodied in the following two aspects: Firstly, in terms of the loose range of surrounding rock, the thickness of the loose circle of surrounding rock caused by the electronic detonator is 1 m less than that of the nonelectric detonator on average, so the length of the anchors can be shortened appropriately when setting the primary support. In terms of the degree of rock fragmentation, the maximum acoustic velocity attenuation caused by an electronic detonator is only half of that of a nonelectric detonator. Therefore, the amount of shotcrete can be reduced, and the spacing of steel arches can be appropriately increased when setting the primary support. This study provided a reference for the optimization design of primary support parameters of the tunnel after blasting, which is helpful in reducing construction costs.

However, the current testing methods cannot accurately divide the failure zone, the crack zone, and the plastic zone, so the size of the loose circle in the test is not uniform. Although specifications [29] provide blasting damage standards, a rock mass is judged to be damaged if the change rate of the rock mass wave velocity before and after blasting is greater than 10%. In some cases, the surrounding rock in the loose circle fractured but still retains high stability. In addition, the influence of initial surrounding rock fissures is not excluded in this study. Therefore, further classification criteria of surrounding rock loose circles and the influence of initial surrounding rock fissures will be considered in future studies.

7. Conclusions

In this study, the damage of protected surrounding rock by using an electronic detonator and a nonelectric detonator in tunnel blasting was compared. The advantages of using an electronic detonator to protect surrounding rock were quantified through field monitoring and numerical simulation. The research results can provide a reliable reference for the design of controlled blasting and the determination of support parameters in tunnel excavation engineering. Some valuable conclusions are drawn from this study:

(1) In the surrounding grade IV rock, when the blasting footage was 3 m, the nonelectric detonators were used for normal construction, and the range of the loose circle was 1.5~2.3 m. The resulting loose circle ranged from 0 m to 1.4 m after the nonelectric detonators were changed to electronic detonators under the condition that the distance between the holes and the amount of charge remained unchanged. The range of the loose circle produced by the blasting of the tunnel electronic detonator was approximately 60~76% of that of the nonelectric detonator blasting, and the surrounding rock in the loose circle was less broken.

(2) When nonelectric detonators were used for blasting, the impact of the blasting load on the loose ring was approximately 33% due to the large blasting load. The small blasting load of the single-hole electronic detonator leads to an approximately 13% effect of the blasting load on the loose circle.

(3) The above studies show that the use of electronic detonators for the protection of surrounding rock has a beneficial effect. The preliminary support parameters of the tunnel can be optimized after excluding the interference of initial surrounding rock fissures. When using electronic detonators, the specific optimization parameters can be expressed as reducing the length of the anchor, increasing the steel arch spacing, and reducing the amount of shotcrete.

Author Contributions: Conceptualization, H.F., X.G. and C.C.; methodology, X.G., J.W. and Q.N.; software, N.Y.; validation, J.W., C.C. and H.F.; formal analysis, J.L ; investigation, Q.N.; resources, H.F.; data curation, Y.L.; writing—original draft preparation, H.F. and N.Y.; writing—review and editing, N.Y., Y.L. and J.L.; visualization, Y.L.; supervision, X.G. All authors have read and agreed to the published version of the manuscript.

Funding: This research was funded by the Service Safety and Intelligent Maintenance of Rail Transit Operation Tunnels (2022JBXT007), the General Program of Shandong Natural Science Foundation of China (ZR2022ME043), and the General Program of National Natural Science Foundation of China (51978356).

Institutional Review Board Statement: Not applicable.

Informed Consent Statement: Not applicable.

Data Availability Statement: The data used to support the findings of this study are included within the article.

Conflicts of Interest: The authors declare no conflict of interest.

References

1. Kwon, S.; Lee, C.; Cho, S.; Jeon, S.; Cho, W. An investigation of the excavation damaged zone at the kaeri underground research tunnel. *Tunn. Undergr. Space Technol.* **2009**, *24*, 1–13. [CrossRef]
2. Hadgu, T.; Kalinina, E.; Wang, Y.F. Investigations of effect of underground excavations using hydrologic and tracer transport modeling. *Tunn. Undergr. Space Technol.* **2022**, *127*, 104577. [CrossRef]
3. Chen, G.H.; Zou, J.F.; Min, Q.; Guo, W.J.; Zhang, T.Z. Face stability analysis of a shallow square tunnel in non-homogeneous soils. *Comput. Geotech.* **2019**, *114*, 103112. [CrossRef]
4. Zhang, X.; Song, H.W.; Yan, X. Analysis of Tunneling Stability Using Classic and Extended Drucker-Prager Model. *Chin. J. Undergr. Space Eng.* **2018**, *6*, 1652–1657.
5. Jiang, F.; Wang, G.; He, P.; Hou, B.; Zhang, S.B.; Sun, S.Q.; Zheng, C.C.; Wu, Y. Mechanical failure analysis during direct shear of double-joint rock mass. *Bull. Eng. Geol. Environ.* **2022**, *81*, 410. [CrossRef]
6. Wang, W.J.; Han, S.; Dong, E.Y. Boundary equation of plastic zone in roadway surrounding rocks considering supporting effect and its application. *J. Min. Saf. Eng.* **2021**, *38*, 749–755. [CrossRef]
7. Wang, R.; Yuan, D.Y.; Zhang, J.Z.; Yang, J. Analysis on broken zone of surrounding rock in tunnel based on Hoek-Brown strength criterion. *J. Saf. Sci. Technol.* **2017**, *13*, 58–63. [CrossRef]
8. Meng, L.; Gao, Z.N.; Meng, X.R. Elasto-plastic analysis of circular roadway surrounding rocks under consideration of rock damage. *J. Saf. Sci. Technol.* **2013**, *9*, 11–16. [CrossRef]
9. Guan, X.M.; Zhang, L.; Wang, Y.W.; Fu, H.X.; An, J.Y. Velocity and stress response and damage mechanism of three types pipelines subjected to highway tunnel blasting vibration. *Eng. Fail. Anal.* **2020**, *118*, 104840. [CrossRef]
10. Guan, X.M.; Zhang, L.; Wang, L.M.; Fu, H.X.; Yu, D.M.; Chen, G.; Ding, Y.; Jiang, W.L. Blasting vibration characteristics and safety standard of pipeline passed down by tunnel in short distance. *J. Cent. South Univ. (Sci. Technol.)* **2019**, *50*, 11. [CrossRef]
11. Ma, Y.C.; Lu, A.Z.; Cai, H.; Zeng, X.T. Analytical solution for determining the plastic zones around two unequal circular tunnels. *Tunn. Undergr. Space Technol.* **2022**, *120*, 104267. [CrossRef]
12. Dong, F.T. *The Supporting Theory Based on Broken Rock Zone and Its Application Technology*; Coal Industry Press: Beijing, China, 2001.
13. Hu, K.H. Application Research on Rapid Surrounding Rock Classification and Loose Circle Test. *Railw. Constr. Technol.* **2022**, *3*, 22–26. [CrossRef]
14. Liu, Y.S.; Zhu, S.Y.; Yang, X.L. Study on Damage Accumulation and Broken Rock Zone Range of Surrounding Rock of Large Span Chamber caused by Multiple Blasts. *Blasting* **2022**, *39*, 9–15, 35. [CrossRef]

15. Zhao, X.; Dai, Z.J.; Li, R.H.; Tao, L.J. Earthquake Mitigation Measures for Tunnels Considering Loosen-zone in Surrounding Rock. *J. Beijing Univ. Technol.* **2022**, *46*, 1027–1038. [CrossRef]
16. Wei, J.B.; Wang, S.M.; Zhao, Z.; Li, D.L. Numerical Study of Damage to Rock Surrounding an Underground Coal Roadway Excavation. *Adv. Civ. Eng.* **2020**, *16*, 8863289. [CrossRef]
17. Han, H.Y.; Fukuda, D.; Liu, H.Y.; Salmi, E.F.; Sellers, E.; Liu, T.J.; Chan, A. Combined finite-discrete element modelling of rock fracture and fragmentation induced by contour blasting during tunnelling with high horizontal in-situ stress. *Int. J. Rock Mech. Min. Sci.* **2020**, *127*, 104214. [CrossRef]
18. Guan, X.M.; Yang, N.; Zhang, W.J.; Li, M.G.; Liu, Z.L.; Wang, X.H.; Zhang, S.L. Vibration response and failure modes analysis of the temporary support structure under blasting excavation of tunnels. *Eng. Fail. Anal.* **2020**, *136*, 106188. [CrossRef]
19. Pan, C.; Li, X.; Li, J.C.; Zhao, J. Numerical investigation of blast-induced fractures in granite: Insights from a hybrid LS-DYNA and UDEC grain-based discrete element method. *Geomech. Geophys. Geo-Energy Geo-Resour.* **2021**, *7*, 1–18. [CrossRef]
20. Siren, T.; Kantia, P.; Rinne, M. Considerations and observations of stress-induced and construction-induced excavation damage zone in crystalline rock. *Int. J. Rock Mech. Min. Sci.* **2015**, *73*, 165–174. [CrossRef]
21. Salum, A.H.; Murthy, V. Optimising blast pulls and controlling blast-induced excavation damage zone in tunnelling through varied rock classes. *Tunn. Undergr. Space Technol.* **2019**, *85*, 307–318. [CrossRef]
22. Zhou, H.; Xiao, M.; Yang, Y.; Liu, G.Q. Seismic Response Analysis Method for Lining Structure in Underground Cavern of Hydropower Station. *KSCE J. Civ. Eng.* **2019**, *23*, 1236–1247. [CrossRef]
23. Li, R.L.; Qu, H.L.; Zhou, T.; Zhou, C.T. An experimental investigation on fatigue characteristics of granite under repeated dynamic tensions. *Int. J. Rock Mech. Min. Sci.* **2022**, *158*, 105185. [CrossRef]
24. Ji, L.; Zhou, C.B.; Lu, S.W.; Jiang, N.; Gutierrez, M. Numerical Studies on the Cumulative Damage Effects and Safety Criterion of a Large Cross-section Tunnel Induced by Single and Multiple Full-Scale Blasting. *Rock Mech. Rock Eng.* **2021**, *54*, 6393–6411. [CrossRef]
25. Yang, J.H.; Lu, W.B.; Li, P.; Yan, P. Evaluation of Rock Vibration Generated in Blasting Excavation of Deep-buried Tunnels. *KSCE J. Civ. Eng.* **2018**, *22*, 2593–2608. [CrossRef]
26. Yang, J.H.; Cai, J.Y.; Yao, C.; Li, P.; Jiang, Q.H.; Zhou, C.B. Comparative Study of Tunnel Blast-Induced Vibration on Tunnel Surfaces and Inside Surrounding Rock. *Rock Mech. Rock Eng.* **2019**, *52*, 1–15. [CrossRef]
27. Fu, H.X.; Zhao, Y.; Xie, J.S. Test Analysis on the Broken Rock Zone of the Surrounding Rock Blasting for Railway Double-Line Tunnel. *China Railw. Sci.* **2010**, *31*, 54–55. [CrossRef]
28. Fu, H.X.; Wang, L.N.Y.; Zhao, Y.; Shen, Z.; Zhang, C.P.; Li, Y.Z. Comparison of Excavation Damage Zones Resulting from Blasting with Nonel Detonators and Blasting with Electronic Detonators. *Rock Mech. Rock Eng.* **2014**, *47*, 809–816. [CrossRef]
29. *DL/T 5389-2007*; Construction Technical Specifications on Rock-Foundation Excavation Engineering of Hydraulic Structures. China Water Resources and Hydropower Press: Beijing, China, 2007.

applied sciences

MDPI

Article

Stability Analysis of Surrounding Rock in the Diversion Tunnel at the Xulong Hydropower Station based on RFPA^{3D} and Microseismic Monitoring

Hongjian Qian [1], Zhou Tan [2] and Biao Li [2,3,*]

1 Guodian Jinshajiang Xulong Hydropower Development Co., Ltd., Chengdu 610041, China
2 School of Geoscience and Technology, Southwest Petroleum University, Chengdu 610500, China
3 State Key Laboratory of Geohazard Prevention and Geoenvironment Protection, Chengdu 610059, China
* Correspondence: libiaoscu@163.com

Abstract: To study the surrounding rock stability of the excavated geologically weak section of the #2 diversion tunnel in the Xulong Hydropower Station, a quasi-3D numerical model was built using the Realistic Failure Process Analysis (RFPA^{3D}) system to simulate the damage and failure process consisting of crack initiation, growth, and penetration in the rock mass after tunnel excavation, and reveal the instability failure mechanism inside the rock mass. Moreover, the microseismic monitoring technology was employed to delineate potential danger areas in the surrounding rock of the tunnel and explore possible instability failure modes. Results indicate that the surrounding rock of the tunnel profile failed as different degrees during the excavation process, most obviously near the vault and corners of the side wall, where tensile failure predominated. As the excavation proceeded, microseismic events increased gradually at the vault and corners of the side wall, and the energy from acoustic emissions accumulated steadily, thus raising the possibility of collapse and rock bursts in this area. The research results can provide technical support for the construction of the diversion tunnel project in the Xulong Hydropower Station and serve as a guide for the construction of similar geologically weak underground projects.

Keywords: surrounding rock stability; RFPA^{3D}; microseismic monitoring; diversion tunnel; Xulong Hydropower Station

Citation: Qian, H.; Tan, Z.; Li, B. Stability Analysis of Surrounding Rock in the Diversion Tunnel at the Xulong Hydropower Station based on RFPA^{3D} and Microseismic Monitoring. *Appl. Sci.* **2022**, *12*, 9939. https://doi.org/10.3390/app12199939

Academic Editors: Zhengzhao Liang, Bei Jiang and Nuwen Xu

Received: 30 August 2022
Accepted: 29 September 2022
Published: 2 October 2022

1. Introduction

China has focused on the development and use of hydropower projects in the lower reaches of the Yalong River and upper reaches of the Jinsha River, with the aim of "achieving carbon neutrality before 2060". Southwest China is located in the high mountain valley area of the Hengduan Mountains on the eastern edge of the Qinghai-Tibet Plateau. It is characterized by its high altitude, high ground stress, and complex rock structure. For hydropower plants in this area, underground caverns are often the most practical and cost-effective option for meeting topographic, spatial, building layout, and construction requirements. The construction of underground caverns for hydropower projects will face great risks and challenges due to the strict requirements for rock quality, construction cycle and safety standards, and complicated construction procedures. The surrounding rock is likely to become unstable and damaged during construction due to poor excavation and delayed support, which could result in casualties and property losses. For example, the Baihetan underground powerhouse has the largest excavation size in the world. During the excavation of underground caverns, rock collapses dominated by high geostress and geological structures posed serious threats to field construction [1]. An unexpectedly powerful rock burst that occurred in November 2009 during the construction of the Jinping-II drainage tunnel destroyed all the support systems in the affected sections, seriously damaged the TBM equipment, and killed seven workers [2]. Tunnel excavation of the Neelum-Jhelum Hydropower Project in Pakistan has been hindered by the ground composition of the Himalayas, faces the project complexities, geological environments involving significant overburden and tectonic stresses, and effects of the excavation method on tunnel stability. This causes time and cost overruns directly [3]. Therefore, it is crucial to assess the stability of surrounding rock during the excavation and unloading of underground caverns, to promptly inform the constructor of unstable areas, and to implement a reasonable construction and support reinforcement plan.

As an important research method, numerical modeling is now widely used by scholars to study the rock excavation and unloading process, and predict the instability failure mode and impacted area in the rock mass. To determine the deformation and stress distribution around the caverns and to analyze the impacts produced by weak structural planes and multiple caverns in the rock mass, Dhawan et al. [4] developed a three-dimensional finite element model for analyzing the stability of underground caverns. Hao et al. [5] examined the impact of faults on the plastic zone and displacement of underground caverns using the Universal Distinct Element Code (UDEC) numerical modeling software. Yazdani et al. [6] performed displacement-based back analysis using finite element and discrete element methods to determine geomechanical properties of the rock in underground caverns, such as stress ratios and joint parameters. Ma et al. [7] investigated the failure mechanisms and movement characteristics of rockfalls through a three-dimensional discontinuous deformation analysis (3D DDA) method. Song et al. [8] conducted pore-scale modeling on the methane hydrate dissociation and transportation in the reconstructed three-dimensional models of the MH-bearing sediment. Cai et al. [9] studied the acoustic emission pattern during the excavation and unloading process of underground plants using the FLAC/PFC coupled numerical modeling method. Jing [10] studied the stability of rock with anisotropic fractures in a tunnel using discontinuous deformation analysis (DDA) in the discrete element method (DEM).

In addition to numerical simulation, Field monitoring is also a common method for studying the mechanical response and stability of the underground rock. Field monitoring data can visually reflect the stress-strain characteristics of the rock mass in underground caverns. Routine monitoring methods include ground stress measurement, bolt stress meters, acoustic testing, and geological radar [11–14]. The monitoring results can better reflect changes in deformation and mechanical parameters on the section, but they have spatial limitations and make it difficult to capture micro-fractures within the rock mass, which are frequently the precursor to macroscopic instability failure in the rock mass. As a 3D monitoring method, microseismic monitoring can detect micro-fractures within the rock mass in advance, allowing for the forecast of risk areas in the rock. Li et al. The authors in [15] analyzed the microseismic b-values associated with rock mass large deformation and their temporal variation. Li et al. [16], based on tempo-spatial characteristics of the microseismic events, analyzed the relation between microseismic activities and field construction and revealed the main damage regions. Leśniak and Isakow [17] analyzed the clustering characteristics of microseismic events and developed a time-based function for high-energy fractures in coal mines. Hudyma and Potvin [18] developed a microseismic monitoring method to manage hazards in underground hard rock mines. Ma et al. [19] investigated the focal mechanism of seismic events induced by underground mines using an optimized moment tensor inversion method. Cai et al. [20] quantified the rock damage in the experimental tunnel using the source radius. Ma et al. [21] examined the stability of the surrounding rock of high-slope tunnels by analyzing the spatial and temporal characteristics of microseismic events. Dai et al. [22] classified the dominant factors of clustering microseismic events and proposed a method for studying the large deformation of rock in underground plant caverns based on source parameters.

In the present study, the incorporated method involving RFPA^{3D} and MS monitoring technique is applied to analyze the stability of the surrounding rock, aiming at the instability failure problem inside the rock mass caused by the redistribution of ground stress after excavation in the geologically weak section of the #2 diversion tunnel in the

Xulong Hydropower Station in the upper reaches of the Jinsha River, and analyzes the progressive instability failure of the tunnel using the RFPA^{3D} numerical modeling software by comparing the spatial and temporal evolution characteristics of microseismic events to reveal the whole process from micro-fracture initiation, growth, and propagation, to penetration within the surrounding rock after tunnel excavation, and identifies potentially unstable areas after tunnel excavation. The research findings can be used to provide technical support for stability evaluation and risk prediction prior to and after excavation of the #2 diversion tunnel in the Xulong Hydropower Station, and can also provide experience for similar hydraulic tunnel excavation.

2. Engineering Background

The Xulong Hydropower Station is located in the upper reaches of the main stream of the Jinsha River at the junction of Deqin County, Yunnan Province and Derong County, Sichuan Province. This primary task of the project development is power generation, which is one of the backbone power points for west-east power transmission. It has an average annual flow of 990 m^3/s at the dam site and an average annual runoff of 31.3 billion m^3. The reservoir has a normal water level of 2302 m, a dead water level of 2294 m, a designed flood level of 2303.42 m, a maximum flood level of 2305.89 m, and a total capacity of about 847 million m^3. The hydropower station has an installed capacity of 2400 MW and can generate 10.319 billion kWh of electricity annually.

According to site exploration data, the larger-scale faults on the left bank of the dam site were F1, F2, and F3, while the rest were III or IV structural planes, primarily steeply dipping faults orthogonal to the Jinsha River. The ground stress increased generally with burial depth, primarily at a medium level, and the concentration area of local stress was dominated by medium and high levels of stress. The riverbed area showed an obvious stress increase, with the maximum horizontal principal stress ranging from 2.9 MPa to 16.7 MPa, and high ground stress phenomena such as core discing occurred locally. The rocks are mainly granite, migmatite, and amphibolite schist. Their uniaxial compressive strength is 80~120 MPa and elastic modulus is 30~50 GPa. Figure 1 shows the geological axis profile of the #2 diversion tunnel on the left bank. Located at the downstream lateral border of the F2 fault, the inlet of the diversion tunnel had an elevation of 2155 m, a length of 1446.67 m, and a thickness of the overlying rock on the vault ranging from 50 m to 280 m. Moreover, the diversion tunnel and auxiliary tunnels were dominated by a medium level of ground stress.

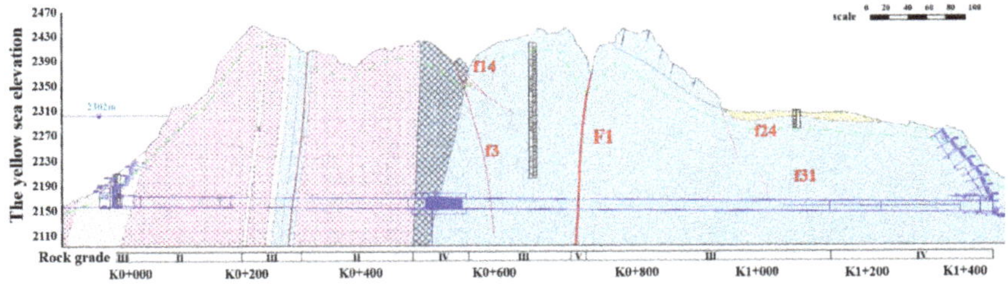

Figure 1. Geological axis profile of the #2 diversion tunnel.

3. Numerical Modeling

3.1. Basic Principles of RFPA

The RFPA is a numerical calculation method for analyzing the material fracture process based on finite element stress analysis and statistical damage theory, as well as a numerical test tool capable of simulating the entire material progressive fracture process until instability [23]. It is a new numerical analysis method for the material fracture process, which can simulate nonlinearity through material non-uniformity and mechanical problems of discontinuous media with the continuum mechanics theory. One significant aspect of this method is that it takes material non-uniformity into account. In 1995, Tang et al. [24] from

Northeastern University developed the RFPA2D system based on the finite element theory and a new material fracture process algorithm thought, which simulates the nonlinearity of material by considering its non-uniformity, simulates material discontinuous behaviors like deformation and failure by weakening elements, and can be used to study the whole process of the rock (rock mass) material from mesoscopic damage to macroscopic failure.

Based on this basic principle, the parallel algorithm was introduced to the RFPA^{3D} system that is a numerical analysis tool [25] for solving the mechanical problems of discontinuous media with the continuum mechanics theory by performing mesoscopic element failure analysis and finite element stress analysis. Combining the statistical damage theory with the numerical calculation method and considering the non-uniformity of rock material, it simulates the nonlinear behavior of rock formation and fracture. The development of the RFPA^{3D} system aimed to build a 3D model with higher theoretical research value, and it has been more widely used in engineering.

3.2. Twin-Shear Unified Strength Criterion

In 1994, Professor Yu [26] from Xi'an Jiaotong University proposed the unified strength theory applicable to geotechnical materials—twin-shear unified strength, which holds that the failure of material begins when the combination of the two larger shear stresses applied to the twin-shear element and the positive stress applied to the element surface reaches a critical value. Its mathematical expression is:

$$F=\begin{cases}\tau_{13}+b\tau_{12}+\beta(\sigma_{13}+b\sigma_{12})=c \\ \qquad(\tau_{12}+\beta\sigma_{12}>\tau_{23}+\beta\sigma_{23}) \\ \tau_{13}+b\tau_{23}+\beta(\sigma_{13}+b\sigma_{23})=c \\ \qquad(\tau_{12}+\beta\sigma_{12}\leqslant\tau_{23}+\beta\sigma_{23})\end{cases} \tag{1}$$

where b is the coefficient reflecting the effect of the intermediate principal shear stress, β is the coefficient reflecting the effect of the positive stress on the failure of the material, and c is the strength parameter of the material.

In geotechnical engineering, the compressive strength parameter σ_c is generally used, and Equation (1) can be expressed as:

$$F=\begin{cases}\frac{1}{\alpha}\sigma_1-\frac{1}{1+b}(b\sigma_2+\sigma_3)=\sigma_c \\ \left(\sigma_2\leqslant\frac{\sigma_1+\alpha\sigma_3}{1+\alpha}\right) \\ \frac{1}{\alpha(1+b)}(\sigma_1+b\sigma_2)-\sigma_3=\sigma_c \\ \left(\sigma_2>\frac{\sigma_1+\alpha\sigma_3}{1+\alpha}\right)\end{cases} \tag{2}$$

where α is the maximum ratio of the tensile to the compressive strength of the material, that is, $\alpha=\sigma_t/\sigma_c$.

In this paper, this strength theory was introduced by simulating the damage to the surrounding rock after tunnel excavation with RFPA^{3D}.

3.3. Modeling

Since stakes K0+765 to K0+775 of the #2 diversion tunnel in the Xulong Hydropower Station pass through the F1 fault, due to the existence of faults, the stress and deformation distribution of surrounding rock are seriously affected. This section is dominated by class V surrounding rock, with high ground stress and high possibility of disaster. [27]. Therefore, this section was selected for the study. Since there are some limitations on the establishment of complex models in RFPA^{3D}, the solid model can be established in the finite element analysis software ANSYS first, and the data transformation interface program AtoR can be prepared using C language to extract and import the node and element information from the ANSYS model into RFPA^{3D}. Modeling and imposing constraints is simple in ANSYS and it is also easy to mesh the model. This will allow for the creation of the tunnel model to be constructed more quickly and conveniently [28–30]. According to site actual conditions, a geological model of 50 m × 50 m × 10 m was firstly established in ANSYS, and a horseshoe-shaped tunnel cavern was excavated in the middle of the model with a height of 16 m and a width of 14 m according to the actual structural and dimensional characteristics of the diversion tunnel. Then, the model was meshed to ensure that all elements were hexahedrons with eight nodes, and the sweep function was recommended in meshing, as shown in Figure 2. Finally, normal constraints were applied to a triple (0, 0, 0), and a stress boundary condition with a lateral pressure coefficient of 1 was applied to a triple (50, 50, 10) according to the actual ground stress environment of the tunnel. After the model with meshes and constraints was imported into RFPA^{3D}, a 3D model considering the non-uniformity of the rock mass was generated according to the Weibull distribution, as shown in Figure 3.

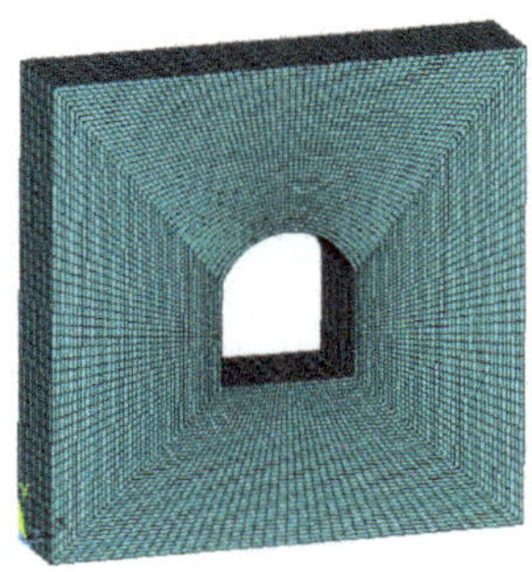

Figure 2. Model meshing.

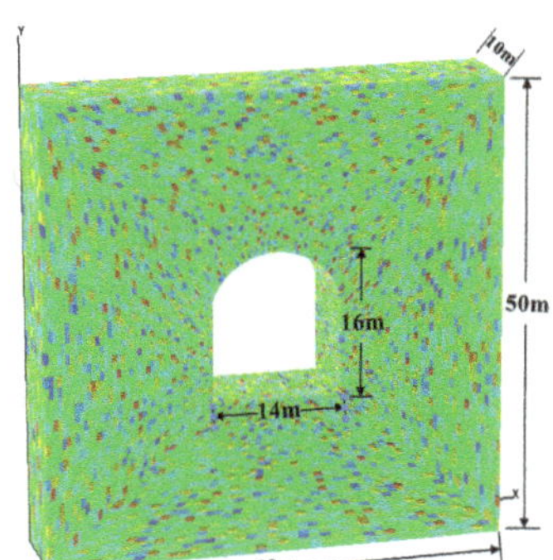

Figure 3. A 3D calculation model.

To reveal the rupture zone, observe acoustic emission and compare with microseismic monitoring, the Mohr–Coulomb model was selected, which is more consistent with reality. The input parameters were determined by consulting the geological data obtained through field measurements; Table 1 lists the selected mechanical parameters of the rock mass. The strength reduction method [31] was employed in the calculation process, which continuously reduced the initial strength f_0 of the material, and f_0 unified the compressive (shear) and tensile strengths of the element material according to the following reduction criterion:

$$f_0{}^{trial} = \frac{f_0}{F_s{}^{\mathrm{trial}}} \tag{3}$$

where $f_0{}^{trial}$ is the test strength and $F_s{}^{trial}$ is the test safety factor.

Table 1. Mechanical parameters of the rock mass.

Elastic Modulus (GPa)	Uniaxial Compressive Strength (MPa)	Poisson's Ratio	Internal Friction Angle (°)	Residual Strength	Coefficient of Confining Pressure	Coefficient of Homogeneity
10	35	0.25	30	0.1	2	5

3.4. Modeling Results and Analysis

Figure 4 shows the distribution of the maximum principal stress. Under the action of the applied confining pressure, the maximum principal stress first started to concentrate at the bottom corner of the side wall of the tunnel, and the stress on the surrounding rock of the tunnel profile also increased immediately. As the loading continued, cracks started to be generated from the tunnel vault and bottom corners of the side wall, forming multiple main cracks that propagated around and bifurcated during the propagation process. The cracks on the vault and bottom developed faster, and those on the side wall developed slower. The tunnel was unstable at this point because the late loading stage caused all the cracks in the surrounding rock of the tunnel profile to be penetrated. Figure 5 shows the distribution of the maximum principal strain, which was constant throughout the loading and failure process of the rock mass. Figure 6 shows the changes in the elastic modulus of the rock, with the red color indicating the failure area. It can be seen that the distribution of the failure is consistent with that of the maximum principal stress.

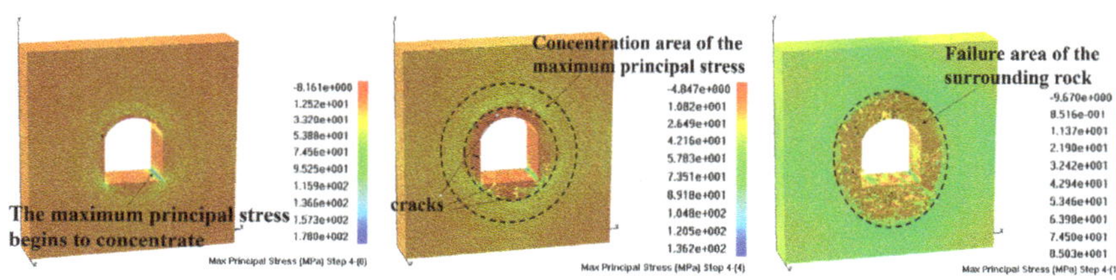

Figure 4. Distribution of the maximum principal stress.

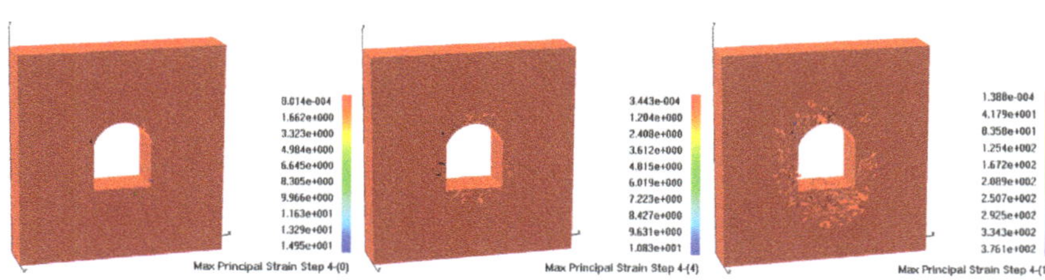

Figure 5. Distribution of the maximum principal strain.

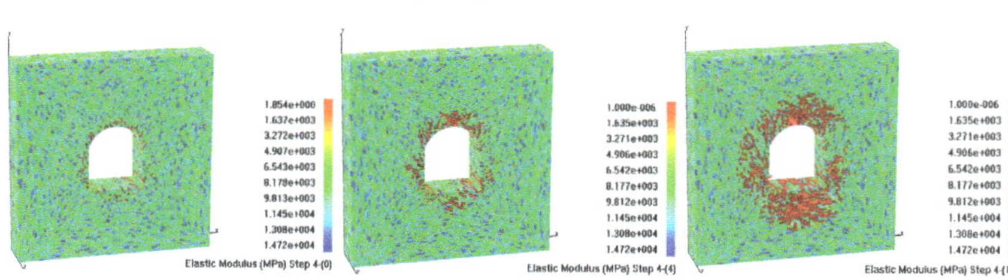

Figure 6. Distribution of the elastic modulus.

In practical engineering, the ground stress field cannot be directly observed, but rock fracture will occur, so the acoustic emission phenomenon can be monitored to understand the failure of the surrounding rock. Figure 7 shows the spatial distribution of the acoustic emission phenomenon during the failure process of rock. The center of the sphere in the figure represents the location of the acoustic emission, and the radius of the sphere indicates the relative magnitude of the energy released from the acoustic emission. In Figure 7, the red sphere and blue sphere represent the acoustic emission events generated from the failure of elements satisfying the tensile criteria or shear criteria, respectively. As can be seen from the figure, in the initial loading stage, a large number of red spheres started to concentrate at the tunnel vault and bottom corners of the side wall, indicating that the rock damage gradually propagated at this moment. As the loading proceeded, the red spheres on both sides of the tunnel also started to increase, and there were red spheres of larger radius gathering at the tunnel vault and corners of the side wall continuously, indicating that the rock fracture intensified and the energy released increased at this time. Meanwhile, blue spheres began to appear at the tunnel vault and bottom, indicating that shear failure occurred locally, but this area was dominated by tensile failure. In the late loading stage, the red spheres kept expanding inward and the blue spheres kept increasing, indicating that the damage to the side wall kept intensifying and that more and more energy was released by micro-fractures. Especially, acoustic emissions were mainly concentrated at the tunnel vault and corners of the side wall. The damage distribution of the vault and the corners of the side walls are influenced by the in situ stress and weak structures. Moreover, the distribution of the initial in situ stress field in the tunnel surrounding rock will influence the damage range and degree. Inevitably, the increase of lateral pressure coefficients will expand the damage range and degree. Therefore, this area had a high probability of rock bursts and required intensive monitoring in the project, so that preventive measures could be taken before the disaster occurred.

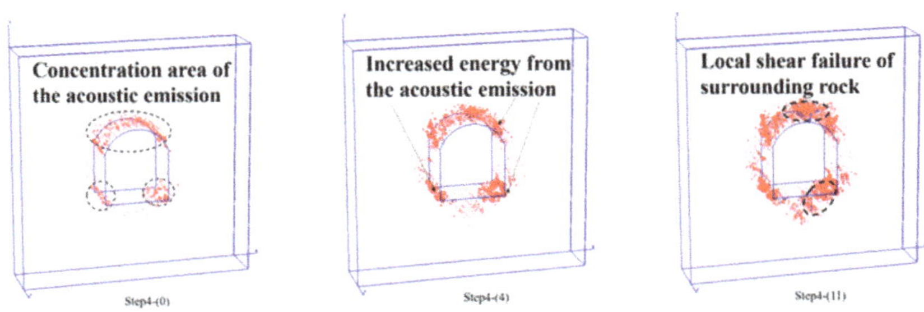

Figure 7. Acoustic emissions during the tunnel failure.

4. Characteristics of Microseismic Events in the Tunnel

4.1. Microseismic Monitoring System of the #2 Diversion Tunnel

The deformation and failure caused by the excavation of underground caverns was essentially the evolution process of rock micro-fractures from initiation, growth, propagation, aggregation, and penetration, to macroscopic failure. Therefore, the aggregation and evolution characteristics of micro-fractures in the rock mass are often used as the precursor of macroscopic instability failure of the rock mass. Microseismic monitoring, as a 3D monitoring method, originated abroad when the first seismic station was established in the Ruhr coalfield in Germany in 1908 to monitor seismic events induced by mining [32]. The microseismic monitoring system can collect micro-fracture information inside the rock mass in real time for assessing the stability of the engineering rock mass. In response to the randomness and multiplicity of deformation, rock bursts, and other issues caused by the excavation and unloading of deep tunnels, the diversion tunnel project in the Xulong Hydropower Station was equipped with the high-sensitivity microseismic monitoring system produced by Canadian ESG, which could provide real-time monitoring. It mainly consists of the acceleration sensor, Paladin digital signal acquisition system, and Hyperion digital signal processing system. Figure 8 shows the network topology of the microseismic monitoring system.

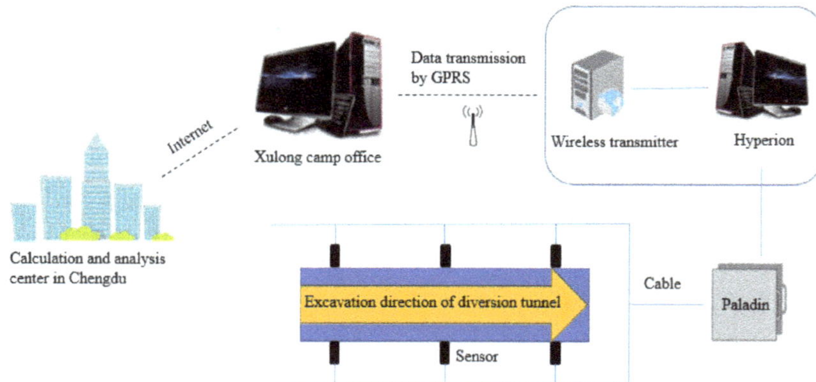

Figure 8. Network topology of the microseismic monitoring system.

The microseismic monitoring system of the #2 diversion tunnel implemented 24 h monitoring. In particular, the microseismic sensor converted the received microseismic signals into electrical signals, which were transmitted to the Paladin digital signal acquisition system through the cable and converted into digital signals with a 24-bit ADC. Finally, the digital signals were recorded and saved automatically by the Hyperion digital signal processing system. Due to the long transmission distance from the Paladin acquisition substation to the Hyperion host, they were connected by optical fiber in order to ensure the stability and integrity of the transmitted signals, and the conversion between digital signals and optical signals was completed by using an optical-to-electrical (O/E) converter. The data collected onsite could be shared through wireless network transmission, so that the Xulong camp office and the calculation and analysis center in Chengdu could download the microseismic data in real time, thus conducting risk warning and stability evaluation for the excavation of the #2 diversion tunnel based on calculation, analysis, and research.

4.2. Spatial and Temporal Distribution of Microseismic Events

The microseismic monitoring system of the diversion tunnel started to run normally on 17 December 2021, and stakes K0+765 to K0+775 of the #2 diversion tunnel were excavated from 3 June to 13 June 2022. During the construction period, a total of 104 valid events were collected after the site interference signals such as mechanical noise were removed, including 92 microseismic events.

Figure 9 shows the temporal distribution of microseismic events. As can be seen from the figure, the microseismic events gradually increased from 3 June to 8 June, reached the peak on 9 June to 10 June, and maintained at about 20, and the surrounding rock was continuously disturbed. Based on the energy dissipation principle, each micro-fracture of the surrounding rock is accompanied by the release of energy. The cumulative energy released from microseismic events suddenly increased on 9 June and 10 June, indicating that frequent microseismic events intensified the release of energy. Then, the number of microseismic events decreased significantly from 11 June to 13 June and maintained at about 5. The moment magnitude was mostly distributed between −1.0 and 0.5, representing small-magnitude events, as shown in Figure 10. The site construction dynamics indicated that the site was under normal construction from 3 June to 8 June, and the construction intensity started to increase on 9 June and 10 June in order to complete the workload, which directly led to the intensification of internal micro-fractures of the rock mass and extremely frequent microseismic events. The site investigation showed that the deformation and failure of the rock mass occurred on the left front of the tunnel face (as shown in Figure 11), and there were risks of rock bursts. For construction safety purposes, the blasting work was stopped on 11 June and 12 June, and support was provided for the surrounding rock. After the support was completed on 13 June, the tunnel excavation was continued, with a much lower intensity than before, so the microseismic events remained at a low level during these days. Thus, the frequency of microseismic events could better reflect the construction intensity of underground plants.

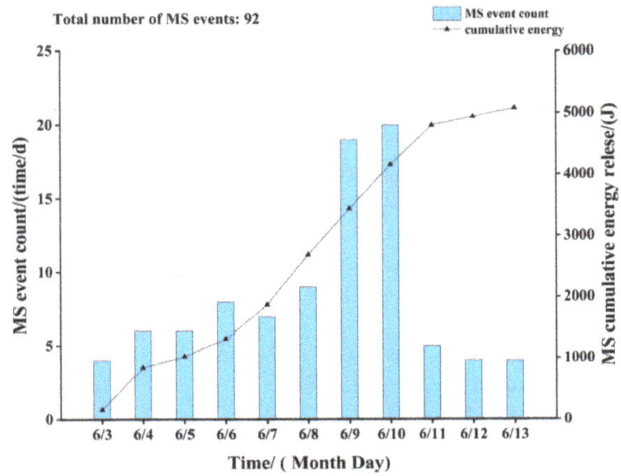

Figure 9. Temporal distribution of microseismic events.

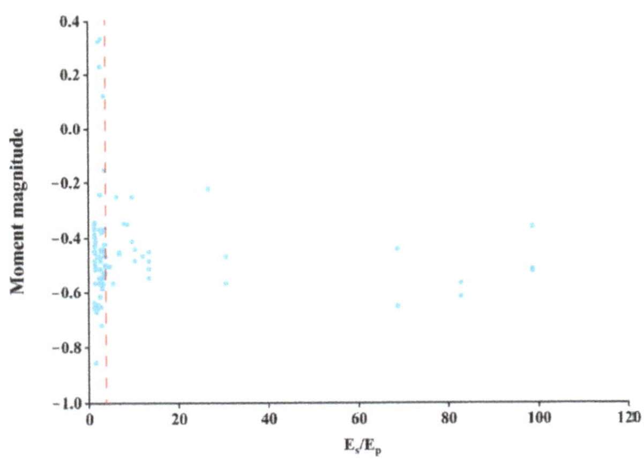

Figure 10. E_S/E_P of microseismic energy.

Figure 11. Deformation and failure of the surrounding rock on the left side of the tunnel face.

Figure 12 shows the spatial distribution of microseismic events, in which the size of the sphere represents the energy level of microseismic events, and different colors represent varying magnitudes of microseismic events. From Figure 12a,b, it can be seen that most of the microseismic events were distributed near the tunnel vault and corners of the side wall, and a few appeared near the side wall of the tunnel, indicating that most of the microseismic events were distributed at the tunnel vault or corners of the side wall. Therefore, more serious failure occurred in this area. The failure area of the surrounding rock of the tunnel can be seen from the spatial distribution of microseismic events. Figure 12c,d are the density cloud maps of the microseismic events, in which the color shades represent varying densities. It can be seen that the energy loss of the rock mass after tunnel excavation was mainly concentrated near the vault and corners of the side wall, which was consistent with the spatial distribution of microseismic events. Therefore, it can be revealed that this area was the main damage zone of the tunnel rock and there were risks of instability failure.

4.3. Comparison between Modeling Results and Microseismic Monitoring Results

A comparative analysis of the spatial evolution of acoustic emissions during the tunnel failure after excavation modeled by RFPA^{3D} and the monitoring data of microseismic events (as shown in Figure 13) shows that the spatial evolution of acoustic emissions based on the twin-shear unified strength theory and the strength reduction method was consistent with the distribution of detected microseismic events. They all prove that the most serious failure of the surrounding rock occurred at the tunnel vault and corners of the side wall, which had high risks of collapse and rock bursts. According to the seismological theory, the ratio of S-wave energy to P-wave energy, E_S/E_P, is often used as one of the indicators to reveal the failure mechanism of the surrounding rock. In particular, slip or shear failure often occurs when $E_S/E_P \geq 10$, and tensile failure often occurs when $E_S/E_P \leq 3$ [33]. As can be seen from Figure 10, the majority of microseismic events had $E_S/E_P \leq 3$ in this area and a fraction of microseismic events had $E_S/E_P \geq 10$, so the excavated section of the tunnel was dominated by tensile failure accompanied by shear failure. This conclusion is consistent with the modeling results in Figure 7, which shows that the integrated method based on the combination of field microseismic monitoring and RFPA^{3D} modeling software can be more accurate and effective for stability analysis of the excavation of the #2 diversion tunnel in the Xulong Hydropower Station.

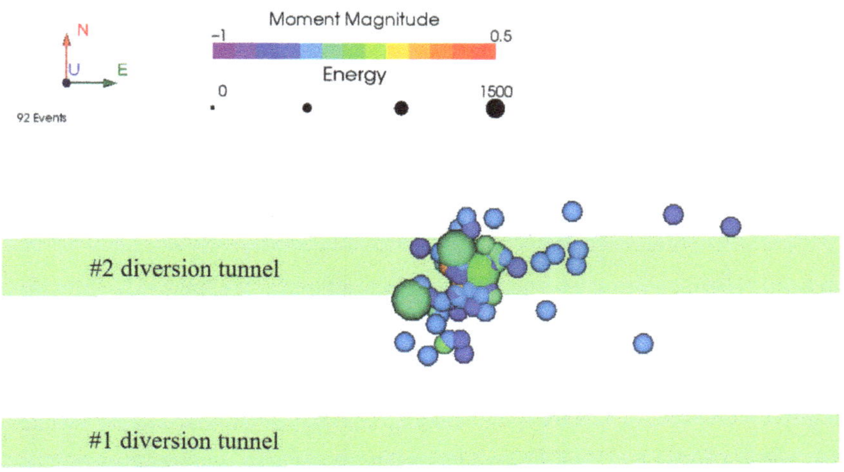

(**a**) Top view of the spatial distribution of microseismic events

Figure 12. *Cont.*

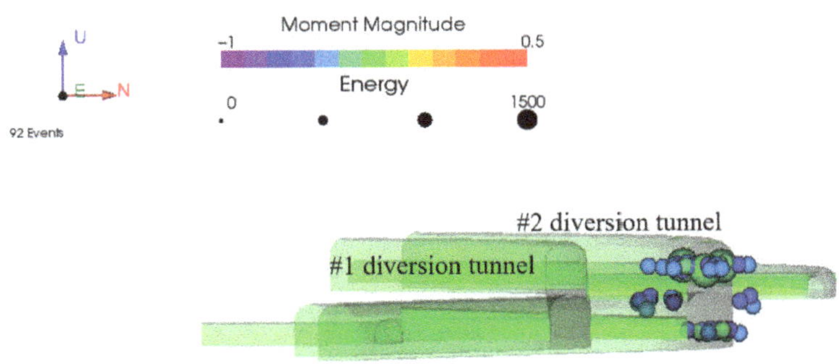

(**b**) Side view of the spatial distribution of microseismic events

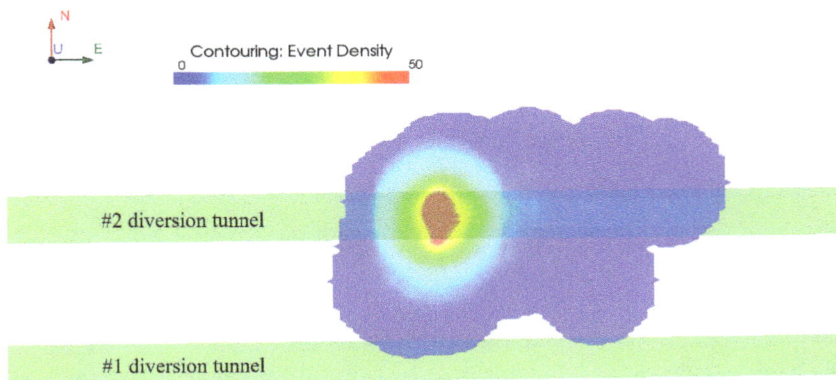

(**c**) Top view of the density of microseismic events

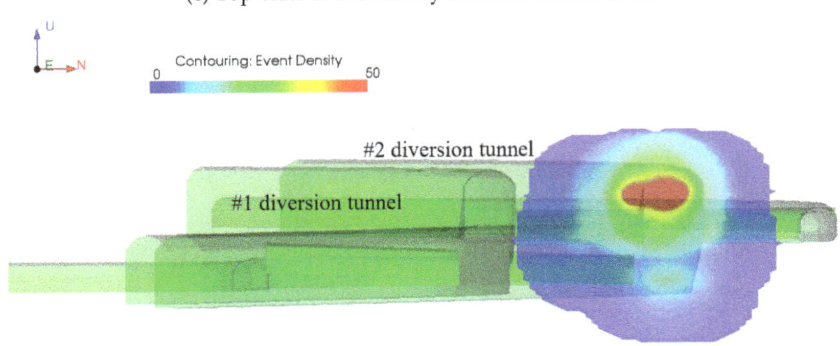

(**d**) Side view of the density of microseismic events

Figure 12. Spatial distribution of microseismic events.

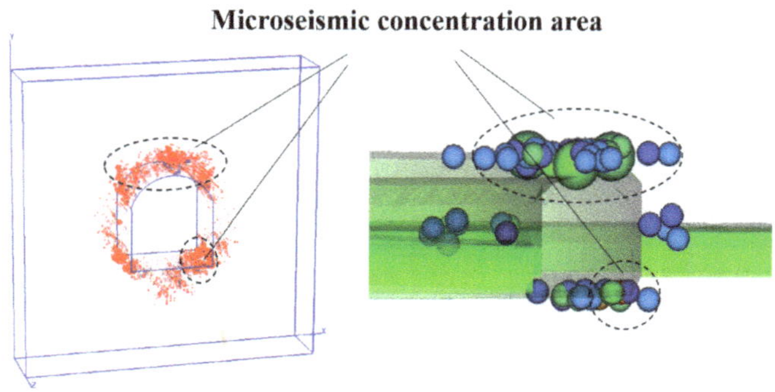

Figure 13. Comparison between numerical modeling and field monitoring.

Tensile failure mainly occurs in the excavation of the geologically weak section of the tunnel, and the failure is concentrated at the tunnel vault and corners of the side wall. The existence of joints or faults may worsen the distribution of stress and deformation of surrounding rock. Fractures tend to occur around the locations of joints or faults due to tress concentration and weak mechanical properties of structures. In addition, the energy and magnitude of microseismic events around the structures are relatively large and the failures are dominated by tensile fractures, accompanied with some shear fractures. Therefore, it is necessary to focus on the areas during excavation.

5. Conclusions

Based on RFPA^{3D} numerical modeling and microseismic monitoring, the stability of the excavated geologically weak section of the #2 diversion tunnel in the Xulong Hydropower Station was analyzed and evaluated, with the following conclusions reached:

(1) Surrounding rock damage after excavation of the geologically weak section of the #2 diversion tunnel was modeled by using RFPA^{3D}. The maximum principal stress concentrated around the tunnel profile and at the corners of the side wall in the study area, and cracks were initiated and grew from the tunnel vault and bottom corner of the side wall until they penetrated the entire profile. Acoustic emissions also started to accumulate at the vault and corners of the side wall and then propagated along the floor and side wall. The acoustic emissions at the vault and corners of the side wall gradually increased in energy, making it a key monitoring target in the study area.

(2) Microseismic events were mainly concentrated near the vault and footwall during the excavation of stakes K0+765 to K0+775. It was dominated by tensile failure accompanied by shear failure. The frequency of microseismic events could better reflect the frequency of excavation activities in underground plants, and microseismic monitoring helped infer the damage evolution of the rock during the instability failure process of the surrounding rock and delineate the potential risk area.

(3) The RFPA^{3D} numerical modeling results and the data collected by the actual microseismic monitoring system were consistent, which verified the damage area of the surrounding rock in the excavated geologically weak section. The vault and corners of the side wall, in particular, sustained the most serious damage, so it is necessary to pay close attention to this area and provide timely support. Microseismic monitoring played a key role in controlling the stability of the tunnel surrounding rock, and it is required to continue to strengthen microseismic monitoring and take effective measures to ensure the safety and stability of the slope during the construction and operation periods in the subsequent construction process.

Author Contributions: H.Q. investigated the geological and construction information of the engineering and wrote the first draft of the manuscript. Z.T. conducted the literature review and carried out the numerical simulation. B.L. developed the overarching research goals and analyzed microseismic data. All authors have read and agreed to the published version of the manuscript.

Funding: The research is supported by the National Natural Science Foundation of China (No. 42177143, 42277461), the Science Foundation for Distinguished Young Scholars of Sichuan Province (Grant No. 2020JDJQ0011) and the opening project (No. SKLGP2021K007) of the State Key Laboratory of Geohazard Prevention and Geoenvironment Protection.

Institutional Review Board Statement: Not applicable.

Informed Consent Statement: Not applicable.

Data Availability Statement: The microseismic data related to this research is confidential and cannot be made public.

Acknowledgments: Thanks are given to colleagues at the hydropower station for their valuable contributions to the project. Finally, the authors would like to thank the editors and reviewers for their valuable comments and constructive suggestions.

Conflicts of Interest: The authors declare no conflict of interest.

References

1. Xiao, Y.-X.; Feng, X.-T.; Feng, G.-L.; Liu, H.-J.; Jiang, Q.; Qiu, S.-L. Mechanism of evolution of stress–structure controlled collapse of surrounding rock in caverns: A case study from the Baihetan hydropower station in China. *Tunn. Undergr. Space Technol.* **2016**, *51*, 56–67. [CrossRef]
2. Zhang, C.Q.; Feng, X.-T.; Zhou, H.; Qiu, S.L.; Wu, W.P. Case Histories of Four Extremely Intense Rockbursts in Deep Tunnels. *Rock Mech. Rock Eng.* **2012**, *45*, 275–288. [CrossRef]
3. Hafeezur, R.; Abdul, M.N.; Kyoungmin, N.; Saeed, A.; Khan, M.; Han, K.Y. Impact of construction method and ground composition on headrace tunnel stability in the Neelum–Jhelum Hydroelectric Project: A case study review from Pakistan. *Appl. Sci.* **2021**, *11*, 1655.
4. Dhawan, K.R.; Singh, D.N.; Gupta, I.D. Three-Dimensional Finite Element Analysis of Underground Caverns. *Int. J. Géoméch.* **2004**, *4*, 224–228. [CrossRef]
5. Hao, Y.; Azzam, R. The plastic zones and displacements around underground openings in rock masses containing a fault. *Tunn. Undergr. Space Technol.* **2005**, *20*, 49–61. [CrossRef]
6. Yazdani, M.; Sharifzadeh, M.; Kamrani, K.; Ghorbani, M. Displacement-based numerical back analysis for estimation of rock mass parameters in Siah Bisheh powerhouse cavern using continuum and discontinuum approach. *Tunn. Undergr. Space Technol.* **2012**, *28*, 41–48. [CrossRef]
7. Ma, K.; Liu, G.Y. Three-dimensional Discontinuous Deformation Analysis of Failure Mechanisms and Movement Characteristics of Slope Rockfalls. *Rock Mech. Rock Eng.* **2022**, *55*, 275–296. [CrossRef]
8. Song, R.; Liu, J.; Yang, C.; Sun, S. Study on the multiphase heat and mass transfer mechanism in the dissociation of methane hydrate in reconstructed real-shape porous sediments. *Energy* **2022**, *254*, 124421. [CrossRef]
9. Cai, M.; Kaiser, P.; Morioka, H.; Minami, M.; Maejima, T.; Tasaka, Y.; Kurose, H. FLAC/PFC coupled numerical simulation of AE in large-scale underground excavations. *Int. J. Rock Mech. Min. Sci.* **2006**, *44*, 550–564. [CrossRef]
10. Jing, L. Formulation of discontinuous deformation analysis (DDA)—An implicit discrete element model for block systems. *Eng. Geol.* **1998**, *49*, 371–381. [CrossRef]
11. Andersson, J.C.; Martin, C.D. The Äspö pillar stability experiment: Part I—Experiment design. *Int. J. Rock Mech. Min. Sci.* **2009**, *46*, 865–878. [CrossRef]
12. Yan, C.-B. Blasting cumulative damage effects of underground engineering rock mass based on sonic wave measurement. *J. Central South Univ. Technol.* **2007**, *14*, 230–235. [CrossRef]
13. Li, S.; Yu, H.; Liu, Y.; Wu, F. Results from in-situ monitoring of displacement, bolt load, and disturbed zone of a powerhouse cavern during excavation process. *Int. J. Rock Mech. Min. Sci.* **2008**, *45*, 1519–1525. [CrossRef]
14. Yan, P.; Lu, W.; Chen, M.; Hu, Y.-G.; Zhou, C.-B.; Wu, X.-X. Contributions of In-Situ Stress Transient Redistribution to Blasting Excavation Damage Zone of Deep Tunnels. *Rock Mech. Rock Eng.* **2014**, *48*, 715–726. [CrossRef]
15. Li, B.; Ding, Q.F.; Xu, N.W.; Dai, F.; Xu, Y.; Qu, H.L. Characteristics of microseismic bvalue associated with rock mass large deformation in underground powerhouse caverns at different stresslevels. *J. Cent. South Univ.* **2022**, *29*, 693–711. [CrossRef]
16. Li, B.; Xu, N.; Dai, F.; Zhang, G.; Xiao, P. Dynamic analysis of rock mass deformation in large underground caverns considering microseismic data. *Int. J. Rock Mech. Min. Sci.* **2019**, *122*, 104078. [CrossRef]
17. Leśniak, A.; Isakow, Z. Space–time clustering of seismic events and hazard assessment in the Zabrze-Bielszowice coal mine, Poland. *Int. J. Rock Mech. Min. Sci.* **2009**, *46*, 918–928. [CrossRef]

18. Hudyma, M.; Potvin, Y.H. An Engineering Approach to Seismic Risk Management in Hardrock Mines. *Rock Mech. Rock Eng.* **2009**, *43*, 891–906. [CrossRef]
19. Ma, J.; Dong, L.; Zhao, G.; Li, X. Focal Mechanism of Mining-Induced Seismicity in Fault Zones: A Case Study of Yongshaba Mine in China. *Rock Mech. Rock Eng.* **2019**, *52*, 3341–3352. [CrossRef]
20. Cai, M.; Kaiser, P.; Martin, C. Quantification of rock mass damage in underground excavations from microseismic event monitoring. *Int. J. Rock Mech. Min. Sci.* **2001**, *38*, 1135–1145. [CrossRef]
21. Ma, K.; Tang, C.-A.; Xu, N.-W.; Liu, F.; Xu, J.-W. Failure precursor of surrounding rock mass around cross tunnel in high-steep rock slope. *J. Central South Univ.* **2013**, *20*, 207–217. [CrossRef]
22. Dai, F.; Li, B.; Xu, N.; Fan, Y.; Zhang, C. Deformation forecasting and stability analysis of large-scale underground powerhouse caverns from microseismic monitoring. *Int. J. Rock Mech. Min. Sci.* **2016**, *86*, 269–281. [CrossRef]
23. Liang, B.; Jiang, H.L.; Sun, W.J.; Jin, J.X. Numerical simulation study on seepage characteristics of multi-fractured rock mass. *J. Water Resour. Water Eng.* **2013**, *6*, 79–81.
24. Tang, C.A.; Hudson, J.A. *Rock Failure Mechanisms: Explained and Illustrated*; CRC Press: Baca Raton, FL, USA, 2010.
25. Liang, Z.Z. *Three-Dimensional Failure Process Analysis of Rock and Associated Numerical Tests*; Northeastern University: Shenyang, China, 2005.
26. Yu, M.H. Unified strength theory for geomaterials and its applications. *Chin. J. Geotech. Eng.* **1994**, *16*, 1–9.
27. Liang, C.; Zhang, G.; Li, L.; Li, B.; Luo, Y.; Li, Y.; Duan, J.; Wu, X. Efficient Fault Surface Grouping in 3-D Seismic Fault Data. *IEEE Trans. Geosci. Remote Sens.* **2022**, *60*, 5916213. [CrossRef]
28. Wu, N.; Liang, Z.; Zhang, Z.; Li, S.; Lang, Y. Development and verification of three-dimensional equivalent discrete fracture network modelling based on the finite element method. *Eng. Geol.* **2022**, *306*, 106759. [CrossRef]
29. Wu, N.; Liang, Z.; Li, Y.; Qian, X.; Gong, B. Effect of confining stress on representative elementary volume of jointed rock masses. *Geomech. Eng.* **2019**, *18*, 627–638.
30. Xue, R.X.; Liang, Z.Z.; Xu, N.W. Rockburst prediction and analysis of activity characteristics within surrounding rock based on microseismic monitoring and numerical simulation. *Int. J. Rock Mech. Min. Sci.* **2021**, *142*, 104750. [CrossRef]
31. Tang, C.A.; Li, L.C.; Li, C.W.; Ma, T.H. RFPA strength reduction method for stability analysis of geotechnical engineering. *Chin. J. Rock Mech. Eng.* **2006**, *25*, 1522–1530.
32. Li, S.Y.; He, X.S.; Zhang, S.Q.; Lu, Q.H.; Jiang, X.Q.; Tong, X.H.; Li, T.; Guan, E.F.; Zuo, Y.; Sun, X.H.; et al. Progress and latest achievements of mine seismic monitoring technology. *Prog. Geophys.* **2004**, *4*, 853–859.
33. Boatwright, J.; Fletcher, J.B. The partition of radiated energy between P and S waves. *Bull. Seismol. Soc. Am.* **1984**, *74*, 361–376. [CrossRef]

applied sciences

MDPI

Article

Sensitivity Analysis of Fracture Geometry Parameters on the Mechanical Behavior of Rock Mass with an Embedded Three-Dimensional Fracture Network

Na Wu [1], Zhengzhao Liang [1], Yan Tao [1], Ting Ai [2,*] and Guijie Li [3]

[1] State Key Laboratory of Coastal and Offshore Engineering, Dalian University of Technology, Dalian 116024, China
[2] Key Laboratory of Deep Earth Science and Engineering, Ministry of Education, Sichuan University, Chengdu 610065, China
[3] School of Aeronautics and Astronautics, Dalian University of Technology, Dalian 116024, China
* Correspondence: aiting@scu.edu.cn

Abstract: The existence of fractures has a significant influence on the mechanical properties of a rock mass. The sensitivity of the rock mass's mechanical properties to the fracture's geometric parameters is conducive to improving the measurement accuracy of fractured rock mass engineering. Firstly, the fracture geometric parameters in the dam site area of Lianghekou Hydropower Station were counted using the ShapeMetriX^{3D} system. Then, the effect of the fracture's geometric parameters on the deformation characteristics, failure mode, and mechanical parameters of the rock mass were investigated based on the RFPA^{3D} under the uniaxial compression test. The results showed that the stress–strain curves of the fractured rock mass mainly exhibited elastic-brittle characteristics. The failure pattern of the fractured rock mass was mainly defined by a compressive-shear composite. Additionally, the influence of the fracture's geometric parameters on the uniaxial compressive strength (UCS) was greater than that of elastic modulus. The sensitivity of the UCS to fracture trace length was more significant.

Keywords: fractured rock mass; ShapeMetriX^{3D} system; fracture geometric parameter; sensitivity analysis; failure mode

Citation: Wu, N.; Liang, Z.; Tao, Y.; Ai, T.; Li, G. Sensitivity Analysis of Fracture Geometry Parameters on the Mechanical Behavior of Rock Mass with an Embedded Three-Dimensional Fracture Network. *Appl. Sci.* **2022**, *12*, 9284. https://doi.org/10.3390/app12189284

Academic Editor: Arcady Dyskin

Received: 26 July 2022
Accepted: 14 September 2022
Published: 16 September 2022

1. Introduction

With the development of the social economy and the increase of national infrastructure investment, the scale and complexity of rock engineering have been increasing. The mechanical characteristics of the rock masses involved are becoming more and more complex. Rock masses differ from rocks in that they contain many discontinuous structural surfaces. Fractures are common discontinuities in rock masses [1,2] that intersect each other to form complex fracture networks. Fracture networks are an important reason for the extremely complex deformation characteristics of rock masses. The nonlinear, size effect, and anisotropic behavior of fractured rock masses are affected by their complex fracture networks [3,4]. The instability of rock masses is usually caused by the deformation, expansion, and penetration of the fractures in the fracture networks [5,6]. Additionally, the strength and deformation parameters are controlled by the complex fracture networks [7]. The complex fracture networks are usually determined by fracture geometry parameters, such as inclination, dip angle, and trace length of the fractures. Therefore, it is of great importance to study the influence of fracture geometry parameters on the mechanical properties of a rock mass.

Identifying and extracting fracture geometry information is essential for analyzing the mechanical and deformation properties of the fractured rock masses. There are two main ways to acquire a fracture's geometric information in an outcrop area, including contact and non-contact measurements. The efficiency of contact measurement is lower, and the measurement accuracy cannot be guaranteed [8,9]. In recent years, non-contact measurement techniques (such as total station and photogrammetry) have been widely used to analyze fracture geometric information in a rock mass. It can effectively capture data from physically inaccessible areas [10]. Laser-scanning techniques are an effective non-contact measurement method. By analyzing the outcrop characteristics of visible fractures on the rock mass surface, point cloud data processing can be used to identify discontinuities and extract information from them [11]. A method to detect and estimate the orientation of plane structures was proposed based on the principal component analysis (PCA) techniques by Gomes et al. [12]. When applied to practical situations, it showed high precision in both the detection and location of fracture planes. A method for generating a high-resolution digital outcrop model (DOM) of the rock exposure based on UAV photogrammetry and the structural self-motion (SFM) technique was introduced and applied to slopes in Greece and China [13]. The results showed that UAV-SFM photogrammetry had higher efficiency and operability, especially in difficult measurement environments. Stereophotogrammetry is a method of extracting the regions of interesting information by constructing stereo images from two or more photographs. ShapeMetriX^{3D} is widely used for discerning the structural surface information statistics of rock slopes and tunnels [14–16]. The close-up photographs turned out useful to obtain detailed images and accurate results.

The distribution of fracture geometry parameters has certain statistical and probability distribution characteristics. Statistical methods are often used to quantitatively describe and analyze fracture geometry parameters. The fracture distribution function is obtained based on a fracture's geometry parameters and statistical methods [17,18]. The discrete fracture network (DFN) model provides an approach to characterize complex fractures based on the fracture distribution function [19]. The sizes, directions, spacings, and spatial locations of fractures are included in the DFN model. The DFN model is also an effective method to study the influence of fracture geometry parameters on the mechanical properties of complex fractured rock mass [20,21]. The model consists of the fractures embedded in the rock. For example, a stochastic DFN model was established based on the two-dimensional finite element method of rock failure process analysis (RFPA^{2D}). The failure modes, strength, and deformation parameters (elastic modulus and peak strength) of fractured rock mass were studied by Wu et al. [22]. Results showed that the fracture geometry parameters had a significant impact on the scale effect of the fractured rock mass. The effect of geometrical parameters of non-persistent fractures on the mechanical behaviors of the rock mass based on PFC^{2D} software was studied by Vaziri et al. [23] under uniaxial loading conditions. The results showed that the effect of random fractures on strength was greater than the effect of the deformation modulus. Amongst the fracture geometrical parameters, the fracture orientation had the most significant effect on the mechanical parameters of the rock mass.

The distribution of fractures in nature is three dimensional and spatial. A 3D equivalent DFN model was proposed based on the three-dimensional rock failure process analysis (RFPA^{3D}) software, and the scale effect and anisotropy of the fractured rock mass were analyzed by Wu et al. [24]. Moreover, the effects of the geometry (orientation, duration, spacing, aperture, and step angle) of parallel fractures on the uniaxial compressive strength (UCS) and deformation modulus were statistically investigated based on PFC^{3D} by Bahaaddini et al. [25]. It was found that the fracture orientation had the greatest influence on the UCS and deformation modulus, followed by the fracture spacing. However, the fractures in the model were not random, which was not consistent with the actual project. The effect of fracture geometry parameters on the equivalent mechanical parameters of the fractured rock mass based on the 3DEC software was studied by Cui et al. [26]. The results showed that the elastic modulus was the most sensitive to the structural density and the UCS was the most sensitive to the trace length of the fracture.

The estimation of the mechanical properties of fractured rock masses has always been a challenging task for practical rock engineers. A review of previous studies clearly showed that the fracture geometry parameters had a strong effect on the mechanical behavior of rock mass. The DFN model is an effective method to study the mechanical behavior of rock masses. However, the DFN model relied heavily on the quality of natural fracture structural data gathered in the field. The purpose of this paper was to evaluate the influence of fracture geometry parameters on the mechanical properties of rock mass using a combination of finite element methods and a fracture network model. First, the ShapeMetriX^{3D} system was used to identify and extract the fracture geometry information in the dam site area of the Lianghekou Hydropower Station. The mean value, variance, and distribution types of fracture geometry parameters were obtained. Then, an orthogonal test scheme was designed based on the influencing factors and factor levels. The corresponding DFN models were established according to the fracture geometry parameters. The effects of fracture geometry parameters, including the inclination, dip angle, trace length and bulk density, on the UCS, elastic modulus, and failure pattern were studied under uniaxial loading conditions. Finally, the degree of influence of the fracture geometry parameters on the mechanical behavior of randomly fractured rock masses was statistically analyzed by using range analysis, variance analysis, and significance test.

2. Acquirement of Fracture Geometry Parameters

2.1. Research Background

The Lianghekou Hydropower Station is located on the main stream of the Yalong River in Yajiang County, Sichuan Province. The dam site is located about 2 km downstream of the confluence of the main stream of the Yalong River and the tributary of Xianshui River. The construction drawing in the dam site area of Lianghekou Hydropower Station is illustrated. The Yalong River flows north to south in the dam site area, crosses the IV exploration line, and returns to S23°E. It then flows from north to south again in the upper reaches of the unnamed ditch. The dam site is located in a lateral valley with a slope of 500–1000 m along the river edge. The left bank is curved and projects towards the right bank with an average topographic slope of 55° and multiple small gullies. The right bank is concave, with an average convex slope of 45°. The geological structure and tectonic environment of the Lianghekou reservoir area are generally simple. There are a few small faults, strong tectonic action, and no large-scale broken rock mass due to the intense tectonic action. The stability of the reservoir bank mainly depends on the stability degree of the existing landslide in the natural state, as well as any potential changes during reservoir storage.

The discontinuous surfaces of the slope rock mass, such as fissures and fractures, control the instability pattern of the slope. Therefore, the slope rock mass is weakened, and the safety of the engineering project is threatened. In this paper, the left bank side slope rock mass downstream of the dam site of Lianghekou Hydropower Station was studied, as shown in Figure 1.

Figure 1. Slope rock mass of the left bank near the dam site of the Lianghekou Hydropower Station.

The geometrical parameters in the fractures of the slope rock mass were counted by the 3D non-contact measurement system (ShapeMetriX^{3D} system). The influence law and sensitivity analysis of the fracture information on the mechanical properties of the fractured rock mass was conducted to provide reference data for the analysis of the slope instability. According to the investigation, the lithology of the rock mass in this area is metamorphic sandstone. The detailed mechanical parameters of the rocks and fractures are shown in Table 1.

Table 1. Mechanical parameters of the rocks and fractures in the study area [24,27].

Material Type	Heterogeneity	UCS/MPa	Elastic Modulus/GPa	Friction Angle/°	Poisson's Ratio
Rock	5	108.9	37.6	56	0.24
Fracture	2	5.45	1.88	30	0.39

2.2. Data Collection from the Study Area

The ShapeMetriX^{3D} system was produced by a 3GSM company in Austria. The system consists of a digital camera capable of calibrating 20 million high-resolution stereo images (Figure 2a), a software package for 3D model reconstruction and visualization analysis (Figure 2b), and a benchmark (Figure 2c). The specified area is imaged from two different angles, and the 3D geometric image is synthesized via technical processing to realize the real 3D model reconstruction of the solid surface.

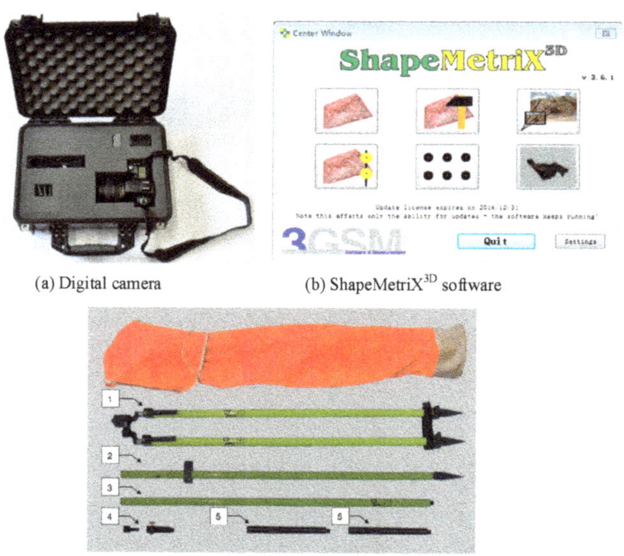

(a) Digital camera (b) ShapeMetriX^{3D} software

(c) Benchmark

Figure 2. Composition and principle of the ShapeMetriX^{3D} system. Research steps 1–5 are the process of data collection based on the equipment.

The steps to obtain the geometric information of 3D rock mass surface fractures are as follows: 1. Select an appropriate camera lens and set the camera parameters according to the working conditions of the shooting site. 2. Place the reference point in front of the test rock mass and make sure it is vertical. 3. Select two shooting points in front of the rock mass on the left and right to photograph the rock mass in the study area. The connection line between the two points is required to be parallel to the rock mass surface and meet the requirements of a certain distance. 4. Find an obvious structural plane on the rock mass surface and measure and record the tendency and inclination of the structural plane with a geological compass. 5. Import the left and right pictures into the Shapemetrix^{3D} software separately, depict the research area, and synthesize. The 3D geometric images are generated. 6. The size and direction of the rock mass are corrected by the size of the benchmark and the inclination and dip angle of the selected structural plane for the reconstruction of a real 3D model of rock mass engineering. 7. The geometric information of the fracture surfaces is described quantitatively, the dominant groups of fracture surfaces are divided according to the production clustering, and the parameters of the production information of the fracture surfaces are obtained. There are three groups of fractures in the study area. The probability distributions and geometrical parameters of fracture groups are listed in Table 2. Refer to Wu et al. [24,27] for more details.

Table 2. Probability distribution and geometrical parameters of the fracture groups [24,27].

Material Type		1# Fracture Group	2# Fracture Group	3# Fracture Group
Inclination/°	Distribution type	Normal	Lognormal	Normal
	Mean value	235.15	351.34	94.73
	Standard deviation	6.48	0.15	13.68
Dip angle/°	Distribution type	Normal	Lognormal	Normal
	Mean value	25.98	28.16	85.80
	Standard deviation	8.41	0.83	9.92
Trace length/m	Distribution type	Lognormal	Normal	Lognormal
	Mean value	3.42	3.12	2.96
	Standard deviation	2.05	1.48	1.39
Spacing/m	Distribution type	Negative exponential	Lognormal	Negative exponential
	Mean value	1.35	1.70	0.69
	Standard deviation	1.21	1.55	0.7
Bulk density/m^{-3}	Mean value	0.01734	0.02020	0.05838

3. Orthogonal Test Scheme and Numerical Model

3.1. Orthogonal Test Scheme

The purpose of this paper was to study the effects of fracture inclination, dip angle, trace length, and bulk density (relating to spacing) on the UCS and elastic modulus of fractured rock masses. Due to a large number of influencing factors and levels, it is quite troublesome to calculate the scheme. The orthogonal test method can not only effectively reduce the workload of calculation, but also obtain satisfactory sample results [28,29]. As a result, an orthogonal test scheme with four influencing factors and three levels was selected, and $L_9(3^4)$ was chosen as the orthogonal table for the design of this test scheme, as shown in Table 3. There are nine group test protocols in the table. The column number indicates the test factor, while the value under that column number indicates the level used for that factor. The level values of each influencing factor based on the range of the main influence factors also are given in Table 3. The level values of the influencing factors of inclination are 47.24°, 94.49°, and 141.73°, respectively. To dip the angle, trace length, and bulk density, the level values were 28.60°, 57.20°, 85.80°, 0.987 m, 1.973 m, 2.960 m, 0.05838 m^{-3}, 0.07784 m^{-3}, and 0.09730 m^{-3} in the order of levels 1, 2, and 3. Therefore, the test protocols for each main control factor were established based on the orthogonal test scheme.

Table 3. Orthogonal arrays $L_9(3^4)$ of the test scheme.

Test Number	Influencing Factors			
	Inclination/°	Dip Angle/°	Trace Length/m	Bulk Density/m^{-3}
T-1	Level 1 (47.24)	Level 1 (28.60)	Level 1 (0.987)	Level 1 (0.05838)
T-2	Level 1	Level 2 (57.20)	Level 2 (1.973)	Level 2 (0.07784)
T-3	Level 1	Level 3 (85.80)	Level 3 (2.960)	Level 3 (0.09730)

Table 3. *Cont.*

Test Number	Influencing Factors			
	Inclination/°	Dip Angle/°	Trace Length/m	Bulk Density/m^{-3}
T-4	Level 2 (94.49)	Level 1	Level 2	Level 3
T-5	Level 2	Level 2	Level 3	Level 1
T-6	Level 2	Level 3	Level 1	Level 2
T-7	Level 3 (141.73)	Level 1	Level 3	Level 2
T-8	Level 3	Level 2	Level 1	Level 3
T-9	Level 3	Level 3	Level 2	Level 1

3.2. The Principle of RFPA^{3D}

A finite element software, RFPA^{3D} (three-dimensional realistic failure process analysis), was applied for the numerical simulation. RFPA^{3D} is an effective method based on the damage mechanics and the statistical theory to simulate the gradual failure process of rock mass. It is worth mentioning that the mechanical parameters (such as the strength and elastic modulus) of the model elements obey the Weibull distribution. Therefore, the distribution characteristics of a rock's mechanical parameters at the mesoscopic level can be considered in the RFPA^{3D}. Additionally, the macroscopic failure is the accumulation process of mesoscopic element failure. For detailed information on the calculational principles and model validity of the RFPA^{3D} software, please refer to papers [24,30,31]. The constitutive equations of the RFPA^{3D} software are briefly summarized in this section.

The elastic-brittle damage constitutive model was introduced into the RFPA^{3D} to describe the stress–strain relationship. The constitutive relationship of the model is expressed as follows:

$$\sigma = E\varepsilon = E_0(1 - D)\varepsilon \tag{1}$$

where σ and ε are the stress applied to the model and the corresponding strain, respectively; E represents the elastic modulus of the damaged element; and E_0 and D denote the initial elastic modulus of the element and the damage variable, respectively. However, with increasing load, the elastic modulus of the element degrades gradually with the increase of the D.

When the model element is in a uniaxial tension state, the maximum tensile strain criterion and the Mohr Coulomb criterion were used in the RFPA^{3D}. The corresponding D was defined as in Equation (2).

$$D = \begin{cases} 0 & \varepsilon > \varepsilon_{to} \\ 1 - \frac{\sigma_{rt}}{\varepsilon E_0} & \varepsilon_{to} \geq \varepsilon \geq \varepsilon_{tu} \\ 1 & \varepsilon < \varepsilon_{tu} \end{cases} \tag{2}$$

where $\sigma_{\rm rt}$ denotes the residual stress of the model element, and $\sigma_{\rm rt} = \lambda\sigma_{\rm t}$. Among them, $\sigma_{\rm t}$ and λ are the tension stress applied on the model and the residual coefficient, respectively; ε_{to} is the failure strain threshold, which is the tensile strain corresponding to the elastic limit; ε_{tu} is the ultimate tensile strain coefficient, and $\varepsilon_{tu} = \eta\varepsilon_{to}$; η denotes the ultimate strain coefficient.

According to the elastic damage theory, the stress–strain relationship can be derived as follows:

$$\sigma_{ij} = \begin{cases} 2G\varepsilon_{ij} + \lambda\delta_{ij}\varepsilon_{kk} & \bar{\varepsilon} > \varepsilon_{to} \\ \frac{\sigma_{rt}}{\bar{\varepsilon}E_0}\left(2G\varepsilon_{ij} + \lambda\delta_{ij}\varepsilon_{kk}\right) = \frac{\sigma_{rt}}{\bar{\varepsilon}}\left(\frac{\varepsilon_{ij}}{1+v} + \frac{\delta_{ij}\varepsilon_{kk}}{(1+v)(1-2v)}\right) & \varepsilon_{to} \geq \bar{\varepsilon} \geq \varepsilon_{tu} \\ 0 & \bar{\varepsilon} < \varepsilon_{tu} \end{cases} \tag{3}$$

where G and v are the shear modulus and the Poisson's ratio of the model element, respectively; when $i = j$, $\delta_{ij} = 1$; otherwise, $\delta_{ij} = 0$.

Moreover, the Mohr–Coulomb criterion is applied when the model element is in a uniaxial compression state. Similarly, the D is expressed as follows:

$$D = \begin{cases} 0 & \varepsilon_1 < \varepsilon_{co} \\ 1 - \frac{\sigma_{rc}}{\varepsilon_1 E_0} & \varepsilon_1 \geq \varepsilon_{co} \end{cases} \tag{4}$$

where σ_{co} and σ_{rc} are the initial and residual UCS of the model element, respectively, and $\sigma_{rc} = \lambda\sigma_{co}$, ε_1, and ε_{co} denote the maximum principal stress and the compressive strain threshold, respectively.

Similarly, the corresponding the stress–strain relationship can be written as follows:

$$\sigma_{ij} = \begin{cases} 2G\varepsilon_{ij} + \lambda\delta_{ij}\varepsilon_{kk} & \varepsilon_1 < \varepsilon_{co} \\ \frac{\sigma_{rc}}{\varepsilon_1 E_0}\left(2G\varepsilon_{ij} + \lambda\delta_{ij}\varepsilon_{kk}\right) = \frac{\sigma_{rc}}{\varepsilon_1}\left(\frac{\varepsilon_{ij}}{1+v} + \frac{\delta_{ij}\varepsilon_{kk}}{(1+v)(1-2v)}\right) & \varepsilon_1 \geq \varepsilon_{co} \end{cases} \tag{5}$$

3.3. Model Establishment Based on RFPA^{3D}

The sensitivity of fracture geometrical parameters to the elastic modulus and UCS of the fractured rock mass was investigated in the sloping rock of the Lianghekou Hydropower Station dam site. However, only the geometric parameters of the 3# fracture group changed during the study. The geometric parameters of the 1# and 2# fracture groups remained unchanged.

The model size of 10 m × 10 m × 10 m and the element number of 120 × 120 × 120 were chosen to build the equivalent rock mass DFN model based on the RFPA^{3D} software [30]. The input mechanical parameters of the RFPA software are the mesoscopic parameters of rock mass, which is related to the element size to reduce the mesh effect of the model. Nine numerical models of the orthogonal arrays of the test scheme were generated based on Tables 1 and 3, as shown in Figure 3. The red lines in the numerical model represent the intersections of the fractures and the rocks.

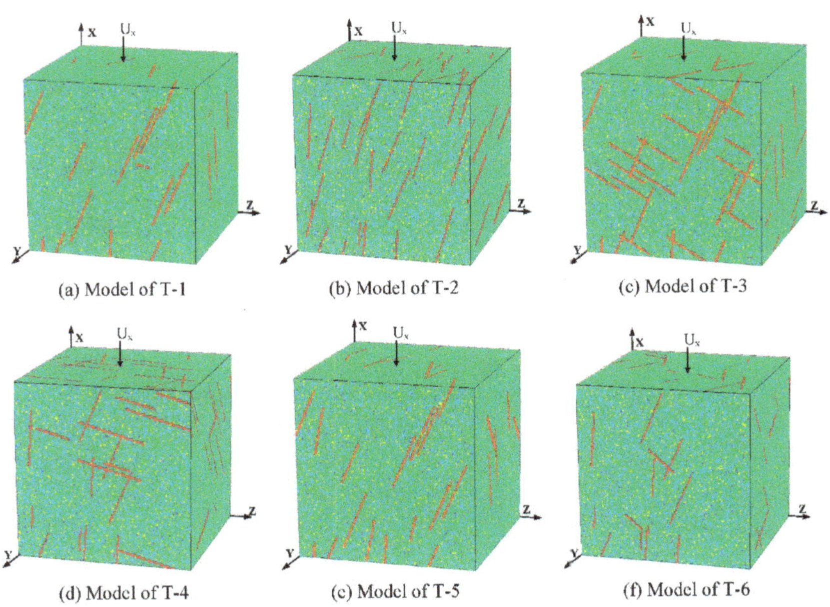

Figure 3. *Cont.*

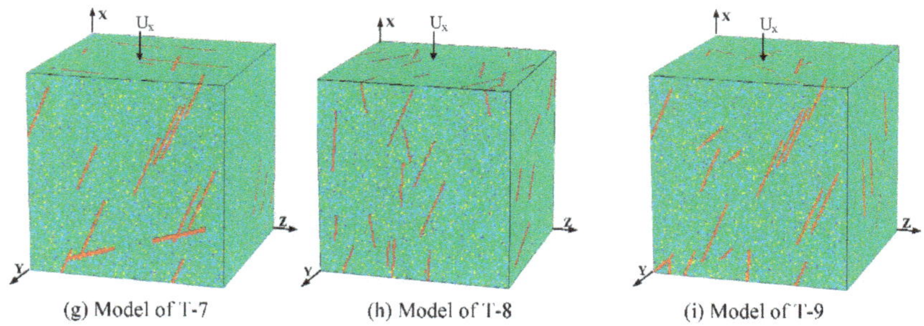

Figure 3. Numerical model of the orthogonal arrays of the test scheme.

Uniaxial, quasi triaxial, and true triaxial compression test of fractured rock mass can be implemented and simulated based on the RFPA3D software. However, the confining pressure may make the calculation results more complicated and diverse. To understand the sensitivity of the mechanical parameters of rock mass to geometric parameters under simple conditions, only the uniaxial compression test was adopted in this section.

During the simulation, the mechanical parameters and boundary conditions were kept constant during the tests except for the factors considered to enhance the reliability of the conclusions. The bottom surface (YOZ) of the model was fixed, and a displacement load (0.2 mm/step) was applied to the upper surface of the model parallel to the X-axis direction until the model failed. The remaining boundaries of the model are free.

4. Results and Analysis

4.1. Deformation Characteristics and Failure Patterns of Numerical Model

The stress–strain curves of the simulation results of the test scheme in Table 3 are shown in Figure 4. The results showed that the fractured rock mass presented brittle deformation, except for T-5, which exhibited elastic–plastic deformation. The deformation characteristics and failure pattern of test number T-1 in Table 3 were investigated as an example, as shown in Figure 5. The numerical model of the case is given in Figure 5a. The X-displacement diagram of the failure fractured rock mass is presented in Figure 5b. Red and blue colors represent the maximum and minimum deformation of the model, respectively. The deformation of the fractured rock mass in the X-axis direction was greatly influenced by DFN, and clear boundaries were formed at the intersections of the fractures and the rock. The deformation or displacement diagram of fractured rock mass decreased from top to bottom in the X-axis direction. Figure 5c shows the acoustic emission diagram of the failure model. Red circles in the acoustic emission diagram represent the damage events induced by compressive under the current load. In Figure 5d, the damage diagram of the failure model is plotted. The results showed that the fractured rock mass was mainly damaged along the fracture surface under the action of the external load. Due to the damage and failure of the fracture, multiple penetration shear faces were generated. The fractured rock mass was primarily defined by compressive-shear composite failure.

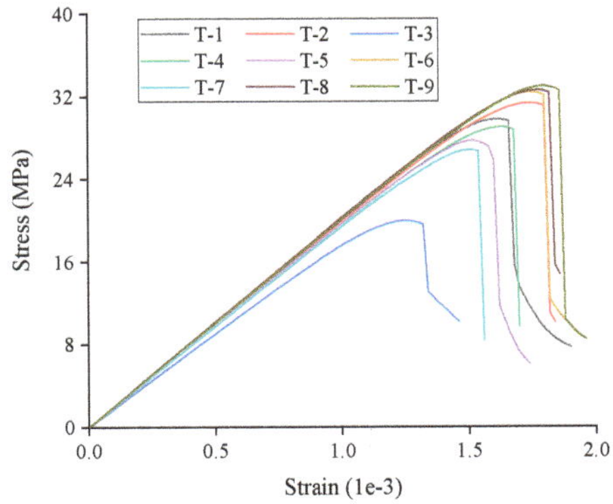

Figure 4. Stress–strain curves of the numerical models.

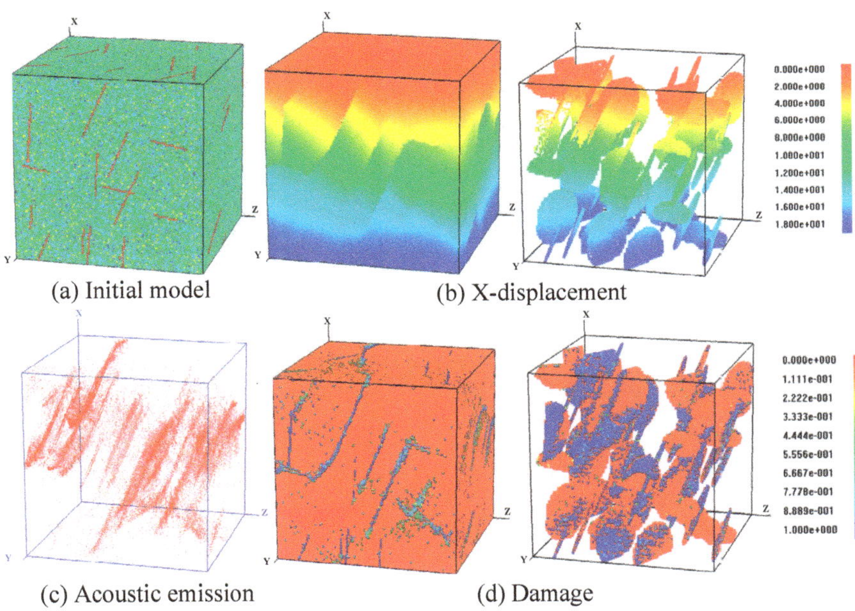

Figure 5. The deformation characteristics and failure pattern for T-1.

4.2. Strength and Deformation Parameters

Meanwhile, the elastic modulus and UCS of the models could be acquired based on the stress–strain curves. The elastic modulus is the slope of the line part of the stress–strain curve in the elastic stage. The UCS is the peak strength in the stress–strain curve. The elastic modulus and UCS provide parameters for sensitivity analysis.

Table 4 lists the test assignment scheme and the numerical results for each factor level based on the orthogonal table of the test scheme. The order of the test scenarios is consistent with Table 3. The values of each influencing factor, and the corresponding elastic modulus and UCS derived from numerical simulation under different schemes are also presented in Table 5. The results showed that the variations in fracture inclination, dip angle, trace length, and bulk density had significant effects on the elastic modulus and UCS of the fractured rock mass.

Table 4. The elastic modulus and UCS under various influencing factors and levels.

Test Number	Influencing Factors				Numerical Results	
	Inclination/°	Dip Angle/°	Trace Length/m	Bulk Density/m^{-3}	Elastic Modulus /GPa	UCS/MPa
T-1	47.24	28.60	1.48	0.029	20.362	29.794
T-2	94.49	28.60	2.96	0.058	20.063	31.378
T-3	141.73	28.60	4.44	0.0871	18.114	19.996
T-4	47.24	57.20	2.96	0.087	19.807	29.094
T-5	94.49	57.20	4.44	0.029	19.881	27.730
T-6	141.73	57.20	1.48	0.058	20.351	32.415
T-7	47.24	85.80	4.44	0.058	19.572	26.869
T-8	94.49	85.80	1.48	0.087	20.356	32.609
T-9	141.73	85.80	2.96	0.029	20.173	32.977

Table 5. Level values of the influencing factors of the Group 2 test.

Influencing Factors	Level 1	Level 2	Level 3
Inclination/°	47.24	94.49	141.73
Dip angle/°	28.60	57.20	85.80
Trace length/m	1.974	3.948	5.92
Bulk density/m^{-3}	0.05838	0.07784	0.09730

4.3. Range Analysis

Range analysis, also known as visual analysis, is one of the most common methods used to analyze the result of orthogonal tests. The problem is analyzed by the difference between the maximum and minimum in the mean effect. The specific formula is as follows:

$$R_i = \max\{k_{i1}, k_{i2}, \cdots, k_{ij}\} - \min\{k_{i1}, k_{i2}, \cdots, k_{ij}\} \quad (6)$$

where k_{ij} presents the mean value of the test results of the i influencing factor on the j level. R_i is refers to the extreme difference in the value of the influencing factor of i, that is, the difference between the maximum and minimum values of the test results at each level of the i influencing factor. The value reflects the variation range in the test results when the influencing factor of i changes. The larger the value of R, the greater the influence of the factor on the overall experiment results, which is a major influence. Conversely, it has a small effect.

Figure 6 shows the data analysis of the influencing factor level on the elastic modulus and UCS, respectively. The results showed that the influence degree of the four influencing factors on the elastic modulus of the fractured rock mass was in the order of track length, bulk density, dip angle, and inclination. However, for the UCS, the track length, inclination, bulk density, and dip angle were arranged in descending order. Therefore, the fracture trace length based on the extreme difference analysis was the main factor affecting the elastic modulus of the fractured rock mass and UCS.

The range analysis is easier to calculate and can prioritize the effect of factors on the calculated results. However, it is impossible to determine whether the influencing factors of the experiment are significant. Furthermore, errors are inevitable during the experiment, and range analysis makes it difficult to distinguish whether the differences corresponding to each factor level are due to experimental error or the factor level itself.

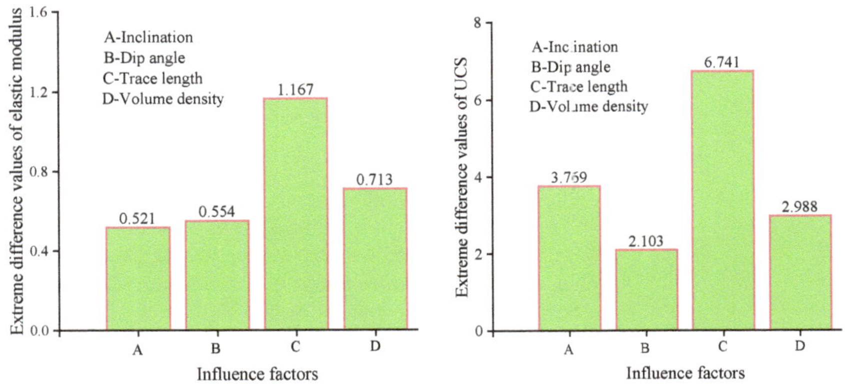

Figure 6. Extreme difference values of the elastic modulus and UCS under various influencing factors.

4.4. Variance Analysis and Significance Test

Variance analysis is widely used in data processing of the orthogonal test and can compensate for the deficiencies of the range analysis. The corresponding sum of squared variance and sum of squared errors need to be calculated for the experiment data when the variance analysis is applied. Based on the calculation results of the elastic modulus and UCS at different influencing factor levels in the orthogonal test, the corresponding sums of deviations squares, errors squares, and degrees of freedom for each factor were calculated with the help of an orthogonal design assistant. Moreover, the mean square is the ratio of the sum of squares of deviation to the degree of freedom. The F ratio is the ratio of the mean square of each factor to the mean square of error, respectively. Therefore, the mean square and the F ratio of each factor could be calculated for the elastic modulus and UCS, as shown in Figure 7. Finally, the F-ratios were compared with the critical values of $F_{0.01}(2, 6)$ and $F_{0.05}(2, 6)$. If $F > F_{0.01}(2, 6)$, the factor was highly significant, denoted by **. If $F_{0.01}(2, 6) > F > F_{0.05}(2, 6)$, the factor was significant, denoted by *. If $F < F_{0.05}(2, 6)$, the factor was highly insignificant. Among them, $F_{0.01}(2, 6) = 10.93$ and $F_{0.05}(2, 6) = 5.14$.

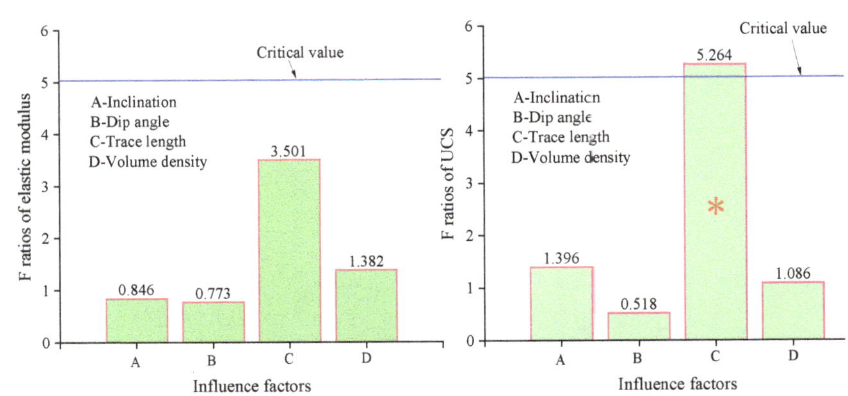

Figure 7. F ratio and significance levels of the elastic modulus and UCS under various influencing factors.

As shown in Figure 7, the results demonstrated that the sensitivity of the rock mass elastic modulus to the four influencing factors was not significant. The main reason is that the elastic modulus is an inherent characteristic of the rock mass, which is mainly related to the chemical composition of material and not its structural state. Besides, the sensitivity of the rock mass elastic modulus to the four influencing factors was in the order of trace length > bulk density > inclination > dip angle. The sensitivity analysis of rock mass UCS to the trace length of fracture was significant, while the sensitivity analysis to the inclination, dip angle, and bulk density of the fracture was insignificant. The sensitivity degree of the rock mass UCS to the four influencing factors was in the order of trace length > inclination > bulk density > dip angle, which is also consistent with the results of the range analysis.

5. Discussion

From the abovementioned research, it could be found that the effect of the fracture on the UCS was greater than the elastic modulus, which is consistent with the results of Vaziri et al. [23]. Additionally, the mechanical parameters of the fractured rock mass were sensitive to the fracture traces, which is in agreement with Cui and Sheng [26]. However, Vaziri et al. [23] and Bahaaddini et al. [24] found that fracture orientation was the most effective parameter among the fracture geometrical parameters, which is inconsistent with the results of this paper. The main reason could be that the fractured rock mass was generated based on a 2D PFC and the fracture orientation (dip angle) was considered in 2D space by Vaziri et al. [23]. Moreover, although a three-dimensional numerical model was applied by Bahaaddini et al. [25], the fractures in the rock mass were assumed to be parallel and only the fracture orientation (dip angle) was changed in the calculation results. Fractures in the rock mass are three-dimensional and their orientation is determined by the inclination and dip angle.

Additionally, the geometric parameters involved in this research do not completely exactly agree with those of other authors. The sensitivity orders of the rock mass mechanical parameters to the rock mass geometry parameters were also different from other authors. To increase the credibility of the conclusions, another orthogonal test was performed. The study methodology, test scheme, and analysis method are consistent with parts III and IV of the paper. Only the level values of the trace length were changed, as shown in Table 5.

The specific calculation results are not described in detail here. Extreme differences and F ratios of the elastic modulus and UCS for various influencing factors in the Group 2 test are presented in Figure 8. The results showed that the influence of the fracture geometry parameters in the UCS was greater than that of the elastic modulus. The results are consistent with the results of the Group 1 test.

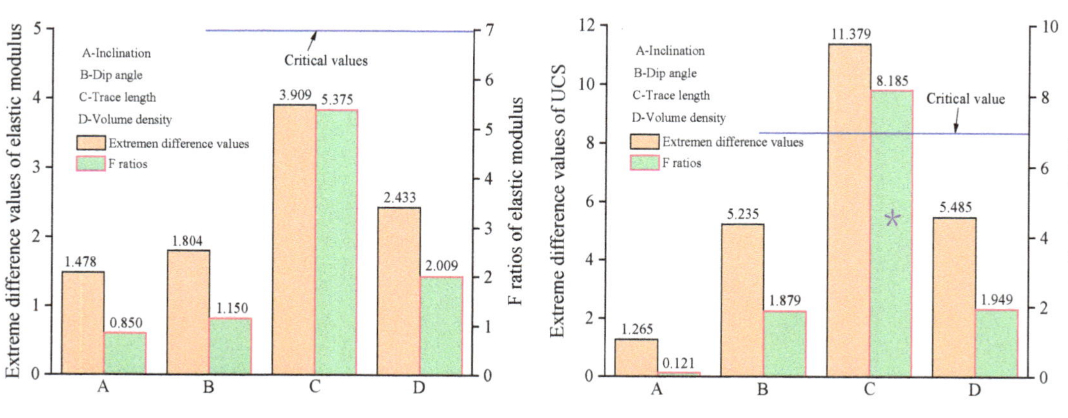

Figure 8. Extreme difference values and F ratios under various influencing factors of the Group 2 test.

Besides, in the Group 2 test, the sensitivity degree of the rock mass elastic modulus and UCS to the four influencing factors was in the order of trace length > bulk density > dip angle > inclination, which are different than the results of the Group 1 test. Therefore, when the influencing factors take different data in the range of values, the order of the sensitivity of the rock mass mechanical parameters to the fracture geometry parameters may be different. To fully understand the problem, the global sensitivity analysis method can be used [32,33]. The Monte Carlo method was applied in the global sensitivity analysis method to generate a random set of data over a reasonable range of parameters to determine the influencing factors. The parameters of different influencing factors were then combined to establish a numerical model and acquire results. Generally, hundreds or even thousands of time-valued simulations are required.

6. Conclusions

The fracture geometry parameters and distribution types of the downstream slope of the left bank of the Lianghekou Hydropower Station were acquired based on the ShapeMetriX3D system. The sensitivity of rock mass mechanical parameters to the 3# group of the fracture geometry parameters was investigated according to the orthogonal test scheme. Moreover, the RFPA3D software was applied.

The behavior characteristics and failure pattern of the fractured rock mass were analyzed. The results showed that the stress–strain curve of the fractured rock mass mainly exhibited an elastic-brittle behavior. The macroscopic failure mode of fractured rock mass is a compressive-shear composite failure pattern.

Range analysis, variance analysis, and significance test were used to analyze the sensitivity of the fractured rock mass mechanical parameters to the fracture geometry. The results indicated that the influence of the fracture geometry parameters in the UCS was greater than that of elastic modulus. The sensitivity of the UCS to fracture trace length was significant, but not to the bulk density, dip angle, and inclination of fracture. The elastic modulus of the fractured rock mass was insensitive to the fracture geometrical parameters.

The research results provide a reference for the measurement and statistics of fracture geometrical parameters in the engineering rock mass, as well as improving the accuracy of the analysis of the mechanical parameters of a fractured rock mass.

Author Contributions: Conceptualization, Z.L.; methodology, N.W.; software, Z.L.; formal analysis, N.W.; investigation, Y.T.; resources, N.W.; data curation, N.W.; writing—original draft preparation, N.W.; writing—review and editing, G.L.; supervision, Z.L.; project administration, T.A.; funding acquisition, Z.L. and T.A. All authors have read and agreed to the published version of the manuscript.

Funding: This research was funded by the National Natural Science Foundation of China (Grant No. 52209123), Open Fund of Key Laboratory of Deep Earth Science and Engineering (Sichuan University), Basic Scientific Research Projects of Dalian University of Technology (DUT21RC(3)071).

Institutional Review Board Statement: Not applicable.

Informed Consent Statement: Not applicable.

Data Availability Statement: Not applicable.

Acknowledgments: The authors thank the anonymous reviewers, who provided valuable suggestions that improved the manuscript.

Conflicts of Interest: The authors declare no conflict of interest.

List of Symbols

U_x	A displacement load parallel to the X-axis direction
k_{ij}	The mean value of test results of the influencing factor (i) on the level (j).
R_i	The extreme difference in the value of the influencing factor of i.
$F_\alpha(x, y)$	F-value with a significant level of α, number of independent variables as x and degrees of freedom as y.

References

1. Mu, W.Q.; Wang, D.Y.; Li, L.C.; Yang, T.H.; Feng, Q.B.; Wang, S.X.; Xiao, F.K. Cement Flow in Interaction Rock Fractures and its Corresponding New Construction Process in Slope Engineering. *Constr. Build. Mater.* **2021**, *303*, 124533. [CrossRef]
2. Gong, B.; Wang, Y.Y.; Zhao, T.; Tang, C.A.; Yang, X.Y.; Chen, T.T. AE Energy Evolution during CJB Fracture Affected by Rock Heterogeneity and Column Irregularity under Lateral Pressure. *Geomat. Nat. Haz. Risk.* **2022**, *13*, 877–907. [CrossRef]

3. Liang, Z.Z.; Wu, N.; Li, Y.C.; Li, H.; Li, W.R. Numerical Study on Anisotropy of the Representative Elementary Volume of Strength and Deformability of Jointed Rock Masses. *Rock Mech. Rock Eng.* **2019**, *52*, 4387–4402. [CrossRef]
4. Wu, N.; Liang, Z.Z.; Li, Y.C.; Li, H.; Li, W.R.; Zhang, M.L. Stress-Dependent Anisotropy Index of Strength and Deformability of Jointed Rock Mass: Insights from a Numerical Study. *Bull. Eng. Geol. Environ.* **2019**, *78*, 5905–5917. [CrossRef]
5. Qiu, L.; Zhu, Y.; Song, D.; He, X.; Wang, W.; Liu, Y.; Xiao, Y.; Wei, M.; Yin, S.; Liiu, Q. Study on the Nonlinear Characteristics of EMR and AE during Coal Splitting Tests. *Minerals* **2022**, *12*, 108. [CrossRef]
6. Strauhal, T.; Zangerl, C. The Impact of Fracture Persistence and Intact Rock Bridge Failure on the In Situ Block Area Distribution. *Appl. Sci.* **2021**, *11*, 3973. [CrossRef]
7. Bonilla-Sierra, V.; Scholtès, L.; Donzé, F.V.; Elmouttie, M.K. Rock Slope Stability Analysis Using Photogrammetric Data and DFN–DEM Modelling. *Acta Geotech.* **2015**, *10*, 497–511. [CrossRef]
8. Li, X.; Chen, Z.; Chen, J.; Zhu, H. Automatic Characterization of Rock Mass Discontinuities Using 3D Point Clouds. *Eng. Geol.* **2019**, *259*, 105131. [CrossRef]
9. Lato, M.J.; Diederichs, M.S.; Hutchinson, D.J. Bias Correction for View-Limited Lidar Scanning of Rock Outcrops for Structural Characterization. *Rock Mech. Rock Eng.* **2010**, *43*, 615–628. [CrossRef]
10. Otoo, J.N.; Maerz, N.H.; Li, X.; Duan, Y. Verification of a 3-D LiDAR Viewer for Discontinuity Orientations. *Rock Mech. Rock Eng.* **2013**, *46*, 543–554. [CrossRef]
11. Pan, D.D.; Li, S.C.; Xu, Z.H.; Zhang, Y.C.; Lin, P.; Li, H.Y. A Deterministic-Stochastic Identification and Modelling Method of Discrete Fracture Networks Using Laser Scanning: Development and Case Study. *Eng. Geol.* **2019**, *262*, 105310. [CrossRef]
12. Gomes, R.K.; de Oliveira, L.P.L.; Gonzaga, L.; Tognoli, F.M.W.; Veronez, M.R.; de Souza, M.K. An Algorithm for Automatic Detection and Orientation Estimation of Planar Structures in LiDAR-Scanned Outcrops. *Comput. Geosci.* **2016**, *90*, 170–178. [CrossRef]
13. Kong, D.; Saroglou, C.; Wu, F.; Sha, P.; Li, B. Development and Application of UAV-SfM Photogrammetry for Quantitative Characterization of Rock Mass Discontinuities. *Int. J. Rock Mech. Min.* **2021**, *141*, 104729. [CrossRef]
14. Cho, J.W.; Jeon, S.; Jeong, H.Y.; Chang, S.-H. Evaluation of Cutting Efficiency during TBM Disc Cutter Excavation within a Korean Granitic Rock Using Linear-Cutting-Machine Testing and Photogrammetric Measurement. *Tunn. Undergr. Sp. Technol.* **2013**, *35*, 37–54. [CrossRef]
15. Sharif, L.K.; Elmo, D.; Stead, D. Improving DFN-Geomechanical Model Integration Using a Novel Automated Approach. *Comput. Geosci.* **2019**, *105*, 228–248.
16. Buyer, A.; Aichinger, S.; Schubert, W. Applying Photogrammetry and Semi-Automated Joint Mapping for Rock Mass Characterization. *Eng. Geol.* **2020**, *264*, 105332. [CrossRef]
17. Yang, T.H.; Wang, P.T.; Xu, T.; Yu, Q.L.; Zhang, P.H.; Shi, W.H.; Hu, G.J. Anisotropic Characteristics of Jointed Rock Mass: A Case Study at Shirengou Iron Ore Mine in China. *Tunn. Undergr. Sp. Technol.* **2015**, *48*, 129–139. [CrossRef]
18. Ma, G.W.; Li, M.Y.; Wang, H.D.; Chen, Y. Equivalent Discrete Fracture Network Method for Numerical Estimation of Deformability in Complexly Fractured Rock Masses. *Eng. Geol.* **2020**, *277*, 105784. [CrossRef]
19. Ren, F.; Ma, G.W.; Fan, L.F.; Wang, Y.; Zhu, H.H. Equivalent Discrete Fracture Networks for Modelling Fluid Flow in Highly Fractured Rock Mass. *Eng. Geol.* **2017**, *229*, 21–30. [CrossRef]
20. Miyoshi, T.; Elmo, D.; Rogers, S. Influence of Data Analysis When Exploiting DFN Model Representation on the Application of Rock Mass Classification Systems. *J. Rock Mech. Geotech.* **2018**, *10*, 1046–1062. [CrossRef]
21. Pakzad, R.; Wang, S.; Sloan, S. Numerical Study of the Failure Response and Fracture Propagation for Rock Specimens with Preexisting Flaws under Compression. *Int. J. Geomech.* **2018**, *18*, 04018070. [CrossRef]
22. Wu, N.; Liang, Z.Z.; Li, Y.C.; Qian, X.K.; Gong, B. Effect of Confining Stress on Representative Elementary Volume of Jointed Rock Masses. *Geomech. Eng.* **2019**, *18*, 627–638.
23. Vaziri, M.R.; Tavakoli, H.; Bahaaddini, M. Statistical Analysis on the Mechanical Behaviour of Non-Persistent Jointed Rock Masses Using Combined DEM and DFN. *B. Eng. Geol. Environ.* **2022**, *81*, 177. [CrossRef]
24. Wu, N.; Liang, Z.Z.; Zhang, Z.H.; Li, S.H.; Lang, Y.X. Development and Verification of Three-Dimensional Equivalent Discrete Fracture Network Modelling Based on the Finite Element Method. *Eng. Geol.* **2022**, *306*, 106759. [CrossRef]
25. Bahaaddini, M.; Hagan, P.; Mitra, R.; Hebblewhite, B.K. Numerical Study of the Mechanical Behavior of Nonpersistent Jointed Rock Masses. *Int. J. Geomech.* **2016**, *16*, 04015035. [CrossRef]
26. Cui, Z.; Sheng, Q. Numerical Modelling of Structural Effect of Equivalent Mechanical Parameters of Fractured Rock Mass. *Rock Soil Mech.* **2018**, *39*, 3830–3841.
27. Wu, N.; Liang, Z.Z.; Song, W.C.; Li, W.R. Construction of a 3D Equivalent Rock Random Fracture Network Model and its Application in the Lianghekou Hydropower Station. *Chin. J. Eng.* **2022**, *44*, 1282–1290.
28. Zhou, J.; Zhao, M.; Wang, C.; Gao, Z. Optimal Design of Diversion Piers of Lateral Intake Pumping Station Based on Orthogonal Test. *Shock. Vib.* **2021**, *2021*, 6616456. [CrossRef]
29. He, Z.; Staples, G.; Ross, M.; Court, I.; Hazzard, K. Orthogonal Software Testing: Taguchi Methods in Software Unit and Subsystem Testing. *Logist. Inf. Manag.* **1997**, *10*, 189–194. [CrossRef]
30. Liang, Z.Z.; Tang, C.A.; Li, H.X.; Zhang, Y.B. Numerical Simulation of 3-D Failure Process in Heterogeneous Rocks. *Int. J. Rock. Mech. Min.* **2004**, *41*, 323–328. [CrossRef]

31. Tang, C.A. Numerical Simulation of Progressive Rock Failure and Associated Seismicity. *Int. J. Rock Mech. Min.* **1997**, *34*, 249–261. [CrossRef]
32. Li, G.; Lu, Z.; Xu, J. Regional Sensitivity Analysis of Aleatory and Epistemic Uncertainties on Failure Probability. *Mech. Syst. Signal. Process.* **2014**, *46*, 209–226. [CrossRef]
33. Li, G.; Xie, C.; Wei, F.; Wang, F. Moment-Independence Global Sensitivity Analysis for the System with Fuzzy Failure State and Its Kriging Method. *J. Syst. Eng. Electron.* **2018**, *29*, 658–666.

MDPI
St. Alban-Anlage 66
4052 Basel
Switzerland
www.mdpi.com

Applied Sciences Editorial Office
E-mail: applsci@mdpi.com
www.mdpi.com/journal/applsci

www.ingramcontent.com/pod-product-compliance
Lightning Source LLC
Chambersburg PA
CBHW040313270926
51850CB00007BC/1158
9783036587004